P9-AFZ-755

SCALE

The Universal Laws of Growth, Innovation,
Sustainability, and the Pace of Life in Organisms,
Cities, Economies, and Companies

GEOFFREY WEST

PENGUIN PRESS | NEW YORK | 2017

PENGUIN PRESS
An imprint of Penguin Random House LLC
375 Hudson Street
New York, New York 10014
penguin.com

Illustration credits appear on page 481.

Library of Congress Cataloging-in-Publication Data
Names: West, Geoffrey B., author.
Title: Scale : the universal laws of growth, innovation, sustainability, and
the pace of life in organisms, cities, economies, and companies / Geoffrey   West.
Description: New York : Penguin Press, [2017] | Includes bibliographical references and index.
Identifiers: LCCN 2016056756 (print) | LCCN 2017008356 (ebook) | ISBN
9781594205583 (hardcover) | ISBN 9781101621509 (ebook)
Subjects: LCSH: Scaling (Social sciences) | Science—Philosophy. | Evolution
(Biology) | Evolution—Molecular aspects. | Urban ecology (Sociology) |
Social sciences—Methodology. | Sustainable development.
Classification: LCC H61.27 .W47 2017 (print) | LCC H61.27 (ebook) | DDC 303.44—dc23
LC record available at https://lccn.loc.gov/2016056756

Printed in the United States of America
1   3   5   7   9   10   8   6   4   2

Designed by Grtechen Achilles

To

Jacqueline

Joshua and Devorah

and

Dora and Alf

With Gratitude and Love

CONTENTS

SCALE

1

THE BIG PICTURE

1. INTRODUCTION, OVERVIEW, AND SUMMARY

Life is probably the most complex and diverse phenomenon in the universe, manifesting an extraordinary variety of forms, functions, and behaviors over an enormous range of scales. It is estimated for instance that there are more than eight million different species of organisms on our planet,[1] ranging in size from the smallest bacterium weighing less than a trillionth of a gram to the largest animal, the blue whale, weighing up to a hundred million grams. If you visited a tropical forest in Brazil you'd find in an area the size of a football field more than a hundred different species of trees and millions of individual insects representing thousands of species. And just think of the amazing differences in how each of these species lives out its life, how differently each is conceived, born, and reproduces and how it dies. Many bacteria live for only an hour and need only a tenth of a trillionth of a watt to stay alive, whereas whales can live for over a century and metabolize at several hundred watts.[2] Add to this extraordinary tapestry of biological life the astonishing complexity and diversity of social life that we humans have brought to the planet, especially in the guise of cities and all of the remarkable phenomena they encompass, ranging from commerce and architecture to the diversity of cultures and the innumerable hidden joys and sorrows of each of their citizens.

Compare any of this complex panoply with the extraordinary simplicity and order of the planets orbiting the sun, or the clockwork regularity of your watch or iPhone, and it's natural to ponder whether there could possibly be any analogous hidden order underlying all of this complexity and diversity. Could there conceivably be a few simple rules that all organisms obey, indeed all complex systems, from plants and animals to cities and companies? Or is all of the drama being played out in the forests, savannahs, and cities across the globe arbitrary and capricious, just one haphazard event after another? Given the random nature of the evolutionary process that gave rise to all of this diversity, it might seem unlikely and counterintuitive that any regularity or systematic behavior would have emerged. After all, each of the multitude of organisms that constitute the biosphere, each of its subsystems, each organ, each cell type, and each genome has evolved by the process of natural selection in its own unique environmental niche following a unique historical path.

Now take a look at the panel of graphs in Figures 1–4. Each represents a well-known quantity that plays an important role in your life and each is plotted against size. The first graph is metabolic rate—how much food is needed each day to stay alive—plotted against the weight or mass of a series of animals. The second is the number of heartbeats in a lifetime, also plotted against the weight or mass of a series of animals. The third is the number of patents produced in a city plotted against its population. And the last is the net assets and income of publicly traded companies plotted against the number of their employees.

You don't have to be a mathematician, a scientist, or an expert in any of these areas to immediately see that although they represent some of the most extraordinarily complex and diverse processes we encounter in our lives, they

Examples of *scaling curves,* which express how quantities *scale* with a change in size: (1) illustrates how metabolic rate[3] and (2) how the number of heartbeats in a lifetime[4] scale with the weight of an animal; (3) illustrates how the number of patents produced in a city[5] scales with its population size; and (4) illustrates how assets and income of companies[6] scale with the number of their employees. Notice that these graphs cover a huge range of scales: for example, both weights of animals and numbers of employees vary by a factor of a million (from mice to elephants and from one-person businesses to the Walmarts and Exxons). In order to be able to put all of these animals, companies, and cities on these plots, the scale on each axis increases by factors of ten.

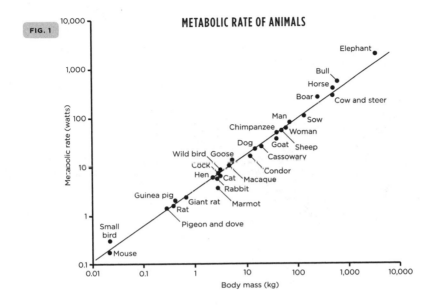

FIG. 1

METABOLIC RATE OF ANIMALS

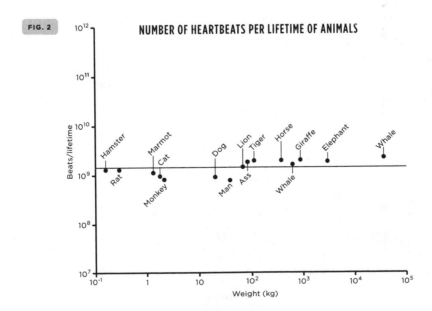

FIG. 2

NUMBER OF HEARTBEATS PER LIFETIME OF ANIMALS

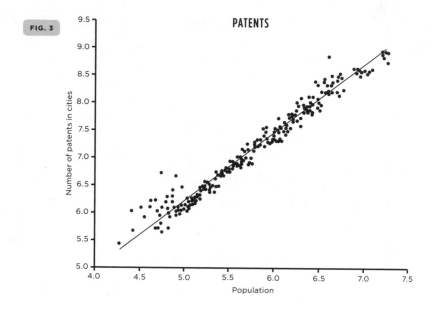

FIG. 3

PATENTS

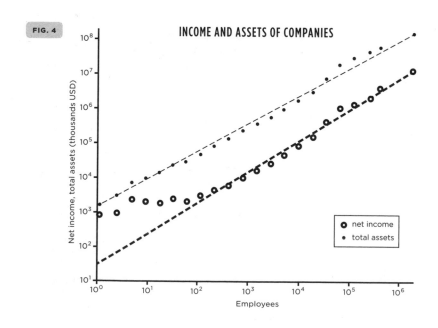

FIG. 4

INCOME AND ASSETS OF COMPANIES

reveal something surprisingly simple, systematic, and regular about each of them. Almost miraculously, the data have lined up in approximately straight lines rather than being arbitrarily distributed across each of these graphs, as might have been anticipated given the unique historical and geographical contingency of each animal, city, or company. Perhaps the most startling of these is Figure 2, which shows that the average number of heartbeats in the lifetime of any mammal is roughly the same, even though small ones like mice live for just a few years whereas big ones like whales can live for a hundred years or more.

The examples shown in Figures 1–4 are just a tiny sampling of an enormous number of such *scaling* relationships that *quantitatively* describe how almost any measurable characteristic of animals, plants, ecosystems, cities, and companies *scales* with size. You will be seeing many more of them throughout this book. The existence of these remarkable regularities strongly suggests that there is a common conceptual framework underlying all of these very different highly complex phenomena and that the dynamics, growth, and organization of animals, plants, human social behavior, cities, and companies arc, in fact, subject to similar generic "laws."

This is the main focus of this book. I will explain the nature and origin of these systematic scaling laws, how they are all interrelated, and how they lead to a deep and broad understanding of many aspects of life and ultimately to the challenge of global sustainability. Taken together, these scaling laws provide us with a window onto underlying principles and concepts that can potentially lead to a quantitative predictive framework for addressing a host of critical questions across science and society.

This book is about a way of thinking, about asking big questions, and about suggesting big answers to some of those big questions. It's a book about how some of the major challenges and issues we are grappling with today, ranging from rapid urbanization, growth, and global sustainability to understanding cancer, metabolism, and the origins of aging and death, can be addressed in an integrated unifying conceptual framework. It is a book about the remarkably similar ways in which cities, companies, tumors, and our bodies work, and how each of them represents a variation on a general theme manifesting surprisingly systematic regularities and similarities in their organization, structure, and dynamics. A common property shared by all of them is that they are highly complex and composed of enormous numbers of individual constituents, whether molecules, cells, or people, connected, interacting, and evolving via networked structures over multiple spatial and temporal scales. Some of these networks are obvious and very physical, like our circulatory system or the roads in a city, but some are more conceptual or virtual, like social networks, ecosystems, and the Internet.

This big-picture framework allows us to address a fascinating spectrum of questions, some of which stimulated my own research interests and some of which will be addressed, sometimes speculatively, in the ensuing chapters. Here's a sampling of some of them:

- Why can we live for up to 120 years but not for a thousand or a million? Why, in fact, do we die and what sets this limit to our life spans? Can life spans be calculated from the properties of cells and complex molecules that make up our bodies? Can they be changed and can life span be extended?

- Why do mice, made of pretty much the same stuff as we are, live for just two to three years whereas elephants live for up to seventy-five? And despite this difference, why is the number of heartbeats in a life span roughly the same for elephants, mice, and all mammals, namely about 1.5 billion?[7]

- Why do organisms and ecosystems ranging from cells and whales to forests scale with size in a remarkably universal, systematic, and predictable fashion? What is the origin of the magic number 4 that seems to control much of their physiology and life history from growth to death?

- Why do we stop growing? Why do we have to sleep for eight hours every day? And why do we get relatively far fewer tumors than mice, but whales get almost none?

- Why do almost all companies live for only a relatively few years whereas cities keep growing and manage to circumvent the apparently inevitable fate that befalls even the most powerful and seemingly invulnerable companies? Can we imagine being able to predict the approximate life spans of companies?

- Can we develop a science of cities and companies, meaning a conceptual framework for understanding their dynamics, growth, and evolution in a quantitatively predictable framework?

- Is there a maximum size of cities? Or an optimum size? Is there a maximum size to animals and plants? Could there be giant insects and giant megacities?

- Why does the pace of life continually increase and why does the rate of innovation have to continue to accelerate in order to sustain socioeconomic life?

- How do we ensure that our human-engineered systems, which evolved only over the past ten thousand years, can continue to coexist with the natural biological world, which evolved over billions of years? Can we maintain a vibrant, innovative society driven by ideas and wealth creation, or are we destined to become a planet of slums, conflict, and devastation?

In addressing questions such as these, I will emphasize conceptual issues and bring together ideas from across the sciences in a transdisciplinary spirit, integrating fundamental questions in biology with those in the social and economic sciences, though shamelessly from the perspective and through the eyes of a theoretical physicist. So much so, in fact, that I will also touch on how the same framework of scaling has played a seminal role in developing a unified picture of the elementary particles and fundamental forces of nature, including their cosmological implications for the evolution of the universe from the Big Bang. In this spirit, I have also tried to be provocative and speculative where appropriate, but in the main, almost all of what is presented is based on established scientific work.

Although many, if not most, of the results and explanations presented in the book have their origins in arguments and derivations couched in the language of mathematics, the book is decidedly nontechnical and pedagogical in spirit and is written for the proverbial "intelligent layperson." This presents quite a challenge and means, of course, that a certain poetic license has to be taken when providing such explanations, and my fellow scientists will have to try to refrain from being overly critical if they find that I have oversimplified the translation from mathematical or technical language into English. For those with a more mathematical inclination, I refer to the technical literature referenced throughout the book.

2. WE LIVE IN AN EXPONENTIALLY EXPANDING SOCIOECONOMIC URBANIZED WORLD

A central topic of the book is the critical role that cities and global urbanization play in determining the future of the planet. Cities have emerged as the source of the greatest challenges the planet has faced since humans became social. The future of humanity and the long-term sustainability of the planet are inextricably linked to the fate of our cities. Cities are the crucible of civilization, the hubs of innovation, the engines of wealth creation and centers of power, the magnets that attract creative individuals, and the stimulant for ideas, growth, and innovation. But they also have a dark side: they are the prime locus of crime, pollution, poverty, disease, and the consumption of energy and resources. Rapid urbanization and accelerating socioeconomic development have generated multiple global challenges ranging from climate change and its environmental impacts to incipient crises in food, energy, and water availability, public health, financial markets, and the global economy.

Given this dual nature of cities as, on the one hand, the origin of many of our major challenges and, on the other, the reservoir of creativity and ideas and therefore the source of their solutions, it becomes a matter of some urgency to ask whether there can be a "science of cities" and by extension a "science of companies," in other words a conceptual framework for understanding their dynamics, growth, and evolution in a quantitatively predictable framework. This is crucial for devising a serious strategy for achieving long-term sustainability, especially as the overwhelming majority of human beings will be urban dwellers by the second half of this century, many in mega-cities of unprecedented size.

Almost none of the problems, challenges, and threats we are facing are new. All of them have been with us since at least the beginnings of the Industrial Revolution, and it is only because of the exponential rate of urbanization that they have now begun to feel like an impending tsunami with the potential to overwhelm us. It is in the very nature of exponential expansion that the immediate future comes upon us increasingly more rapidly, potentially presenting us with unforeseen challenges whose threat we recognize only after it's too late. Consequently, it is only relatively recently that we have become conscious of global warming, long-term environmental changes, limitations on energy, water, and other resources, health and pollution issues, stability of financial markets, and so on. And even as we have become concerned, it has been implicitly presumed that these are temporary aberrations that will eventually be solved and disappear. Not surprisingly, most politicians, economists, and policy makers have continued to take a fairly optimistic long-term view that our innovation and ingenuity will triumph, as indeed they have in the past. As will be elaborated on later, I am not so sure.

For almost the entire time span of human existence most human beings have resided in nonurban environments. Just two hundred years ago the United States was predominantly agricultural, with barely 4 percent of the population living in cities, compared with more than 80 percent today. This is typical of almost all developed countries such as France, Australia, and Norway, but it is also true for many that are considered as "developing," such as Argentina, Lebanon, and Libya. Nowadays, no country on the planet comes close to being just 4 percent urban; even Burundi, perhaps the poorest and least developed of all nations, is over 10 percent urbanized. In 2006 the planet crossed a remarkable historical threshold, with more than half of the world's population residing in urban centers, compared with just 15 percent a hundred years ago and still only 30 percent by 1950. It is now expected to rise above 75 percent by 2050, with more than two billion more people moving to cities, mostly in China, India, Southeast Asia, and Africa.[8]

This is an enormous number. It means that, when averaged over the next thirty-five years, about a million and a half people will be urbanized *each week*. To get an idea of what this implies, consider the following: today is August 22; by October 22 there will be the equivalent of another New York metropolitan area on the planet, and by Christmas another one, and by February 22 yet another, and so on. . . . Inextricably, from now to well into the middle of the century another New York metropolitan area is being added to the planet every couple of months. And note that we are talking of a New York metropolitan area consisting of 15 million people, not just New York City, which has only 8 million.

Perhaps the most astonishing and ambitious urbanization program on the planet is being carried out by China, where the government is on a fast track to build up to three hundred new cities each in excess of a million people over the next twenty to twenty-five years. Historically, China was slow to urbanize and industrialize but is now making up for lost time. In 1950, China was not much more than 10 percent urbanized but will very likely cross the halfway mark this year. At the present rate it will be moving the equivalent of the entire U.S. population (more than 300 million people) to cities in the next twenty to twenty-five years. And not far behind are India and Africa. This will be by far the largest migration of human beings to have ever taken place on the planet and will very likely never be equaled in the future. The resulting challenges to the availability of energy and resources and the enormous stress on the social fabric across the globe are mind-boggling . . . and the timescales to address them are very short. Everyone will be affected; there is no hiding place.

3. A MATTER OF LIFE AND DEATH

The open-ended exponential growth of cities stands in marked contrast to what we see in biology: most organisms, like us, grow rapidly when young but then slow down, cease growing, and eventually die. Most companies follow a similar pattern, with almost all of them eventually disappearing, whereas most cities don't. Nevertheless, biological imagery is routinely used when writing about cities as well as companies. Typical phrases include "the DNA of the company," "the metabolism of the city," "the ecology of the marketplace," and so on. Are these just metaphors or do they encode something of real scientific substance? To what extent, if any, are cities and companies very large organisms? They did, after all, evolve from biology and consequently share many features in common.

There are clearly characteristics of cities that are not biological, and these will be discussed in detail later. But if cities are indeed some sort of superorganism, then why do almost none of them ever die? There are, of course, classic examples of cities that have died, especially ancient ones, but they tend to be special cases due to conflict and the abuse of the immediate environment. Overall, they represent only a tiny fraction of all those that have ever existed. Cities are remarkably resilient and the vast majority persist. Just think of the awful experiment that was done seventy years ago when atom bombs were dropped on two cities, yet just thirty years later they were thriving. It's extremely difficult to kill a city! On the other hand, it's relatively easy to kill animals and companies—overwhelmingly, almost all of them eventually die, even the most powerful and seemingly invulnerable. Despite the continuing increase in the average life span of human beings over the last 200 years, our maximum life span has remained unchanged. No human being has ever lived for more than 123 years, and very few companies have lived for much longer—most have disappeared after 10 years. So why *do* almost all cities remain viable, whereas the vast majority of companies and organisms die?

Death is integral to all biological and socioeconomic life: almost all living things are born, live, and eventually die, yet death as a serious focus of study and contemplation tends to be suppressed and neglected, both socially and scientifically, relative to birth and life. At a personal level, it wasn't until I reached my fifties that I started thinking seriously about aging and dying. I had gone through my twenties, thirties, and forties and into my fifties without being much concerned about my own mortality, unconsciously maintaining the myth common among the "young" that I was immortal. However, I come from a long line of short-lived males, so perhaps it was inevitable that at some stage in my fifties it would begin to dawn on me that I might be dead in five to ten years and that it would be prudent to start contemplating what that would mean.

I suppose that one could view all religion and philosophical reflection as having its origins in how we integrate the inevitable imminence of death into our daily lives. So I started thinking and reading about aging and death, first in personal, psychological, religious, and philosophical terms, which though extremely engaging, left me with more questions than answers. And then, because of other events that I shall relate later in the book, I started thinking about them in scientific terms, which serendipitously led me on a path that changed both my personal and professional life.

As a physicist thinking about aging and death it was natural not only to ask about possible mechanisms for why we age and why we die but, equally important, to ask where the *scale* of human life span comes from. Why hasn't anybody lived for more than 123 years? What is the origin of the mysterious threescore years and ten deemed to be the scale of human life span in the Old Testament? Could we possibly live for a thousand years like the mythical Methuselah? Most companies, on the other hand, live for only a few years. Half of all U.S. publicly traded companies have disappeared within ten years of entering the market. Although a small minority live for considerably longer, almost all seem destined to go the way of Montgomery Ward, TWA, Studebaker, and Lehman Brothers. Why? Can we develop a serious mechanistic theory for understanding not only our own mortality but also that of companies? Can we imagine being able to quantitatively understand the processes of aging and death of companies and thereby "predict" their approximate life spans? And what is it about cities that they manage to circumvent this apparently inevitable fate?

4. ENERGY, METABOLISM, AND ENTROPY

Addressing these questions naturally leads to asking where all the other scales of life come from. Why, for instance, do we sleep approximately eight hours a night whereas mice sleep fifteen and elephants just four? Why are the tallest trees a few hundred feet high and not a mile? Why do the largest companies stop growing when their assets reach half a trillion dollars? And why are there roughly five hundred mitochondria in each of your cells?

To answer such questions, and to understand quantitatively and mechanistically processes such as aging and mortality, whether for humans, elephants, cities, or companies, we must first come to terms with how each of these systems grew and how each stays alive. In biology these are controlled and maintained by the process of metabolism. Quantitatively, this is expressed in terms of *metabolic rate,* which is the amount of energy needed per second to keep an organism alive; for us it's about 2,000 food calories a day, which, surprisingly, corresponds to a rate of only about 90 watts, the equivalent of a standard incandescent lightbulb. As can be seen from Figure 1, our metabolic rate has the "correct" value for a mammal of our size. This is our *biological* metabolic rate living as naturally evolved animals. As social animals now living in cities we still need just a lightbulb equivalent of food to stay alive but, in addition, we now require homes, heating, lighting, automobiles, roads, airplanes, computers, and so on. Consequently, the amount of energy needed to support an average person living in the United States has risen to an astounding 11,000 watts. This *social* metabolic rate is equivalent to the entire needs of about a dozen elephants. Furthermore, in making this transition from the biological to the social our overall population has increased from just a few million to more than seven billion. No wonder there's a looming energy and resource crisis.

None of these systems, whether "natural" or man-made, can operate without a continuous supply of energy and resources that have to be transformed into something "useful." Appropriating the concept from biology, I shall refer to all such processes of energy transformation as *metabolism.* Depending on the sophistication of the system, these outputs of useful energy are allocated between doing physical work and fueling maintenance, growth, and reproduction. As social human beings and in marked contrast to all other creatures, the major portion of our metabolic energy has been devoted to forming communities and institutions such as cities, villages, companies, and collectives, to the manufacture of an extraordinary array of artifacts, and to the creation of an astonishing litany of ideas ranging from airplanes, cell phones, and cathedrals to symphonies, mathematics, and literature, and much, much more.

However, it's not often appreciated that without a continuous supply of energy and resources, not only can there be no manufacturing of any of these things but, perhaps more important, there can be no ideas, no innovation, no growth, and no evolution. Energy is primary. It underlies everything that we do and everything that happens around us. As such, its role in all of the questions addressed will be another continuous thread that runs throughout the book. This may seem self-evident, but it is surprising how small a role, if any, the generalized concept of energy plays in the conceptual thinking of economists and social scientists.

There is always a price to pay when energy is processed; there is no free lunch. Because energy underlies the transformation and operation of literally everything, no system operates without consequences. Indeed, there is a fundamental law of nature that cannot be transgressed, called the *Second Law of Thermodynamics*, which says that whenever energy is transformed into a useful form, it also produces "useless" energy as a degraded by-product: "unintended consequences" in the form of inaccessible disorganized heat or unusable products are inevitable. There are no perpetual motion machines. You need to eat to stay alive and maintain and service the highly organized functionality of your mind and body. But after you've eaten, sooner or later you will have to go to the bathroom. This is the physical manifestation of your personal entropy production.

This fundamental, universal property resulting from how all things interact by interchanging energy and resources was called *entropy* by the German physicist Rudolf Clausius in 1855. Whenever energy is used or processed in order to make or maintain order within a closed system, some degree of disorder is inevitable—entropy always increases. The word *entropy,* by the way, is the literal Greek translation of "transformation" or "evolution." Lest you think there might be some loophole in this law, it is worth quoting Einstein on the subject: "It is the only physical theory of universal content which I am convinced will never be overthrown" . . . and he included his own laws of relativity in this.

Like death, taxes, and the Sword of Damocles, the Second Law of Thermodynamics hangs over all of us and everything around us. Dissipative forces, analogous to the production of disorganized heat by friction, are continually and inextricably at work leading to the degradation of all systems. The most brilliantly designed machine, the most creatively organized company, the most beautifully evolved organism cannot escape this grimmest of grim reapers. To maintain order and structure in an evolving system requires the continual supply and use of energy whose by-product is disorder. That's why to stay alive we need to continually eat so as to combat the inevitable, destructive forces of entropy production. Entropy kills. Ultimately, we are all subject to the forces of "wear and tear" in its multiple forms. The battle to combat entropy by continually having to supply more energy for growth, innovation, maintenance, and repair, which becomes increasingly more challenging as the system ages, underlies any serious discussion of aging, mortality, resilience, and sustainability, whether for organisms, companies, or societies.

5. SIZE REALLY MATTERS: SCALING AND NONLINEAR BEHAVIOR

In addressing these diverse and seemingly unrelated questions, the lens I shall use will predominantly be that of *Scale* and the conceptual framework that of *Science*. Scaling and scalability, that is, how things change with size, and the fundamental rules and principles they obey are central themes that run throughout the book and are used as points of departure for developing almost all of the arguments presented. Viewed through this lens, cities, companies, plants, animals, our bodies, and even tumors manifest a remarkable similarity in the ways that they are organized and function. Each represents a fascinating variation on a general universal theme that is manifested in surprisingly systematic mathematical regularities and similarities in their organization, structure, and dynamics. These will be shown to be consequences of a broad, big-picture conceptual framework for understanding such disparate systems in an integrated unifying way, and with which many of the big issues can be addressed, analyzed, and understood.

Scaling simply refers, in its most elemental form, to how a system responds when its size changes. What happens to a city or a company if its size is doubled? Or to a building, an airplane, an economy, or an animal if its size is halved? If the population of a city is doubled, does the resulting city have approximately twice as many roads, twice as much crime, and produce twice as many patents? Do the profits of a company double if its sales double, and does an animal require half as much food if its weight is halved?

Addressing such seemingly innocuous questions concerning how systems respond to a change in their size has had remarkably profound consequences across the entire spectrum of science, engineering, and technology and has affected almost every aspect of our lives. Scaling arguments have led to a deep understanding of the dynamics of tipping points and phase transitions (how, for example, liquids freeze into solids or vaporize into gases), chaotic phenomena (the "butterfly effect" in which the mythical flapping of a butterfly's wings in Brazil leads to a hurricane in Florida), the discovery of quarks (the building blocks of matter), the unification of the fundamental forces of nature, and the evolution of the universe after the Big Bang. These are but a few of the more spectacular examples where scaling arguments have been instrumental in illuminating important universal principles or structure.[9]

In a more practical context, scaling plays a critical role in the design of increasingly large human-engineered artifacts and machines, such as buildings, bridges, ships, airplanes, and computers, where extrapolating from the small to the large in an efficient, cost-effective fashion is a continuing challenge. Even more challenging and of perhaps greater urgency is the need to understand how to scale organizational structures of increasingly large and complex social organizations such as companies, corporations, cities, and governments, where the underlying principles are typically not well understood because these are continuously evolving complex adaptive systems.

A greatly underappreciated case in point is the hidden role that scaling plays in medicine. Much of the research and development on diseases, new drugs, and therapeutic procedures is undertaken using mice as "model" systems. This immediately raises the critical question of how to scale up the findings and experiments on mice to humans. For instance, huge resources are spent each year on investigating cancer in mice, yet a typical mouse develops many more tumors per gram of tissue per year than we do, whereas whales get almost none, begging the question as to the relevance of such research for humans. To put it slightly differently: if we are to gain a deep understanding and solve the challenge of human cancer from such studies we need to know how to reliably scale up from mice to humans, and conversely, down from whales. Dilemmas such as this will be discussed in chapter 4 when addressing scaling issues inherent in biomedicine and health.

To introduce some of the language that will be used throughout the book and ensure that we're all on the same page as we begin this exploration, I want to review some commonly used concepts and terms that most people have some familiarity with—because they are used colloquially—but about which there are often misconceptions.

So let us return to the simple question posed above: does an animal require half as much food if its weight is halved? You might expect the answer to this to be yes, because halving its weight halves the number of cells that need to be fed. This would imply that "half as big requires half as much" and, conversely, that "twice as big requires twice as much" and so on. This is a simple example of classic *linear* thinking. Surprisingly, it is not always easy to recognize linear thinking, despite its apparent simplicity, because it often tends to be implicit rather than explicit.

For instance, it is not usually appreciated that the ubiquitous use of per capita measures as a way of characterizing and ranking countries, cities, companies, or economies is a subtle manifestation of this. Let me give a simple example. The gross domestic product (GDP) of the United States was estimated to be about $50,000 per capita in 2013, meaning that, averaged over the entire economy, each person can effectively be thought of as having produced $50,000 worth of "goods." Metropolitan Oklahoma City, with a population of about 1.2 million people, has a GDP of about $60 billion, so its per capita GDP ($60 billion divided by 1.2 million) is indeed close to the average for the United States, namely $50,000. Extrapolating this to a city with a population ten times larger, having 12 million people, would predict its GDP to be $600 billion (obtained by multiplying the $50,000 per capita by the 12 million people), ten times larger than Oklahoma City. However, metropolitan Los Angeles, which is indeed ten times larger than Oklahoma City with 12 million inhabitants, has a GDP that is actually more than $700 billion, which is more than 15 percent larger than the "predicted" value obtained by the linear extrapolation implicit in using a per capita measure.

This, of course, is just a single example which you might think is a special case—Los Angeles is simply a richer city than Oklahoma City. While that is indeed true, it turns out that the underestimation from comparing Oklahoma City with Los Angeles is not a special case but on the contrary is, in fact, an example of a general systematic trend across all cities across the globe which shows that simple linear proportionality, implicit in using per capita measures, is almost *never* valid. GDP, like almost any other quantifiable characteristic of a city, or indeed of almost any complex system, typically scales *nonlinearly*. I will be much more precise about what this means and what it implies later but, for the time being, *nonlinear* behavior can simply be thought of as meaning that measurable characteristics of a system generally do *not* simply double when its size is doubled. In the example given here, this can be restated as saying that there is a systematic increase in per capita GDP, as well as in average wages, crime rates, and many other urban metrics, as city size increases. This reflects an essential feature of all cities, namely that social activity and economic productivity are *systematically* enhanced with increasing size of the population. This systematic "value-added" bonus as size increases is called *increasing returns to scale* by economists and social scientists, whereas physicists prefer the more sexy term *superlinear scaling*.

An important example of nonlinear scaling arises in the biological world when we look at the amount of food and energy consumed each day by animals (including us) in order to stay alive. Surprisingly, an animal that is twice the size of another, and therefore composed of about twice as many cells, requires only about 75 percent more food and energy each day, rather than 100 percent more, as might naively have been expected from a linear extrapolation. For example, a 120-pound woman typically requires about 1,300 food calories a day just to stay alive without doing any activity or performing any tasks. This is called her *basal metabolic rate* by biologists and doctors and is to be distinguished from her *active metabolic rate*, which includes all of the additional daily activities of living. Her big English sheepdog, on the other hand, which weighs half as much as she does (60 pounds) and therefore has approximately half as many cells, would therefore be expected to require only about half as much food energy each day just to stay alive, namely about 650 food calories. In fact, her dog requires about 880 food calories each day.

Although a dog is not a small woman, this example is a special case of the general scaling rule for how metabolic rate scales with size. It operates across all mammals ranging from tiny shrews, weighing just a few grams, to giant blue whales, weighing greater than a hundred million times more. A profound consequence of this rule is that on a per gram basis, the larger animal (the woman in this example) is actually more efficient than the smaller one (her dog) because less energy is required to support each gram of her tissue (by about 25 percent). Her horse, by the way, would be even more efficient. This systematic savings with increasing size is known as an *economy of scale*. Put succinctly, this states that the bigger you are, the less you need per capita (or, in the case of animals, per cell or per gram of tissue) to stay alive. Notice that this is the opposite behavior to the case of increasing returns to scale, or superlinear scaling, manifested in the GDP of cities: in that case, the bigger you are, the more there is per capita, whereas for economics of scale, the bigger you are, the less there is per capita. This kind of scaling is referred to as *sublinear scaling*.

Size and scale are major determinants of the generic behavior of highly complex, evolving systems, and much of the book is devoted to explaining and understanding the origins of such nonlinear behavior and how it can be used to address a broad range of questions with examples drawn from across the entire spectrum of science, technology, economics, and business, as well as from daily life, science fiction, and sports.

6. SCALING AND COMPLEXITY: EMERGENCE, SELF-ORGANIZATION, AND RESILIENCE

I have already used the term *complexity* several times in just these few short pages and have cavalierly referred to systems as being *complex* as if this designation were both well understood and well defined. Neither, in fact, is the case, and I want to make a short detour here to discuss this much-overworked concept because almost all of the systems that I'm going to be talking about are usually thought of as being "complex."

I am hardly unique in my casual use of the word or its many derivatives without defining it. Over the past quarter of a century, terms like complex adaptive systems, the science of complexity, emergent behavior, self-organization, resilience, and adaptive nonlinear dynamics have begun to pervade not just the scientific literature but also that of the business and corporate world as well as the popular media.

To set the stage, I'd like to quote two distinguished thinkers, one a scientist, the other a lawyer. The first is the eminent physicist Stephen Hawking, who in an interview[10] at the turn of the millennium was asked the following question:

> Some say that while the 20th century was the century of physics, we are now entering the century of biology. What do you think of this?

To which he responded:

> I think the next century will be the century of complexity.

I wholeheartedly agree. As I hope I have already made clear, we urgently need a science of complex adaptive systems to address the host of extraordinarily challenging societal problems we face.

The second is a well-known quote from the eminent U.S. Supreme Court justice Potter Stewart, who when discussing the concept of pornography and its relationship to free speech in a landmark decision of 1964 made the following marvelous comment:

> I shall not today attempt further to define the kinds of material I understand to be embraced within that shorthand description ["hard-core pornography"]; and perhaps I could never succeed in intelligibly doing so. But I know it when I see it.

Just substitute the word "complexity" for "hard-core pornography" and that's pretty much what many of us would say: we may not be able to define it but we know it when we see it!

Unfortunately, however, while "knowing it when we see it" may be good enough for the U.S. Supreme Court, it's not considered good enough for science. Science has progressed famously by being concise and accurate about the objects that it studies and the concepts it invokes. We typically demand them to be precise, unambiguous, and operationally measurable. Momentum, energy, and temperature are classic examples of quantities that are precisely defined in physics but are used colloquially or metaphorically in everyday language. Having said that, however, there are a sizable number of really big concepts whose precise definitions still engender significant debate. These include life, innovation, consciousness, love, sustainability, cities, and, indeed, complexity. So rather than trying to give a scientific definition of complexity, I am going to resort to a middle ground and describe what I view as some of the essential features of typical complex systems so that we can recognize them when we see them and distinguish them from systems we might describe as *simple* or "just" very *complicated,* though not necessarily *complex.* This discussion is by no means complete but is intended to help clarify the more salient features of what we mean when we call a system complex.[11]

A typical complex system is composed of myriad individual constituents or agents that once aggregated take on collective characteristics that are usually not manifested in, nor could easily be predicted from, the properties of the individual components themselves. For example, you are much more than the totality of your cells and, similarly, your cells are much more than the totality of all of the molecules from which they are composed. What you think of as *you*—your consciousness, your personality, and your character—is a collective manifestation of the multiple interactions among the neurons and synapses in your brain. These are themselves exchanging continuous interactions with the rest of the cells of your body, many of which are constituents of semiautonomous organs, such as your heart or liver. In addition, all of these are, to varying degrees, continuously interacting with the external environment. Furthermore, and somewhat paradoxically, none of the 100 trillion or so cells that constitute your body have properties that you would recognize or identify as being *you,* nor do any of them have any consciousness or knowledge that they are a part of *you.* Each, so to speak, has its own specific characteristics and follows its own local rules of behavior and interaction, and in so doing, almost miraculously integrates with all the other cells of your body to be *you.* This, despite the huge range of scales, both spatial and temporal, that are operating within your body from the microscopic molecular level up to the macroscopic scales associated with living your daily life for up to a hundred years. You are a complex system par excellence.

In a similar fashion, a city is much more than the sum of its buildings, roads, and people, a company much more than the sum of its employees and products, and an ecosystem much more than the plants and animals that inhabit it. The economic output, the buzz, the creativity and culture of a city or a company all result from the nonlinear nature of the multiple feedback mechanisms embodied in the interactions between its inhabitants, their infrastructure, and the environment.

A wonderful example of this that we're all very familiar with is a colony of ants. In a matter of days, they literally build their cities from the ground up, one grain at a time. These remarkable edifices are constructed with multilevel networks of tunnels and chambers, ventilation systems, food storage and incubation units, all supplied by complex transportation routes. Their efficiency, resilience, and functionality would be considered major award-winning accomplishments by our very best engineers, architects, and urban planners had they been the designers and builders. Yet there are no tiny brilliant (or for that matter even mediocre) ant engineers, architects, or urban planners, and there never have been. No one is in charge.

Ant colonies are built without forethought and without the aid of any single mind or any group discussion or consultation. There is no blueprint or master plan. Just thousands of ants working mindlessly in the dark moving millions of grains of earth and sand to create these impressive structures. This feat is accomplished by each individual ant obeying just a few simple rules mediated by chemical cues and other signals, resulting in an extraordinarily coherent collective output. It is almost as if they were programmed to be microscopic operations in a giant computer algorithm.

Speaking of algorithms, computer simulations of such processes have successfully modeled this kind of outcome in which complex behavior emerges from a continuous iteration of very simple rules operating between individual agents. These simulations have given credence to the idea that the bewildering dynamics and organization of highly complex systems have their origin in very simple rules governing the interaction between their individual constituents. This discovery was only possible beginning about thirty years ago once computers were sufficiently powerful for such large calculations to be carried out. Nowadays, these computations can readily be done on your laptop. These computer investigations were very important in providing strong support for the idea that there might actually be a *simplicity* underlying the *complexity* that we observe in many such systems and that they might therefore be amenable to scientific analysis. Thus was conceived the conceptual possibility of developing a serious quantitative *science of complexity*, to which we shall return later.

In general, then, a universal characteristic of a complex system is that the whole is greater than, and often significantly different from, the simple linear sum of its parts. In many instances, the whole seems to take on a life of its own, almost dissociated from the specific characteristics of its individual building blocks. Furthermore, even if we understood how the individual constituents, whether cells, ants, or people, interact with one another, predicting the systemic behavior of the resulting whole is not usually possible. This collective outcome, in which a system manifests significantly different characteristics from those resulting from simply adding up all of the contributions of its individual constituent parts, is called an *emergent behavior*. It is a readily recognizable characteristic of economies, financial markets, urban communities, companies, and organisms.

The important lesson that we learn from these investigations is that in many such systems there is no central control. So, for example, in building an ant colony, no individual ant has any sense of the grand enterprise to which he is contributing. Some ant species even go so far as to use their own bodies as building blocks to construct sophisticated structures: army ants and fire ants assemble themselves into bridges and rafts for use in crossing waterways and overcoming impediments during foraging expeditions. These are examples of what is called *self-organization*. It is an emergent behavior in which the constituents themselves agglomerate to form the emergent whole, as in the formation of human social groups, such as book clubs or political rallies, or your organs, which can be viewed as the self-organization of their constituent cells, or a city as a manifestation of the self-organization of its inhabitants.

Closely related to the concepts of emergence and self-organization is another critical characteristic of many complex systems, namely their ability to adapt and evolve in response to changing external conditions. The quintessential example of such a *complex adaptive system* is, of course, life itself in all of its extraordinary manifestations from cells to cities. The Darwinian theory of natural selection is the scientific narrative that has been developed for understanding and describing how organisms and ecosystems continuously evolve and adapt to changing conditions.

The study of complex systems has taught us to be wary of naively breaking the system down into independently acting component parts. Furthermore, a small perturbation in one part of the system may have giant consequences elsewhere. The system can be prone to sudden and seemingly unpredictable changes—a market crash being a classic example. One or more trends can reinforce other trends in a positive feedback loop until things swiftly spiral out of control and cross a tipping point beyond which behavior radically changes. This was spectacularly manifested by the 2008 meltdown of financial markets across the globe with potentially devastating social and commercial consequences worldwide, stimulated by misconceived dynamics in the parochial and relatively localized U.S. mortgage industry.

It is only over the last thirty years or so that scientists have started to seriously investigate the challenges of understanding complex adaptive systems in their own right and seeking novel ways of addressing them. A natural outcome has been the emergence of an integrated systemic transdisciplinary approach involving a broad spectrum of techniques and concepts derived from diverse areas of science ranging from biology, economics, and physics to computer science, engineering, and the socioeconomic sciences. An important lesson from these investigations is that, while it is not generally possible to make detailed predictions about such systems, it is sometimes possible to derive a coarse-grained quantitative description for the average salient features of the system. For example, although we will never be able to predict precisely when a particular person will die, we ought to be able to predict why the life span of human beings is on the order of one hundred years. Bringing such a quantitative perspective to the challenge of sustainability and the long-term survival of our planet is critical because it inherently recognizes the kinds of interconnectedness and interdependencies so frequently ignored in current approaches.

Scaling up from the small to the large is often accompanied by an evolution from simplicity to complexity while maintaining basic elements or building blocks of the system unchanged or conserved. This is familiar in engineering, economies, companies, cities, organisms, and, perhaps most dramatically, evolutionary processes. For example, a skyscraper in a large city is a significantly more complex object than a modest family dwelling in a small town, but the underlying principles of construction and design, including questions of mechanics, energy and information distribution, the size of electrical outlets, water faucets, telephones, laptops, doors, et cetera, all remain approximately the same independent of the size of the building. These basic building blocks do not significantly change when scaling up from my house to the Empire State Building; they are shared by all of us. Similarly, organisms have evolved to have an enormous range of sizes and an extraordinary diversity of morphologies and interactions, which often reflect increasing complexity, yet fundamental building blocks like cells, mitochondria, capillaries, and even leaves do not appreciably change with body size or increasing complexity of the class of systems in which they are embedded.

7. YOU ARE YOUR NETWORKS: GROWTH FROM CELLS TO WHALES

I began this chapter by pointing out the very surprising and counterintuitive fact that, despite the vagaries and accidents inherent in evolutionary dynamics, almost all of the most fundamental and complex measurable characteristics of organisms scale with size in a remarkably simple and regular fashion. This is explicitly illustrated, for example, in Figure 1, where metabolic rate is plotted against body mass for a sequence of animals.

This systematic regularity follows a precise mathematical formula which, in technical parlance, is expressed by saying that "metabolic rate scales as a *power law* whose *exponent* is very close to the number ¾." I'll explain this in much greater detail later but here I want to give a simple illustration of what it means colloquially. So consider the following: elephants are roughly 10,000 times (*four* orders of magnitude, 10^4) heavier than rats; consequently, they have roughly 10,000 times as many cells. The ¾ power scaling law says that, despite having 10,000 times as many cells to support, the metabolic rate of an elephant (that is, the amount of energy needed to keep it alive) is only 1,000 times (*three* orders of magnitude, 10^3) larger than a rat's; note the ratio of 3:4

in the powers of ten. This represents an extraordinary economy of scale as size increases, implying that the cells of elephants operate at a rate that is about a tenth that of rat cells. Parenthetically, it's worth pointing out that the subsequent decrease in the rates of cellular damage from metabolic processes underlies the greater longevity of elephants and provides the framework for understanding aging and mortality. The scaling law can be expressed in the slightly different way that I used earlier: if an animal is twice the size of another (whether 10 lbs. vs. 5 lbs. or 1,000 lbs. vs. 500 lbs.) we might naively expect metabolic rate to be twice as large, reflecting classic linear thinking. The scaling law, however, is *nonlinear* and says that metabolic rates don't double but, in fact, increase by only about 75 percent, representing a whopping 25 percent savings with every doubling of size.[12]

Notice that the ¾ ratio is just the slope of the graph in Figure 1, where the quantities (metabolic rate and mass) are plotted *logarithmically*—meaning that they increase by factors of ten along both axes. When plotted this way, the slope of the graph is just the exponent of the power law.

This scaling law for metabolic rate, known as Kleiber's law after the biologist who first articulated it, is valid across almost all taxonomic groups, including mammals, birds, fish, crustacea, bacteria, plants, and cells. Even more impressive, however, is that similar scaling laws hold for essentially all physiological quantities and life-history events, including growth rate, heart rate, evolutionary rate, genome length, mitochondrial density, gray matter in the brain, life span, the height of trees and even the number of their leaves. Furthermore, when plotted logarithmically this dizzying array of scaling laws all look like Figure 1 and therefore have the same mathematical structure. They are all "power laws" and are typically governed by an exponent (the slope of the graph), which is a simple multiple of ¼, the classic example being the ¾ for metabolic rate. So, for example, if the size of a mammal is doubled, its heart rate decreases by about 25 percent. The number 4 therefore plays a fundamental and almost magically universal role in all of life.[13]

How do such surprising regularities emerge from the statistical processes and historical contingencies inherent in natural selection? The universality and predominance of ¼ power scaling strongly suggests that natural selection has been constrained by other general physical principles that transcend specific design. Highly complex, self-sustaining structures, whether cells, organisms, ecosystems, cities, or corporations, require the close integration of enormous numbers of their constituent units that need efficient servicing at all scales. This has been accomplished in living systems by evolving fractal-like, hierarchical branching network systems presumed optimized by the continuous "competitive" feedback mechanisms implicit in natural selection. It is the generic physical, geometric, and mathematical properties of these network systems that underlie the origin of these scaling laws, including the prevalence of the one-quarter exponent. As an example, Kleiber's law follows from requiring that the energy needed to pump blood through mammalian circulatory systems, including ours, is minimized so that the energy we devote to reproduction is maximized. Examples of other such networks include the respiratory, renal, neural, and plant and tree vascular systems. These ideas, as well as the concepts of *space filling* (the need to feed all cells in the body) and *fractals* (the geometry of the networks), will be elaborated upon in some detail.

The same underlying principles and properties operate across the networks of mammals, fish, birds, plants, cells, and ecosystems even though they have evolved different designs. When expressed in mathematical language, they lead to the explanation for the origin of universal ¼ power scaling laws but they also predict many quantitative results that capture essential features of these systems, including, for example, the size of the smallest and largest mammal (the shrew and whale), blood flow and pulse rate in any vessel of the circulatory system of any mammal, the height of the tallest tree anywhere in the United States, how long elephants or mice sleep, and the vascular structure of tumors.[14]

They also lead to a theory of *growth*. Growth can be viewed as a special case of a scaling phenomenon. A mature organism is essentially a *nonlinearly* scaled-up version of the infant—just compare the various proportions of your body with those of a baby. Growth at any stage of development is accomplished by apportioning the metabolic energy being delivered through networks to existing cells to the production of new cells that build up new tissue. This process can be analyzed using the network theory to predict a universal quantitative theory of growth curves applicable to any organism, including tumors. A growth curve is simply a graph of the size of the organism plotted as a function of its age. You are probably familiar with growth curves if you have had children, as pediatricians routinely show them to parents so that they can see how their child's development compares with the expectations for the average infant. The growth theory also explains a curious paradoxical phenomenon that you might have pondered, namely, why we eventually stop growing even though we continue to eat. This turns out to be a consequence of the sublinear scaling of metabolic rate and the economies of scale embodied in the network design. In a later chapter, the same paradigm will be applied to the growth of cities, companies, and economies to understand the fundamental question as to the origins of open-ended growth and its possible sustainability.

Because networks determine the rates at which energy and resources are delivered to cells, they set the pace of all physiological processes. Because cells are constrained to operate systematically slower in larger organisms relative to smaller ones, *the pace of life* systematically decreases with increasing size. Thus, large mammals live longer, take longer to mature, have slower heart rates, and cells that don't work as hard as those of small mammals, all to the same predictable degree. Small creatures live life in the fast lane while large ones move ponderously, though more efficiently, through life; think of a scurrying mouse relative to a sauntering elephant.

Having established this way of thinking, the scene will shift to ask how the network and scaling paradigm, successfully established in the biological arena, can be fruitfully applied to ask similar questions about the dynamics, growth, and structure of cities and companies with a view to developing an analogous mechanistic *science of cities and companies.* This will in turn be used as a point of departure for addressing the big questions of global sustainability and the challenge of continuous innovation and the increasing pace of life.

8. CITIES AND GLOBAL SUSTAINABILITY: INNOVATION AND CYCLES OF SINGULARITIES

Scaling as a manifestation of an underlying network theory implies that despite external appearances and habitats, a whale is to a good approximation a scaled-up elephant, an elephant is a scaled-up dog, and a dog is, in turn, a scaled-up mouse, when viewed in terms of their measurable characteristics and traits. At an 80 to 90 percent level they are scaled versions of one another following predictable nonlinear mathematical rules. Put slightly differently, all mammals that have ever existed including you and me are, on average, approximately scaled versions of a single idealized mammal. Could this be true of cities and companies? Is New York a scaled-up San Francisco, which is a scaled-up Boise, which is a scaled-up Santa Fe? Is Tokyo a scaled-up Osaka, which is a scaled-up Kyoto, which is a scaled-up Tsukuba? Even within their own national urban systems all of these cities surely look different from one another, and each has a different history, geography, and culture. However, the same could have been said of whales, horses, dogs, and mice. The only way to answer such questions seriously is to look at data.

Remarkably, analyses of such data show that, as a function of population size, city infrastructure—such as the length of roads, electrical cables, water pipes, and the number of gas stations—scales in the same way whether in the United States, China, Japan, Europe, or Latin America. As in biology, these quantities scale sublinearly with size, indicating a systematic economy of scale but with an exponent of about 0.85 rather than 0.75. So, for example, across the globe, fewer roads and electrical cables are needed per capita the bigger the city. Like organisms, cities are indeed approximately scaled versions of one another, despite their different histories, geographies, and cultures, at least as far as their physical infrastructure is concerned.

Perhaps even more remarkably they are also scaled socioeconomic versions of one another. Socioeconomic quantities such as wages, wealth, patents, AIDS cases, crime, and educational institutions, which have no analog in biology and did not exist on the planet before humans invented cities ten thousand years ago, also scale with population size but with a *superlinear* (meaning bigger than one) exponent of approximately 1.15. An example of this is the number of patents produced in a city shown in Figure 3. Thus, on a per capita basis, all of these quantities systematically *increase* to the same degree as city size increases and, at the same time, there are equivalent savings from economies of scale in all infrastructural quantities. Despite their amazing diversity and complexity across the globe, and despite localized urban planning, cities manifest a surprising coarse-grained simplicity, regularity, and predictability.[15]

To put it in simple terms, scaling implies that if a city is twice the size of another city in the same country (whether 40,000 vs. 20,000 or 4 million vs. 2 million), then its wages, wealth, number of patents, AIDS cases, violent crime, and educational institutions all increase by approximately the same degree (by about 15 percent above mere doubling), with similar savings in all of its infrastructure. The bigger the city, the more the average individual systematically owns, produces, and consumes, whether goods, resources, or ideas. The good, the bad, and the ugly are integrated in an approximately predictable package: a person may move to a bigger city drawn by more innovation, a greater sense of "action," and higher wages, but she can also expect to confront an equivalent increase in the prevalence of crime and disease.

The fact that the same scaling laws are observed for diverse urban metrics in cities and urban systems that evolved independently across the globe strongly suggests that, as in biology, there are underlying generic principles transcending history, geography, and culture and that a fundamental, coarse-grained theory of cities is possible. In chapter 8 I will discuss how the inextricable tension between benefits and costs of social and infrastructural networks has its origins in the underlying universal dynamics of social network structures and group clustering of human interactions. Cities provide a natural mechanism for reaping the benefits of high social connectivity among very different people conceiving and solving problems in a diversity of ways. I will discuss the nature and dynamics of these social network structures and show how scaling laws emerge, including the intriguing link between the 15 percent enhancement of all socioeconomic activities, whether good or bad, and the equivalent 15 percent savings on physical infrastructure.

When humans began forming sizable communities they brought a fundamentally new dynamic to the planet. With the invention of language and the consequent exchange of information in social network space we discovered how to innovate and create wealth and ideas, ultimately manifested in super-linear scaling. In biology, network dynamics constrains the pace of life to decrease systematically with increasing size following the ¼ power scaling laws. In contrast, the dynamics of social networks underlying wealth creation and innovation leads to the opposite behavior, namely, the *systematically increasing pace of life* as city size increases: diseases spread faster, businesses are born and die more often, commerce is transacted more rapidly, and people even walk faster, all following the approximate 15 percent rule. We all sense that life is faster in the big city than in the small town and that it has ubiquitously accelerated during our lifetimes as cities and their economies grew.

Resources and energy are the necessary fuel for growth. In biology, growth is driven by metabolism whose sublinear scaling leads to a predictable, approximately stable size at maturity. Such a behavior would be considered a disaster in traditional economic thinking where healthy economies, whether for cities or nations, are characterized by continuous open-ended exponential expansion of at least a few percent per annum ad infinitum. Just as bounded growth in biology follows from the sublinear scaling of metabolic rate, the superlinear scaling of wealth creation and innovation (as measured by patent production, for example) leads to unbounded, often faster-than-exponential growth consistent with open-ended economies. This is satisfyingly consistent, but there's a big catch, which goes under the forbidding technical name of a *finite time singularity*. In a nutshell, the problem is that the theory also predicts that unbounded growth cannot be sustained without having either infinite resources or inducing major paradigm shifts that "reset" the clock before potential collapse occurs. We have sustained open-ended growth and avoided collapse by invoking continuous cycles of paradigm-shifting innovations such as those associated on the big scale of human history with discoveries of iron, steam, coal, computation, and, most recently, digital information technology. Indeed, the litany of such discoveries both large and small is testament to the extraordinary ingenuity of the collective human mind.

Unfortunately, however, there is another serious catch. Theory dictates that such discoveries must occur at an increasingly accelerating pace; the time between successive innovations must systematically and inextricably get shorter and shorter. For instance, the time between the "Computer Age" and the "Information and Digital Age" was perhaps twenty years, in contrast to the thousands of years between the Stone, Bronze, and Iron ages. If we therefore insist on continuous open-ended growth, not only does the pace of life inevitably quicken, but we must innovate at a faster and faster rate. We are all too familiar with its short-term manifestation in the increasingly faster pace at which new gadgets and models appear. It's as if we are on a succession of accelerating treadmills and have to jump from one to another at an ever-increasing rate. This is clearly not sustainable, potentially leading to the collapse of the entire urbanized socioeconomic fabric. Innovation and wealth creation that fuel social systems, if left unchecked, potentially sow the seeds of their inevitable collapse. Can this be avoided or are we locked into a fascinating experiment in natural selection that is doomed to fail?

9. COMPANIES AND BUSINESSES

It is natural to extend these ideas to ask how they might relate to companies. Could there possibly be a quantitative, predictive *science of companies*? Do companies manifest systematic regularities that transcend their size and business character? For example, in terms of sales and assets, are Walmart and Exxon, whose revenues exceed half a trillion dollars, approximately scaled-up versions of smaller companies with sales of less than $10 million? Amazingly, the answer to this is yes, as can be seen from Figure 4: like organisms and cities, companies also scale as simple power laws. Equally surprising is that they scale sublinearly as functions of their size, rather than superlinearly like socioeconomic metrics in cities. In this sense, companies are much more like organisms than cities. The scaling exponent for companies is around 0.9, to be compared with 0.85 for the infrastructure of cities and 0.75 for organisms. However, there is considerably more variation around precise scaling among companies than for organisms or cities. This is especially so in their early stages of development as they jostle for a place in the market. Nevertheless, the surprising regularity manifested in their average behavior suggests that, despite their broad diversity and apparent individuality, companies grow and function under general constraints and principles that transcend their size and business sector.

For organisms, the sublinear scaling of metabolic rate underlies their cessation of growth and a size at maturity that remains approximately stable until death. A similar life-history trajectory is at work for companies. They grow rapidly in their early years but taper off as they mature and, if they survive, eventually stop growing relative to the GDP. In their youth, many are dominated by a spectrum of innovative ideas as they seek to optimize their place in the market. However, as they grow and become more established, the spectrum of their product space inevitably narrows and, at the same time, they need to build a significant administration and bureaucracy. Relatively quickly, economies of scale and sublinear scaling, reflecting the challenge of efficiently administering a large and complex organization, dominate innovation and ideas encapsulated in superlinear scaling, ultimately leading to stagnation and to mortality. Half of all the companies in any given cohort of U.S. publicly traded companies disappear within ten years, and a scant few make it to fifty, let alone a hundred years.[16]

As they grow companies tend to become more and more unidimensional, driven partially by market forces but also by the inevitable ossification of the top-down administrative and bureaucratic needs perceived as necessary for operating a traditional company in the modern era. Change, adaptation, and reinvention become increasingly difficult to effect, especially as the external socioeconomic clock is continually accelerating and conditions change at a faster and faster rate. Cities, on the other hand, become increasingly multidimensional as they grow in size. Indeed, in stark contrast to almost all companies, the diversity of cities, as measured by the number of different kinds of jobs and businesses that comprise their economic landscape, continually and systematically increases in a predictable way with increasing city size. From this perspective it comes as no surprise that the growth and mortality curves of companies closely resemble the corresponding growth and mortality curves of organisms. Both cases exhibit systematic sublinear scaling, economies of scale, bounded growth, and finite life spans. Furthermore, in both cases, the probability of dying, usually referred to as the mortality rate, which is the rate at which deaths are occurring relative to the number still alive, is the same no matter the age of either the animal or the company. Publicly traded companies die through acquisitions, mergers, and bankruptcies at the same rate regardless of how well established they are or what they actually do. The mechanistic basis for understanding the growth, mortality, and organizational dynamics of companies, comparing and contrasting them with the growth and mortality of organisms and the unbounded growth and apparent "immortality" of cities, will be discussed in greater detail in chapter 9.

2

THE MEASURE OF ALL THINGS

An Introduction to Scaling

Before turning to the many issues and questions raised in the opening chapter I want to devote this chapter to a broad introduction to some of the basic concepts used throughout the rest of the book. Although some readers may well be familiar with some of this material, I want to make sure we're all on the same page.

The overview is primarily from a historical perspective, beginning with Galileo explaining why there can't be giant insects and ending with Lord Rayleigh explaining why the sky is blue. In between, I will touch upon Superman, LSD and drug dosages, body mass indices, ship disasters and the origin of modeling theory, and how all of these are related to the origins and nature of innovation and limits to growth. Above all, I want to use these examples to convey the conceptual power of thinking quantitatively in terms of *scale*.

1. FROM GODZILLA TO GALILEO

From time to time, like many scientists I receive requests from journalists asking for an interview, usually about some question or problem related to cities, urbanization, the environment, sustainability, complexity, the Santa Fe

Institute, or occasionally even about the Higgs particle. Imagine my surprise, then, when I was contacted by a journalist from the magazine *Popular Mechanics* informing me that Hollywood was going to release a new blockbuster version of the classic Japanese film *Godzilla* and that she was interested in getting my views on it. You may recall that Godzilla is an enormous monster that mostly roams around cities (Tokyo, in the original 1954 version) causing destruction and havoc while terrorizing the populace.

The journalist had heard that I knew something about scaling and wanted to know "in a fun, goofy, nerdy sort of way, about the biology of Godzilla (to tie in with the release of the new movie) . . . how fast such a large animal would walk . . . how much energy his metabolism would generate, how much he would weigh, etc." Naturally, this new twenty-first-century all-American Godzilla was the biggest incarnation of the character yet, reaching a lofty height of 350 feet (106 meters), more than twice the height of the original Japanese version, which was "only" 164 feet (50 meters). I immediately responded by telling the journalist that almost any scientist she contacted would tell her that no such beast as Godzilla could actually exist because, if it were made of pretty much the same basic stuff as we are (meaning all of life), it could not function because it would collapse under its own weight.

The scientific argument upon which this is based was articulated more than four hundred years ago by Galileo at the dawn of modern science. It is in its very essence an elegant scaling argument: Galileo asked what happens if you try to indefinitely scale up an animal, a tree, or a building, and with his response discovered that there are limits to growth. His argument set the basic template for all subsequent scaling arguments right up to the present day.

For good reason, Galileo is often referred to as the "Father of Modern Science" for his many seminal contributions to physics, mathematics, astronomy, and philosophy. He is perhaps best known for his mythical experiments dropping objects of different sizes and compositions from the top of the Leaning Tower of Pisa to show that they all reached the ground at the same time. This nonintuitive observation contradicted the accepted Aristotelian dogma that heavy objects fall faster than lighter ones in direct proportion to their weight, a fundamental misconception that was universally believed for almost two

GALILEVS
GALILEVS
MATHVS:

Galileo at ages thirty-five and sixty-nine; he died less than ten years later. Aging and impending mortality, vividly exhibited in these portraits, will be discussed in some detail in chapter 4.

thousand years before Galileo actually tested it. It is amazing in retrospect that until Galileo's investigations no one seems to have thought of, let alone bothered, testing this apparently "self-evident fact."

Galileo's experiment revolutionized our fundamental understanding of motion and dynamics and paved the way for Newton to propose his famous laws of motion. These laws led to a precise quantitative mathematical, predictive framework for understanding *all* motion whether here on Earth or across the universe, thereby uniting the heavens and Earth under the same natural laws. This not only redefined man's place in the universe, but provided the gold standard for all subsequent science, including setting the stage for the coming of the age of enlightenment and the technological revolution of the past two hundred years.

Galileo is also famous for perfecting the telescope and discovering the moons of Jupiter, which convinced him of the Copernican view of the solar system. By continuing to insist on a heliocentric view derived from his observations, Galileo was to end up paying a heavy price. At the age of sixty-nine and in poor health, he was brought before the Inquisition and found guilty of heresy. He was forced to recant and after a brief imprisonment spent the rest of his life under house arrest (nine more years during which he went blind). His books were banned and put on the Vatican's infamous *Index Librorum Prohibitorum*. It wasn't until 1835, more than two hundred years later, that his works were finally dropped from the Index, and until 1992—almost four hundred years later—and for Pope John Paul II to publicly express regret for how Galileo had been treated. It is sobering to realize that words written long ago in Hebrew, Greek, and Latin, based on opinion, intuition, and prejudice, can so overwhelmingly outweigh scientific observational evidence and the logic and language of mathematics. Sad to say, we are hardly free from such misguided thinking today.

Despite the terrible tragedy that befell Galileo, humanity reaped a wonderful benefit from his incarceration. It may very well have happened anyway, but it was while he was under house arrest that he wrote what is perhaps his finest work, one of the truly great books in the scientific literature, titled *Discourses and Mathematical Demonstrations Relating to Two New Sciences.*[1] The book is basically his legacy from the preceding forty years during which he grappled with how to systematically address the challenge of understanding the natural world around us in a logical, rational framework. As such, it laid the groundwork for the equally monumental contribution of Isaac Newton and pretty much for all of the science that followed. Indeed, in praising the book, Einstein was not exaggerating when he called Galileo "the Father of Modern Science."[2]

It's a great book. Despite its forbidding title and somewhat archaic language and style, it's surprisingly readable and a lot of fun. It is written in the style of a "discourse" between three men (Simplicio, Sagredo, and Salviati) who meet over four days to discuss and debate the various questions, big and small, that Galileo is seeking to answer. Simplicio represents the "ordinary" layperson who is curious about the world and asks a series of apparently naive questions. Salviati is the smart fellow (Galileo!) with all of the answers, which are presented in a compelling but patient manner, while Sagredo is the middleman who alternates between challenging Salviati and encouraging Simplicio.

On the second day of their discourse they turn their attention to what appears to be a somewhat arcane discussion on the strength of ropes and beams, and just as you are wondering where this somewhat tedious, detailed discussion is going, the fog clears, the lights flash, and Salviati makes the following pronouncement:

> *From what has already been demonstrated, you can plainly see the impossibility of increasing the size of structures to vast dimensions either in art or in nature; likewise the impossibility of building ships, palaces, or temples of enormous size in such a way that their oars, yards, beams, iron-bolts, and, in short, all their other parts will hold together; nor can nature produce trees of extraordinary size because the branches would break down under their own weight; so also it would be impossible to build up the bony structures of men, horses, or other animals so as to hold together and perform their normal functions if these animals were to be increased enormously in height . . . for if his height be increased inordinately he will fall and be crushed under his own weight.*

There it is: our paranoid fantasies of giant ants, beetles, spiders, or for that matter, Godzillas, so graphically displayed by the comic and film industries, had already been conjectured nearly four hundred years ago by Galileo, who then brilliantly demonstrates that they are a physical impossibility. Or, more precisely, that there are fundamental constraints as to how large they can actually get. So many such science-fiction images are indeed just that: fiction.

Galileo's argument is elegant and simple, yet it has profound implications. Furthermore, it provides an excellent introduction to many of the concepts we'll be investigating in the following chapters. It consists of two parts: a geometrical argument showing how areas and volumes associated with any object scale as its size increases (Figure 5) and a structural argument showing that the strength of pillars holding up buildings, limbs supporting animals, or trunks supporting trees is proportional to their cross-sectional areas (Figure 6).

In the accompanying box, I present a nontechnical version of the first of these, showing that if the shape of an object is kept fixed, then when it is scaled up, all of its areas increase as the *square* of its lengths while all of its volumes increase as the *cube*.

GALILEO'S ARGUMENT ON HOW AREAS AND VOLUMES SCALE

To begin, consider one of the simplest possible geometrical objects, namely, a floor tile in the shape of a square, and imagine scaling it up to a larger size; see Figure 5. To be specific let's take the length of its sides to be 1 ft. so that its area, obtained by multiplying the length of two adjacent sides together, is 1 ft. × 1 ft. = 1 sq. ft. Now, suppose we double the length of all of its sides from 1 to 2 ft., then its area increases to 2 ft. × 2 ft. = 4 sq. ft. Similarly, if we were to triple the lengths to 3 ft., then its area would increase to 9 sq. ft., and so on. The generalization is clear: *the area increases with the square of the lengths*.

This relationship remains valid for *any* two-dimensional geometric shape, and not just for squares, provided that the shape is kept fixed when all of its linear dimensions are increased by the same factor. A simple example is a circle: if its radius is doubled, for instance, then its area increases by a factor of 2 × 2 = 4. A more general example is that of doubling the dimensions of every length in your house while keeping its

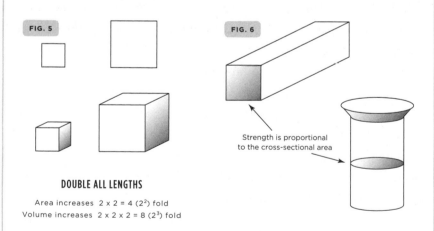

FIG. 5

DOUBLE ALL LENGTHS

Area increases 2 x 2 = 4 (2^2) fold
Volume increases 2 x 2 x 2 = 8 (2^3) fold

FIG. 6

Strength is proportional to the cross-sectional area

(5) Illustration of how areas and volumes scale for the simple case of squares and cubes. (6) The strength of a beam or limb is proportional to its cross-sectional area.

shape and structural layout the same, in which case the area of all of its surfaces, such as its walls and floors, would increase by a factor of four.

This argument can be straightforwardly extended from areas to volumes. Consider first a simple cube: if the lengths of its sides are increased by a factor of two from, say, 1 ft. to 2 ft., then its volume increases from 1 cubic foot to $2 \times 2 \times 2 = 8$ cubic. Similarly, if the lengths are increased by a factor of three, the volume increases by a factor of $3 \times 3 \times 3 = 27$. As with areas, this can straightforwardly be generalized to any object, regardless of its shape, provided we keep it fixed, to conclude that if we scale it up, *its volume increases with the cube of its linear dimensions.*

Thus, when an object is scaled up in size, its volumes increase at a much faster rate than its areas. Let me give a simple example: if you double the dimensions of every length in your house keeping its shape the same, then its volume increases by a factor of $2^3 = 8$ while its floor area increases by only a factor of $2^2 = 4$. To take a much more extreme case, suppose all of its linear dimensions were increased by a factor of 10, then all surface areas such as floors, walls, and ceilings would increase by a factor of $10 \times 10 = 100$ (that is, a hundredfold), whereas the volumes of its rooms would increase by the much larger factor of $10 \times 10 \times 10 = 1,000$ (a thousandfold).

This has huge implications for the design and functionality of much of the world around us, whether it's the buildings we live and work in or the structure of the animals and plants of the natural world. For instance, most heating, cooling, and lighting is proportional to the corresponding surface areas of the heaters, air conditioners, and windows. Their effectiveness therefore increases much more slowly than the volume of living space needed to be heated, cooled, or lit, so these need to be disproportionately increased in size when a building is scaled up. Similarly, for large animals, the need to dissipate heat generated by their metabolism and physical activity can become problematic because the surface area through which it is dissipated is proportionately much smaller relative to their volume than for smaller ones. Elephants, for example, have solved this challenge by evolving disproportionately large ears to significantly increase their surface area so as to dissipate more heat.

This essential difference in the way areas and volumes scale was very likely appreciated by many people before Galileo. His additional new insight was to combine this geometric realization with his realization that the strength of pillars, beams, and limbs is determined by the size of their cross-sectional areas and not by how long they are. Thus a post whose rectangular cross-section is 2 inches by 4 inches (= 8 sq. in.) can support *four* times the weight of a similar post of the same material whose cross-sectional dimensions are only half as big, namely 1 inch by 2 inches (= 2 sq. in.), regardless of the length of *either* post. The first could be 4 feet long and the second 7 feet, it doesn't matter. That's why builders, architects, and engineers involved in construction classify wood by its cross-sectional dimensions, and why lumber yards at Home Depot and Lowe's display them as "two-by-twos, two-by-fours, four-by-fours," and so on.

Now, as we scale up a building or an animal, their weights increase in direct proportion to their volumes provided, of course, that the materials they're made of don't change so that their densities remain the same: so doubling the volume doubles the weight. Thus, the weight being supported by a pillar or a limb increases much faster than the corresponding increase in strength, because weight (like volume) scales as the cube of the linear dimensions whereas strength increases only as the square. To emphasize this point, consider increasing the height of a building or tree by a factor of 10 keeping its shape the same; then the weight needed to be supported increases a *thousand*fold (10^3) whereas the strength of the pillar or trunk holding it up increases by only a *hundred*fold (10^2). Thus, the ability to safely support the additional weight is only a tenth of what it had previously been. Consequently, if the size of the structure, whatever it is, is arbitrarily increased it will eventually collapse under its own weight. There are limits to size and growth.

To put it slightly differently: *relative* strength becomes progressively weaker as size *increases*. Or, as Galileo so graphically put it: "the smaller the body the greater its relative strength. Thus a small dog could probably carry on his back two or three dogs of his own size; but I believe that a horse could not carry even one of his own size."

2. MISLEADING CONCLUSIONS AND
MISCONCEPTIONS OF SCALE: SUPERMAN

Superman made his earthly debut in 1938 and still remains one of the great icons of the sci-fi/fantasy world. I have reproduced the first page of the original *Superman* comic from 1938 in which his origins are explained.[3] He had arrived as a baby from the planet Krypton "whose inhabitants' physical structure was millions of years advanced of our own. Upon reaching maturity the people of his race became gifted with titanic strength." Indeed, upon maturity Superman "could easily leap ⅛th of a mile; hurdle a twenty-story building . . . raise tremendous weights . . . run faster than an express train . . ." all triumphantly summed up in the famous introduction to the radio serials and subsequent TV series and films: "Faster than a speeding bullet. More powerful than a locomotive. Able to leap tall buildings in a single bound. . . . It's Superman."

All of which may well be true. However, in the last frame of that first page there is another bold pronouncement, so important that it warranted being put in capital letters: A SCIENTIFIC EXPLANATION OF CLARK KENT'S AMAZING STRENGTH . . . incredible? *No!* For even today on our world exist creatures with *super-strength!*" To support this, two examples are given: "The lowly ant can support weights hundreds of times its own" and "the grasshopper leaps what to man would be the space of several city blocks."

As persuasive as these examples might appear to be, they represent a classic case of misconceived and misleading conclusions drawn from correct facts. Ants appear to be, at least superficially, much stronger than human beings. However, as we have learned from Galileo, *relative* strength systematically *increases* as size decreases. Consequently, scaling down from a dog to an ant following the simple rules of how strength scales with size will show that if "a small dog could probably carry on his back two or three dogs of his own size," then an ant can carry on his back a hundred ants of his size. Furthermore, because we are about 10 million times heavier than an average ant, the same argument shows that we are capable of carrying only about one other person on ours. Thus, ants have, in fact, the correct strength appropriate for a

The origin myth of Superman and an explanation for his superstrength;
from the opening page of the first Superman comic book in 1938.

creature of their size, just as we do, so there's nothing extraordinary or surprising about their lifting one hundred times their own weight.

The misconception arises because of the natural propensity to think *linearly*, as encapsulated in the implicit presumption that doubling the size of an
animal leads to a doubling of its strength. If this were so, then we would be

10 million times stronger than ants and be able to lift about a ton, corresponding to our being able to lift more than ten other people, just like Superman.

3. ORDERS OF MAGNITUDE, LOGARITHMS, EARTHQUAKES, AND THE RICHTER SCALE

We just saw that if the lengths of an object are increased by a factor of 10 without changing its shape or composition, then its areas (and therefore strengths) increase by a factor of 100, and its volumes (and therefore weights) by a factor of 1,000. Successive powers of ten, such as these, are called *orders of magnitude* and are typically expressed in a convenient shorthand notation as 10^1, 10^2, 10^3, et cetera, with the exponent—the little superscript on the ten—denoting the number of zeros following the one. Thus, 10^6 is shorthand for a million, or 6 orders of magnitude, because it is 1 followed by 6 zeros: 1,000,000.

In this language Galileo's result can be expressed as saying that for every order of magnitude increase in length, areas and strengths increase by *two* orders of magnitude, whereas volumes and weights increase by *three* orders of magnitude. From which it follows that for a single order of magnitude increase in area, volumes increase by 3/2 (that is, one and a half) orders of magnitude. A similar relationship therefore holds between strength and weight: for every order of magnitude increase in strength, the weight that can be supported increases by one and a half orders of magnitude. Conversely, if the weight is increased by a single order of magnitude, then the strength only increases by ⅔ of an order of magnitude. This is the essential manifestation of a *nonlinear* relationship. A *linear* relationship would have meant that for every order of magnitude increase in area, the volume would have also increased by one order of magnitude.

Even though many of us may not be aware of it, we have all been exposed to the concept of orders of magnitude, including fractions of orders of magnitude, through the reporting of earthquakes in the media. Not infrequently we hear news announcements along the lines that "there was a moderate-size earthquake today in Los Angeles that measured 5.7 on the Richter scale which shook many buildings but caused only minor damage." And occasionally we hear of earthquakes such as the one in the Northridge region of Los Angeles in 1994, which was only a single unit larger on the Richter scale but caused enormous amounts of damage. The Northridge earthquake, whose magnitude was 6.7, caused more than $20 billion worth of damage including sixty fatalities, making it one of the costliest natural disasters in U.S. history, whereas a 5.7 earthquake caused only negligible damage. The reason for this vast difference in impact despite an apparently small increase in magnitude is that the Richter scale expresses size in terms of *orders of magnitude*.

So an increase of one unit actually means an increase of one order of magnitude, so that a 6.7 earthquake is actually 10 times the size of a 5.7 earthquake. Likewise, a 7.7 earthquake, such as the Sumatra one of 2010, is 10 times bigger than the Northridge quake and 100 times bigger than a 5.7 earthquake. The Sumatra earthquake was in a relatively unpopulated area but still caused widespread destruction via a tsunami that displaced more than twenty thousand people and killed almost five hundred. Sadly, five years earlier, Sumatra had suffered an even more destructive earthquake whose magnitude was 8.7 and therefore yet another 10 times larger. Obviously, in addition to its size, the destruction wreaked by an earthquake depends a great deal on the local conditions such as population size and density, the robustness of buildings and infrastructure, and so on. The 1994 Northridge earthquake and the more recent 2011 Fukushima one, both of which caused huge amounts of damage, were "only" 6.7 and 6.6, respectively.

The Richter scale actually measures the "shaking" amplitude of the earthquake recorded on a seismometer. The corresponding amount of energy released scales nonlinearly with this amplitude in such a way that for every order of magnitude increase in the measured amplitude the energy released increases by one and a half (that is $\frac{3}{2}$) orders of magnitude. This means that a difference of *two orders* of magnitude in the amplitude, that is a change of 2.0 on the Richter scale, is equivalent to a factor of *three* orders of magnitude (1,000) in the energy released, while a change of just 1.0 is equivalent to a factor of the square root of $1,000 = 31.6$.[4]

Just to give some idea of the enormous amounts of energy involved in earthquakes, here are some numbers to peruse: the energy released by the detonation of a pound (or half a kilogram) of TNT corresponds roughly to a magnitude of 1 on the Richter scale; a magnitude of 3 corresponds to about 1,000 pounds (or about 500 kg) of TNT, which was roughly the size of the 1995 Oklahoma City bombing; 5.7 corresponds to about 5,000 tons, 6.7 to about 170,000 tons (the Northridge and Fukushima earthquakes), 7.7 to about 5.4 million tons (the 2010 Sumatra earthquake), and 8.7 to about 170 million tons (the 2005 Sumatra earthquake). The most powerful earthquake ever recorded was the Great Chilean Earthquake of 1960 in Valdivia, which registered 9.5, corresponding to 2,700 million tons of TNT, almost a thousand times larger than Northridge or Fukushima.

For comparison, the atomic bomb ("Little Boy") that was dropped on Hiroshima in 1945 released the energy equivalent of about 15,000 tons of TNT. A typical hydrogen bomb releases well over 1,000 times more, corresponding to a major earthquake of magnitude 8. These are enormous amounts of energy when you realize that 170 million tons of TNT, the size of the 2005 Sumatra earthquake, can fuel a city of 15 million people, equivalent to the entire New York City metropolitan area, for an entire year.

This kind of scale where instead of increasing linearly as in 1, 2, 3, 4, 5 . . . we increase by factors of 10 as in the Richter scale: 10^1, 10^2, 10^3, 10^4, 10^5 . . . is called *logarithmic*. Notice that it's actually linear in terms of the numbers of orders of magnitude, as indicated by the exponents (the superscripts) on the tens. Among its many attributes, a logarithmic scale allows one to plot quantities that differ by huge factors on the same axis, such as those between the magnitudes of the Valdivia earthquake, the Northridge earthquake, and a stick of dynamite, which overall cover a range of more than a billion (10^9). This would be impossible if a linear plot was used because almost all of the events would pile up at the lower end of the graph. To include all earthquakes, which range over five or six orders of magnitude, on a linear plot would require a piece of paper several miles long—hence the invention of the Richter scale.

Because it conveniently allows quantities that vary over a vast range to be represented on a line on a single page of paper such as this, the logarithmic technique is ubiquitously used across all areas of science. The brightness of stars, the acidity of chemical solutions (their pH), physiological characteristics of animals, and the GDPs of countries are all examples where this technique is commonly utilized to cover the entire spectrum of the variation of the quantity being investigated. The graphs shown in Figures 1–4 in the opening chapter are plotted this way.

4. PUMPING IRON AND TESTING GALILEO

An essential component of science that often distinguishes it from other intellectual pursuits is its insistence that hypothesized claims be verified by experiment and observation. This is highly nontrivial, as evidenced by the fact that it took more than two thousand years before Aristotle's pronouncement that objects falling under gravity do so at a rate proportional to their weight to actually be tested—and when carried out, to be found wanting. Sadly, many of our present-day dogmas and beliefs, especially in the nonscientific realm, remain untested yet rigidly adhered to without any serious attempt to verify them—sometimes with unfortunate or even devastating consequences.

So following our detour into powers of ten, I want to use what we've learned about orders of magnitude and logarithms to address the issue of checking Galileo's predictions about how strength should scale with weight. Can we show that in the "real world" strength really does increase with weight according to the rule that it should do so in the ratio of two to three when expressed in terms of orders of magnitude?

In 1956, the chemist M. H. Lietzke devised a simple and elegant confirmation of Galileo's prediction. He realized that competitive weight lifting across different weight classes provides us with a data set of how maximal strength scales with body size, at least among human beings. All champion weight lifters try to maximize how big a load they can lift, and to accomplish this they have all trained with pretty much the same intensity and to the same degree, so if we compare their strengths we do so under approximately similar conditions. Furthermore, championships are decided by three different kinds of lifts—the press, the snatch, and the clean and jerk—so taking the total of these effectively averages over individual variation of specific talents. These totals are therefore a good measure of maximal strength.

Using the totals of these three lifts from the weight lifting competition in the 1956 Olympic Games, Lietzke brilliantly confirmed the ⅔ prediction for how strength should scale with body weight. The totals for the individual gold medal winners were plotted *logarithmically* versus their body weight, where each axis represented increases by factors of ten. If strength, which is plotted along the *vertical* axis, increases by *two* orders of magnitude for every *three* orders of magnitude increase in body weight, which is plotted along the *horizontal* axis, then the data should exhibit a straight line whose slope is ⅔. The measured value found by Lietzke was 0.675, very close to the prediction of ⅔ = 0.667. His graph is reproduced in Figure 7.[5]

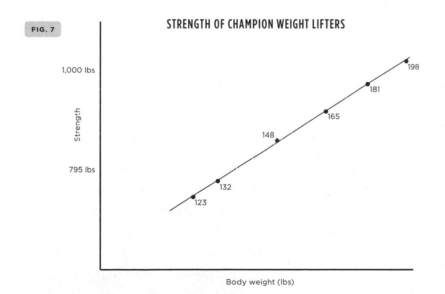

FIG. 7

STRENGTH OF CHAMPION WEIGHT LIFTERS

1,000 lbs

Strength

795 lbs

198

181

165

148

132

123

Body weight (lbs)

Total weights lifted by champion weight lifters in the 1956 Olympic Games plotted logarithmically versus their body weights, confirming the ⅔ prediction for the slope. Who was the strongest and who was the weakest?

5. INDIVIDUAL PERFORMANCE AND DEVIATIONS FROM
SCALING: THE STRONGEST MAN IN THE WORLD

The regularity exhibited by the weight lifting data and its close agreement with the ⅔ prediction for the scaling of strength might seem surprising given the simplicity of the scaling argument. After all, each of us has a slightly different shape, different body characteristics, different history, slightly different genetics, and so on, none of which is accounted for in the derivation of the ⅔ prediction. Using the total sum of the weights lifted by champions who have trained to approximately the same degree helps to average over some of these individual differences. On the other hand, all of us are to a good approximation made of pretty much the same stuff with very similar physiologies. We function very similarly and, as manifested in Figure 7, approximate scaled versions of one another at least as far as strength is concerned. Indeed, by the end of the book, I hope to convince you that this broad similarity extends to almost every aspect of your physiology and life history. So much so in fact that when I talk about "we" being approximately scaled versions of one another, I will mean not just all human beings, but all mammals and, to varying degrees, all of life.

Another way of viewing these scaling laws is that they provide an idealized baseline that captures the dominant, essential features that unite us not only as human beings but also as variations on being an organism and an expression of life. Each individual, each species, and even each taxonomic group deviates to varying degrees from the idealized norms manifested by the scaling laws, with deviations reflecting the specific characteristics that represent individuality.

Let me illustrate this using the weight lifting example. If you look carefully at the graph of Figure 7 you can clearly see that four of the points lie almost exactly on the line, indicating that these weight lifters are lifting almost precisely what they should for their body weights. However, notice that the remaining two, the heavyweight and the middleweight, both lie just a little off the line, one below and one above. Thus, the heavyweight, even though he has lifted more than anyone else, is actually *under*performing relative to what he should be lifting given his weight, whereas the middleweight is *over*performing relative to his weight. In other words, from the egalitarian level playing field perspective of a physicist, the strongest man in the world in 1956 was actually the middleweight champion because he was overperforming relative to his weight. Ironically, the weakest of all of the champions from this scientific scaling perspective is the heavyweight, despite the fact that he lifted more than anyone else.

6. MORE MISLEADING CONCLUSIONS AND MISCONCEPTIONS OF SCALE: DRUG DOSAGES FROM LSD AND ELEPHANTS TO TYLENOL AND BABIES

The role of size and scale pervades medicine and health even though the ideas and conceptual framework inherent in scaling laws are not explicitly integrated into the biomedical professions. For example, we are all very familiar with the idea that there are standard charts showing how height, growth rate, food intake, and even the circumference of our waists should correlate with our weight, or how these metrics should change during our early development. These charts are none other than representations of scaling laws that have been deemed applicable to the "average healthy human being." Indeed, doctors are trained to recognize how such variables should on average be correlated with the weight and age of their patients.

Also well known is the related concept of an *invariant* quantity, such as our pulse rate or body temperature, which does not change systematically with the weight or height of the average healthy individual. Indeed, substantial deviations from these invariant averages are routinely used in the diagnosis of disease or ill health. Having a body temperature of 101°F or a blood pressure of 275/154 is a signal that something's wrong. These days, a standard physical examination results in a plethora of such metrics that your doctor uses to assess the state of your health. A major challenge of the medical and health industry is to ascertain the quantifiable baseline scale of life, and consequently the extended suite of metrics for the average, healthy human being, including how large a variation or deviation from them can be tolerated.

Not surprisingly, then, many critical problems in medicine can be addressed in terms of scaling. In later chapters several important health issues that concern us all, ranging from aging and mortality to sleep and cancer, will be addressed using this framework. Here, however, I first want to whet the appetite by considering some equally important medical issues that involve ideas stemming from Galileo's insight into the tension between the way in which areas and volumes scale. These will reveal how easy it is to form misconceptions that lead to seriously misleading conclusions arising from unconsciously using linear extrapolation.

In the development of new drugs and in the investigation of many diseases, much of the research is conducted on so-called model animals, typically standard cohorts of mice that have been bred and refined precisely for research purposes. Of fundamental importance to medical and pharmaceutical research is how results from such studies should be scaled up to human beings in order to prescribe safe and effective dosages or draw conclusions regarding diagnoses and therapeutic procedures. A comprehensive theory of how this can be accomplished has not yet been developed, although the pharmaceutical industry devotes enormous resources to addressing it when developing new drugs.

A classic example of some of the challenges and pitfalls is an early study investigating the potentially therapeutic effects of LSD on humans. Although the term "psychedelic" was already coined by 1957, the drug was almost unknown outside of a specialized psychiatric community in 1962, when the psychiatrist Louis West (no relation), together with Chester Pierce at the University of Oklahoma and Warren Thomas, a zoologist at the Oklahoma City zoo, proposed investigating its effects on elephants.

Elephants? Yes, elephants and, in particular, Asiatic elephants. Although it may sound somewhat eccentric to use elephants rather than mice as the "model" for studying the effects of LSD, there were some not entirely implausible reasons for doing so. It so happens that Asiatic elephants periodically undergo an unpredictable transition from their normal placid obedient state to one in which they become highly aggressive and even dangerous for periods of up to two weeks. West and his collaborators speculated that this bizarre and often destructive behavior, known as *musth*, was triggered by the autoproduction of LSD in elephants' brains. So the idea was to see if LSD would induce this curious condition and, if so, thereby gain insight into LSD's effects on humans from studying how they react. Pretty weird, but maybe not entirely unreasonable.

However, this immediately raises an intriguing question: how much LSD should you give an elephant?

At that time little was known about safe dosages of LSD. Although it had not yet entered the popular culture it was known that even dosages of less than a quarter milligram would induce a typical "acid trip" for a human being and that a safe dose for cats was about one tenth of a milligram per kilogram of body weight. The investigators chose to use this latter number to estimate how much LSD they should give to Tusko the elephant, their unsuspecting subject that resided at the Lincoln Park Zoo in Oklahoma City.

Tusko weighed about 3,000 kilograms, so using the number known to be safe for cats, they estimated that a safe and appropriate dose for Tusko would be about 0.1 milligram per kilogram multiplied by 3,000 kilograms, which comes out to 300 milligrams of LSD. The amount they actually injected was 297 milligrams. Recall that a good hit of LSD for you and me is less than a quarter milligram. The results on Tusko were dramatic and catastrophic. To quote directly from their paper: "Five minutes after the injection he [the elephant] trumpeted, collapsed, fell heavily onto his right side, defecated, and went into status epilepticus." Poor old Tusko died an hour and forty minutes later. Perhaps almost as disturbing as this awful outcome was that the investigators concluded that elephants are "proportionally very sensitive to LSD."

The problem, of course, is something we've already stressed several times, namely the seductive trap of linear thinking. The calculation of how big a dosage should be used on Tusko was based on the implicit assumption that effective and safe dosages scale *linearly* with body weight so that the dosage per kilogram of body weight was presumed to be the same for all mammals. The 0.1 milligram per kilogram of body weight obtained from cats was therefore naively multiplied by Tusko's weight, resulting in the outlandish estimate of 297 milligrams, with disastrous consequences.

Exactly how doses should be scaled from one animal to another is still an open question that depends, to varying degrees, on the detailed properties of the drug and the medical condition being addressed. However, regardless of details, an understanding of the underlying mechanism by which drugs are transported and absorbed into specific organs and tissues needs to be considered in order to obtain a credible estimate. Among the many factors involved, metabolic rate plays an important role. Drugs, like metabolites and oxygen, are typically transported across surface membranes, sometimes via diffusion and sometimes through network systems. As a result, the dose-determining factor is to a significant degree constrained by the scaling of surface areas rather than the total volume or weight of an organism, and these scale *nonlinearly* with weight. A simple calculation using the ⅔ scaling rule for areas as a function of weight shows that a more appropriate dose for elephants should be closer to a few milligrams of LSD rather than the several hundred that were actually administered. Had this been done, Tusko would no doubt have lived and a vastly different conclusion about the effects of LSD would have been drawn.

The lesson is clear: the scaling of drug dosages is nontrivial, and a naive approach can lead to unfortunate results and mistaken conclusions if not done correctly with due attention being paid to the underlying mechanism of drug transport and absorption. It is clearly an issue of enormous importance, sometimes even a matter of life and death. This is one of the main reasons it takes so long for new drugs to obtain approval for their general use.

Lest you think this was some fringe piece of research, the paper on elephants and LSD was published in one of the world's most highly regarded and prestigious journals, namely *Science*.[6]

Many of us are very familiar with the problem of how drug doses should be scaled with body weight from having dealt with children with fevers, colds, earaches, and other such vagaries of parenting. I recall many years ago being quite surprised when trying to console a screaming infant struggling in the middle of the night with a high fever to discover that the recommended dose of baby Tylenol, printed on the label of the bottle, scaled *linearly* with body weight. Being familiar with the tragic story of Tusko I felt a certain degree of concern. The label had a small chart on it showing how big a dose should be given to a baby of a given age and weight. For example, for a 6-pound baby, the recommended dose was ¼ teaspoon (40 mg) whereas for a baby of 36 pounds (six times heavier) the dose was 1½ teaspoons (240 mg), exactly six times larger. However, if the nonlinear ⅔ power scaling law is followed, the dosage should only have been increased by a factor of $6^{\frac{2}{3}} \approx 3.3$, corresponding to 132 milligrams, which is just over *half* of the recommended dose! So if the ¼ teaspoon recommended for the 6-pound baby is correct, then the dose of 1½ teaspoons recommended for the 36-pound baby was almost twice too large.

Hopefully this has not been putting children at risk, but I have noticed in more recent years that such a chart no longer appears on the bottle or on the pharmaceutical company Web site. However, the Web site still shows such a chart indicating linear scaling for the recommended dosages for infants from 36 to 72 pounds, although they now wisely recommend that a physician be consulted for babies less than 36 pounds (less than two years old). Nevertheless, other reputable Web sites still recommend linear scaling for babies younger than this.[7]

7. BMI, QUETELET, THE AVERAGE MAN, AND SOCIAL PHYSICS

Another important medical issue related to scale is the use of the *body-mass index* (BMI) as a proxy for body fat content and, by extrapolation, as an important metric of health. This has become very topical in recent years because of its ubiquitous use in the diagnosis of obesity and its association with many deleterious health issues including hypertension, diabetes, and heart disease. Although introduced more than 150 years ago by the Belgian mathematician Adolphe Quetelet as a simple means of classifying sedentary individuals, BMI has taken on a powerful authority among physicians and the general public despite its somewhat murky theoretical underpinnings.

Until the 1970s and the rise of its popularity, the BMI was actually known as the Quetelet index. Although trained as a mathematician, Quetelet was a classic polymath who contributed to a wide range of scientific disciplines including meteorology, astronomy, mathematics, statistics, demography, sociology, and criminology. His major legacy is the BMI, but this was just a very small part of his passion for bringing serious statistical analysis and quantitative thinking to problems of societal interest.

Quetelet's goal was to understand the statistical laws underlying social phenomena such as crime, marriage, and suicide rates and to explore their interrelationships. His most influential book, published in 1835, was *On Man and the Development of His Faculties, or Essays on Social Physics.* The title was shortened for the English translation to the much more grandiose-sounding *Treatise on Man.* In the book, he introduces the term *social physics* and describes his concept of the "average man" (*l'homme moyen*). This concept is very much in the spirit of our earlier discussion concerning Galileo's argument on how the strength of the mythical "average person" scales with his or her weight and height, or the idea that there are meaningful average baseline values for physiological characteristics such as our body temperature and blood pressure.

The "average man" (and woman!) is characterized by the values of the various measurable physiological and social metrics averaged over a sufficiently large population sample. These include everything from heights and life spans to the numbers of marriages, the amount of alcohol consumed, and the rates of disease. However, Quetelet brought something new and important to these analyses, namely the statistical variation of these quantities around their mean values, including estimates of their associated probability distributions. He found, though sometimes he merely assumed, that these variances mostly followed a so-called normal, or Gaussian, distribution, popularly known as the *bell curve.* Thus, in addition to measuring average values of these various quantities, he analyzed the distribution of how much they varied around that mean. So health, for example, would be defined not just as having specific values of these metrics (such as having a body temperature of 98.6°F) but that these should fall within well-defined bounds determined by the variation from the mean value for healthy individuals in the entire population.

Quetelet's ideas, and his use of the term *social physics,* were somewhat controversial at the time because they were interpreted as implying a deterministic framework for social phenomena and therefore contradicted the concepts of free will and freedom of choice. In retrospect this is surprising given that Quetelet was obsessed with statistical variation, which we now might view as providing a quantitative measure of how much "freedom of choice" we have to deviate from the norm. This tension between the role of underlying "laws" that constrain the structure and evolution of a system, whether social or biological, and the extent to which they can be "violated" is a recurring theme that will be returned to later. How much freedom do we have in shaping our destiny, whether collectively or individually? At a detailed, high-resolution level, we may have great freedom in determining events into the near future, but at a coarse-grained, bigger-picture level where we deal with very long timescales life may be more deterministic than we think.

Although the term *social physics* faded from the scientific landscape it has been resurrected more recently by scientists from various backgrounds who have started to address social science questions from a more quantitative analytic viewpoint typically associated with the paradigmatic framework of traditional physics. Much of the work that my colleagues and I have been involved in and which will be elucidated in some detail in later chapters could be described as social physics, although it is not a term any of us uses with ease. Ironically, it has been picked up primarily by computer scientists, who are neither social scientists nor physicists, to describe their analysis of huge data sets on social interactions. As they characterize it: "Social Physics is a new way of understanding human behavior based on analysis of Big Data."[8] While this body of research is very interesting, it is probably safe to say that few physicists would recognize it as "physics," primarily because it does not focus on underlying principles, general laws, mathematical analyses, and mechanistic explanations.

Quetelet's body-mass index is defined as your body weight divided by the square of your height and is therefore expressed in terms of pounds per square inch or kilograms per square meter. The idea behind it is that weights of healthy individuals, and in particular those with a "normal" body shape and proportion of body fat, are presumed to scale with the square of their heights. So dividing the weight by the square of the height should lead to a quantity that is approximately the same for all healthy individuals, with values falling within a relatively narrow range (between 18.5 and 25.0 kg/sq. m). Values outside of this range are interpreted as an indication of a potential health problem associated with being either over- or underweight relative to height.[9]

The BMI is therefore presumed to be approximately *invariant* across idealized healthy individuals, meaning that it has roughly the same value regardless of body weight and height. However, this implies that body weight should increase as the *square* of height, which seems to be seriously at odds with our earlier discussion of Galileo's work where we concluded that body weight should increase much faster as the *cube* of height. If this is so, then BMI, as defined, would *not* be an invariant quantity but would instead increase linearly with height, thereby consistently overdiagnosing taller people as overweight while underdiagnosing shorter ones. Indeed, there is evidence that tall people have uncharacteristically high values compared with their actual body fat content.

So how in fact does weight scale with height for human beings? Various statistical analyses of data have led to varying conclusions, ranging from confirmation of the cubic law to more recent analyses suggesting exponents of 2.7 and values that are even smaller and closer to two.[10] To understand why this might be so, we have to remind ourselves of a major assumption that was made in deriving the cubic law, namely that the shape of the system, our bodies in this case, should remain the same when its size increases. However, shapes change with age, from the extreme case of a baby, with its large head and chunky limbs, to a mature "well-proportioned" adult, and finally to the sagging bodies of people my age. In addition, shapes also depend on gender, culture, and other socioeconomic factors that may or may not be correlated with health and obesity.

Many years ago I analyzed a data set on the heights of men and women as a function of their weights and found excellent agreement with the classic cubic law. Serendipitously, the data I had analyzed came from a relatively narrow cohort of U.S. males aged fifty to fifty-nine and U.S. females aged forty to forty-nine. Because these were analyzed separately by gender and within a similar, rather narrow age group, these cohorts represented meaningful "average" healthy men and women having similar characteristics. Ironically, this is in contrast to much more serious and comprehensive studies in which the averaging was performed over *all* age groups with diverse characteristics, making the interpretation much less clear. It is therefore not surprising that they resulted in exponents that differ from the idealized value of three. This suggests that a more reasonable approach would be to deconstruct the entire data set into cohorts with similar characteristics, such as age, and develop metrics for the resulting subgroups.

Unlike the cubic scaling law, the conventional definition of the BMI has no theoretical or conceptual underpinning and is therefore of dubious statistical significance. In contrast, the cubic law does have a conceptual basis and if we control for the characteristics of the cohort is supported by data. It is not surprising, therefore, that an alternative definition of the BMI has been suggested in which the BMI is defined as body weight divided by the *cube* of the height; it is known as the *Ponderal index*. Although it does somewhat better than the Quetelet definition in being meaningfully correlated with body fat content, it nevertheless suffers from similar problems because it has not been deconstructed into cohorts having similar characteristics.

Of course, good physicians use a range of BMI values to assess health, thereby mitigating gross misinterpretations with the exception perhaps of individuals with BMIs near the boundaries. In any case, it is clear that the classic BMI as presently used should not be taken too seriously without further investigation and the development of more subtle detailed indices that recognize, for example, age and cultural differences, especially for those who might appear to be at risk.

I have used these examples to illustrate how the conceptual framework of scaling underlies the use of critical metrics in our health care repertoire and in so doing have revealed potential pitfalls and misconceptions. As with drug doses, this is a complex and very important component of medical practice whose underlying theoretical framework has not yet been fully developed or realized.[11]

8. INNOVATION AND LIMITS TO GROWTH

Galileo's deceptively simple argument for why there are limits to the heights of trees, animals, and buildings has profound consequences for design and innovation. Earlier, when explaining his argument I concluded with the remark: *Clearly, the structure, whatever it is, will eventually collapse under its own weight if its size is arbitrarily increased. There are limits to size and growth.* To which should have been added the critical phrase *"unless something changes."* Change and, by implication, *innovation,* must occur in order to continue growing and avoid collapse. Growth and the continual need to be adapting to the challenges of new or changing environments, often in the form of "improvement" or increasing efficiency, are major drivers of innovation.

Galileo, like most physicists, did not concern himself with adaptive processes. We had to wait for Darwin to learn how important these are in shaping the world around us. As such, adaptive processes are primarily the domain of biology, economics, and the social sciences. However, in the mechanical examples he considered, Galileo introduced the fundamental concept of scale and by implication growth, both of which play an integral role in complex adaptive systems. Because of the conflicting scaling laws that constrain different attributes of a system—for example, the strengths of structures supporting a system scale differently from the way the weights being supported scale—growth, as manifested by an open-ended increase in size, cannot be sustained forever.

Unless, of course, an innovation occurs. A crucial assumption in the derivation of these scaling laws was that the system maintains the same physical characteristics, such as shape, density, and chemical composition, as it changes size. Consequently, in order to build larger structures or evolve larger organisms beyond the limits set by the scaling laws, innovations must occur that either change the material composition of the system or its structural design, or both.

A simple example of the first kind of innovation is to use a stronger material such as steel in place of wood for bridges or buildings, while a simple example of the second kind is to use arches, vaults, or domes in their construction rather than just horizontal beams and vertical pillars. The evolution of bridges is, in fact, an excellent example of how innovations in both materials and design were stimulated by the desire, or perceived requirement, to meet new challenges: in this case, to traverse wider and wider rivers, canyons, and valleys in a safe, resilient manner.

The most primitive kind of bridge is simply a log that has fallen across a stream or has been purposely placed there by humans—already an act of innovation. Perhaps the first significant act of engineering innovation in constructing bridges was to use purposely cut wooden logs or planks. Driven by the challenges of safety, stability, resilience, convenience, *and* the desire to span wider rivers, this was extended to incorporate stone structures as simple support systems on each bank, forming what is known as a beam bridge. Given the limited tensile strength of wood there is clearly a limit as to how long a span can be traversed in this way. This was solved by the simple design innovation of introducing stone support piers in the middle of the river that effectively extended the bridge to be a succession of individual beam bridges.

An alternative strategy was the much more sophisticated innovation of constructing bridges entirely of stone and using the physical principles of the arch, thereby changing both the materials and design. Such bridges had the great advantage of being able to withstand conditions that would damage or destroy earlier designs. Remarkably, arched stone bridges go back more than three thousand years to the Greek Bronze Age (thirteenth century BC), with some still in use today. The greatest stone arch bridge builders of antiquity were the Romans, who built vast numbers of beautiful bridges and aqueducts throughout their empire, many of which still stand.

To cross ever wider and deeper chasms such as the Avon Gorge in England or the entrance to San Francisco Bay in the United States required new technology, new materials, and new design. In addition, increases of traffic density and the need to support larger loads, especially with the coming of the railway, led to the development of arched cast iron bridges, truss systems of wrought iron, and eventually to the use of steel and the development of the modern suspension bridge. There are many variants of these designs, such as cantilevered bridges, tied arch bridges (Sydney Harbor being the most famous), and movable bridges like Tower Bridge in London. In addition, modern bridges are now constructed from a plethora of different materials, including combinations of concrete, steel, and fiber-reinforced polymers. All of these represent innovative responses to a combination of generic engineering challenges, including the constraints of scaling laws that transcend the individuality of each bridge, and the multiple local challenges of geography, geology, traffic, and economics that define the uniqueness and individuality of each bridge.

All of these innovative variations, driven by the perceived need to traverse wider and ever more challenging chasms, eventually hit a limit. Innovation in this context can then be viewed as the response to the challenge of continually scaling up the width of space to be crossed, beginning with a tiny stream and ending up with the widest expanses of water and the deepest and broadest canyons and valleys. You cannot cross San Francisco Bay with a long plank of wood. To bridge it you need to embark on a long evolutionary journey across many levels of innovation to the discovery of iron and the invention of steel and their integration with the design concept of a suspension bridge.

This way of thinking about innovation, which relates it to the drive or need to grow bigger, to expand horizons and compete in ever-larger markets with its inevitable confrontation with potential limitations imposed by physical constraints, will form the paradigm later in the book for addressing similar kinds of innovation in the larger context of biological and socioeconomic adaptive systems.

In the following sections, this will be extended to show how the idea of *modeling* a system arose. Modeling is now so commonplace and so taken for granted that we don't usually recognize that it is a relatively modern development. We can hardly countenance a time when it was not an essential and inseparable feature of industrial processes or scientific activity. Models of various kinds have been built for centuries, especially in architecture, but these were primarily to illustrate the aesthetic characteristics of the real thing rather than as scale models to test, investigate, or demonstrate the dynamical or physical principles of the system being constructed. And most important, they were almost always "made to scale," meaning that each detailed part was in some fixed proportion to the full size—1:10, for example—just like a map. Each part of the model was a *linearly* scaled representation of the actual-size ship, cathedral, or city being "modeled." Fine for aesthetics and toys but not much good for learning how the real system works.

Nowadays, every conceivable process or physical object, from automobiles, buildings, airplanes, and ships to traffic congestion, epidemics, economies, and the weather, is simulated on computers as "models" of the real thing. I discussed earlier how specially bred mice are used in biomedical research as scaled-down "models" of human beings. In all of these cases, the big question is how do you realistically and reliably scale up the results and observations of the model system to the real thing? This entire way of thinking has its origins in a sad failure in ship design in the middle of the nineteenth century and the marvelous insights of a modest gentleman engineer into how to avoid it in the future.

9. THE *GREAT EASTERN,* WIDE-GAUGE RAILWAYS, AND THE REMARKABLE ISAMBARD KINGDOM BRUNEL

Failure and catastrophe can provide a huge impetus and opportunity in stimulating innovation, new ideas, and inventions whether in science, engineering, finance, politics, or one's personal life. Such was the case in the history of shipbuilding and the origins of modeling theory and the role played by an extraordinary man with an extraordinary name: Isambard Kingdom Brunel.

In 2002 the BBC conducted a nationwide poll to select the "100 Greatest Britons." Perhaps predictably, Winston Churchill came in first with Princess Diana third (she had only been dead for five years at that time), followed by Charles Darwin, William Shakespeare, and Isaac Newton, a pretty impressive triumvirate. But who was second? None other than the remarkable Isambard Kingdom Brunel!

On occasions when I mention Brunel's name in lectures outside of the United Kingdom I usually ask the audience if they've ever heard of him. At best, there is a small smattering of hands, usually by people from Britain. I then inform them that according to a BBC poll, Brunel is the second-greatest Briton of all time ahead of Darwin, Shakespeare, Newton, and even John Lennon and David Beckham. It gets a good laugh but more important provides a natural segue into some provocative issues related to science, engineering, innovation, and scaling.

So who was Isambard Kingdom Brunel, and why is he famous? Many consider him the greatest engineer of the nineteenth century, a man whose vision and innovations, particularly concerning transport, helped make Britain the most powerful and richest nation in the world. He was a true engineering polymath who strongly resisted the trend toward specialization. He typically worked on all aspects of his projects beginning with the big-picture concept through to the detailed preparation of the drawings, carrying out surveys in the field and paying attention to the minutiae of design and manufacture. His accomplishments are numerous and he left an extraordinary legacy of remarkable structures ranging from ships, railways, and railway stations to spectacular bridges and tunnels.

Brunel was born in Portsmouth in the south of England in 1806 and died relatively young in 1859. His father, Sir Marc Brunel, was born in Normandy, France, and was also a highly accomplished engineer. They worked together when Isambard was only nineteen years old building the first ever tunnel under a navigable river, the Thames Tunnel at Rotherhithe in East London. It was a pedestrian tunnel that became a major tourist attraction with almost two million visitors a year paying a penny apiece to traverse it. Like many such underground walkways it sadly became the haunt of the homeless, muggers, and prostitutes and in 1869 was eventually transformed into a railway tunnel, becoming part of the London Underground system still in use to this day.

In 1830 at age twenty-four Brunel won a very stiff competition to build a suspension bridge over the River Avon Gorge in Bristol. It was an ambitious design and, upon its eventual completion in 1864, five years after his death, it had the longest span of any bridge in the world (702 feet, and 249 feet above the river). Brunel's father did not believe that a single span of such a length was physically possible and recommended that Isambard include a central support for the bridge, which he duly ignored.

Brunel subsequently became the chief engineer and designer for what was considered the finest railway of its time, the Great Western Railway, running from London to Bristol and beyond. In this role he designed many spectacular bridges, viaducts, and tunnels—the Box Tunnel, near Bath, was the longest railway tunnel in the world at the time—and even stations. Familiar to many, for example, is London's Paddington Station with its marvelous wrought-iron work.

One of his most fascinating innovations was the unique introduction of a broad gauge of 7 feet ¼ inch for the width between tracks. The standard gauge of 4 feet 8½ inches, which was used in all other British railways at that time, was adopted worldwide and is used on almost all railways today. Brunel pointed out that the standard gauge was an arbitrary carryover from the mine

A rakish-looking Isambard Kingdom Brunel posing in front of the chains he designed for the launching of the *Great Fastern* in 1858. Also shown is the giant ship under construction and the Clifton suspension bridge over the River Avon, which he designed in 1830 when only twenty-four years old.

railways built before the invention of the world's first passenger trains in 1830. It had simply been determined by the width needed to fit a cart horse between the shafts that pulled carriages in the mines. Brunel rightly thought that serious consideration should be given to determining what the optimum gauge should be and tried to bring some rationality to the issue. He claimed that his calculations, confirmed by a series of trials and experiments, showed that his broader gauge was the optimum size for providing higher speeds, greater stability, and a more comfortable ride to passengers. Consequently, the Great Western Railway was unique in having a gauge that was almost twice as wide as every other railway line. Unfortunately, in 1892, following the evolution of a national railway system, the British Parliament forced the Great Western Railway to conform to the standard gauge, despite its acknowledged inferiority.

The parallels with similar issues we are facing today regarding the inevitable tension and trade-offs between innovative optimization and the uniformity and fixing of standards determined by historical precedence, especially in our fast-developing high-tech industry, are clear. The battle over railway track gauges provides an informative case study of how innovative change may not always lead to the optimum solution.

Though Brunel's projects were not always completely successful, they typically contained inspired innovative solutions to long-standing engineering problems. Perhaps his grandest achievements—and failures—were in shipbuilding. As global trade was developing and competitive empires were being established, the need to develop rapid, efficient ocean transport over long distances was becoming increasingly pressing. Brunel formulated a grand vision of a seamless transition between the Great Western Railway and his newly formed Great Western Steamship Company so that a passenger could buy a ticket at Paddington Station in London and get off in New York City, powered the entire way by steam. He whimsically called this the Ocean Railway. However, it was widely believed that a ship powered purely by steam would not be able to carry enough fuel for the trip and still have room for sufficient commercial cargo to be economically viable.

Brunel thought otherwise. His conclusions were based on a simple scaling argument. He realized that the volume of cargo a ship could carry increases as the *cube* of its dimensions (like its weight), whereas the strength of the drag forces it experiences as it travels through water increases as the cross-sectional area of its hull and therefore only as the *square* of its dimensions. This is just like Galileo's conclusions for how the strength of beams and limbs scale with body weight. In both cases the strength increases more slowly than the corresponding weight following a ⅔ power scaling law. Thus the strength of the hydrodynamic drag forces on a ship relative to the weight of the cargo it can carry *decreases* in direct proportion to the length of the ship. Or to put it the other way around: the weight of its cargo relative to the drag forces its engines need to overcome systematically *increases* the bigger the ship. In other words, *a larger ship requires proportionately less fuel to transport each ton of cargo than a smaller ship.* Bigger ships are therefore more energy efficient and cost effective than smaller ones—another great example of an economy of scale and one that had enormous consequences for the development of world trade and commerce.[12]

Although these conclusions were nonintuitive and not generally believed, Brunel and the Great Western Steamship Company were convinced. Brunel boldly proceeded to design the company's first ship, the *Great Western*, which was the first steamship purposely built for crossing the Atlantic. She was a paddle-wheel ship constructed of wood (with four sails as backup, just in case) and, when completed in 1837, was the largest and fastest ship in the world.

Following the success of the *Great Western* and confirmation of the scaling argument that bigger ships were more efficient than smaller ones, Brunel moved to build an even larger one, brazenly combining new technologies and materials never before incorporated into a single design. The *Great Britain*, launched in 1843, was built of iron rather than wood and driven by a screw propeller in the rear rather than paddle wheels on the sides. In so doing, the *Great Britain* became the prototype for all modern ships. She was longer than any previous ship and was the first iron-hulled, propeller-driven ship to cross the Atlantic. You can still see her today fully renovated and preserved in the dry dock built in Bristol by Brunel for her original construction.

Having conquered the Atlantic, Brunel turned his attention to the biggest challenge of all, namely, connecting the far-flung reaches of the burgeoning British Empire to consolidate its position as the dominant global force. He wanted to design a ship that could sail nonstop from London to Sydney and back without refueling using only a single load of coal (and this was before the opening of the Suez Canal). This meant that the ship would have to be more than twice the length of the *Great Britain* at almost 700 feet and have a displacement (effectively its weight) almost ten times bigger. It was named the *Great Eastern* and launched in 1858. It took almost fifty years and into the twentieth century before another ship approached its size. Just to give a sense of scale, the huge oil supertankers plying the world's oceans today more than 150 years later are still only a little more than twice as long as the *Great Eastern*.

Sadly, however, the *Great Eastern* was not a success. Though an extraordinary engineering accomplishment that raised the bar to a level not again reached until well into the twentieth century, it suffered like many of Brunel's achievements from construction delays and budget overruns. But more pointedly, the

Great Eastern was not a technical success either. She was ponderous and ungainly, rolled too much even in moderately heavy waves, and, most pertinent, could barely move her gargantuan mass at even moderate speeds. Nor, surprisingly, was she very efficient and as a result was never used for her original grandiose purpose of serving the Empire by shipping large loads of cargo and large numbers of passengers to and from India and Australia. She made a small number of transatlantic crossings before being ignominiously transformed into a ship for laying cables. The first resilient transatlantic telegraph cable was laid by the *Great Eastern* in 1866, enabling reliable telecommunication between Europe and North America, thereby revolutionizing global communication.

The *Great Eastern* ended up as a floating music hall and advertising billboard in Liverpool before being broken up in 1889. Such was the sad ending to a glorious vision. A bizarre footnote to this tale that is probably of interest only to ardent soccer fans is that in 1891 when the famous British football club Liverpool was being founded, they searched for a flagpole for their new stadium and purchased the top mast of the *Great Eastern* for that purpose. It still proudly stands there today.

How did all of this happen? How could such a marvelous vision overseen by one of the most brilliant and innovative practitioners of all time end in such a shambles? The *Great Eastern* was hardly the first ship to be poorly designed, but its sheer size, its innovative vision, and its huge cost relative to its serious underperformance made it a spectacular failure.

10. WILLIAM FROUDE AND THE ORIGINS OF MODELING THEORY

When systems fail or designs don't meet their expectations there are usually a plethora of reasons that could be the problem. These include poor planning and execution, faulty workmanship or materials, poor management, and even a lack of conceptual understanding. However, there are key examples like that of the *Great Eastern* where the major reason for failure was that they were designed without a deep understanding of the underlying science and of the basic principles of scale. Indeed, until the last half of the nineteenth century, neither science nor scale played any significant role in the manufacture of most artifacts, let alone ships.

There were some significant exceptions to this, the most salient being in the development of steam engines where understanding the relationship between pressure, temperature, and volume of steam helped advance the design of very large, efficient boilers, the kind that allowed engineers to contemplate building great ships the size of the *Great Eastern* that could sail across the globe. More significant, investigations into understanding the fundamental principles and characteristics of efficient engines and the nature of different forms of energy, whether heat, chemical, or kinetic, led to the development of the fundamental science of *thermodynamics*. And of even greater significance, the laws of thermodynamics and the concepts of *energy* and *entropy* extend far beyond the narrow confines of steam engines and are universally applicable to *any* system where energy is being exchanged, whether for a ship, an airplane, a city, an economy, your body, or the entire universe itself.

Even at the time of the *Great Eastern,* there was very little, if any, such "real" science in shipbuilding. Success in designing and building ships had been achieved by the gradual accumulation of knowledge and technique via a process of trial and error, resulting in well-established rules of thumb passed on, to a large extent, by the mechanisms of apprenticeship and learning on the job. Typically, each new ship was a minor variant on a previous one, with small changes here and there as demanded by the projected needs and uses of the vessel. Small errors resulting from simple extrapolation from what had worked before to the new situation usually had a relatively small impact. So increasing the length of a ship by 5 percent, for example, might produce a vessel that didn't quite meet design expectations or one that didn't behave quite as expected, but these "errors" could be readily corrected and even improved upon by an appropriate adjustment or inspired innovation in future versions. Thus, to a large extent, shipbuilding, like almost all other developments in artifactual manufacturing, evolved almost organically, mimicking a process akin to natural selection.

Superimposed on this incremental and essentially linear process was an occasional inspired innovative nonlinear leap that changed something significant in the design or in the materials used, such as the introduction of sails, propellers, or the use of steam and iron. Although such innovative leaps still built on previous designs, they required a rethinking and oftentimes major readjustments before a new successful prototype emerged.

The tried and tested process of simply extrapolating from previous design worked well when designing and building new ships, provided the changes were incremental. There was no need for a deep scientific understanding of why something worked the way it did because the long succession of previous successful vessels effectively ensured that most of the problems to be addressed had already been solved. This paradigm is succinctly summarized in a comment about the shipwrights who built a much earlier catastrophic failure, the Swedish warship *Vasa:* "The trouble at that time was that the science of ship design wasn't fully understood. There was no such thing as construction drawings and ships were designed by 'rule of thumb'; largely based on previous experience."[13] Shipwrights were given the overall dimensions and used their own experience to produce a ship with good sailing qualities.

Sounds pretty straightforward, and it might well have been had the *Vasa* been just a minor extension of previous ships built by the Stockholm shipyard. However, King Gustav Adolf had demanded a ship that was 30 percent longer than previous ships with an extra deck carrying much heavier artillery than usual. With such radical demands, no longer would a small mistake in design lead to a small error in performance. A ship of this size is a complicated structure and its dynamics, especially regarding its stability, are inherently nonlinear. A small error in design could, and did, result in macroscopic errors in performance, resulting in catastrophic consequences. Unfortunately, the shipwrights did not have the scientific knowledge to know how to correctly scale up a ship by such a large amount. In fact, they didn't have the scientific knowledge to know how to correctly scale up a ship by a small amount either, but this hardly mattered. Consequently, the ship ended up being too narrow and too top-heavy so that even a light breeze was sufficient to capsize her—and it did, even before she left the harbor in Stockholm on her maiden voyage, resulting in the loss of many lives.[14]

The same may be said of the *Great Eastern* where the increase in size was even greater, with the length being doubled and the weight increased by almost a factor of ten. Brunel and his colleagues simply didn't have the scientific knowledge to correctly scale up a ship by such a large factor. Fortunately, in this case, there were no human catastrophes, only economic ones. In such a fiercely competitive economic market, underperformance is the kiss of death.

It was only in the decade prior to the building of the *Great Eastern* that the underlying science governing the motion of ships was developed. The field of *hydrodynamics* was first formalized independently by the French engineer Claude-Louis Navier and the great Irish mathematical physicist George Stokes. The fundamental equation, which is universally known as the *Navier-Stokes equation,* arises from applying Newton's laws to the motion of fluids, and by extension to the dynamics of physical objects moving through fluids, such as ships through water or airplanes through air.

This may sound pretty arcane, and it is very likely that you've never heard of the Navier-Stokes equation, but it has played, and continues to play, a major role in almost all aspects of your life. Among many other things, it underlies the design of airplanes, automobiles, hydroelectric power stations, and artificial hearts, the understanding of blood flow in your circulatory system, and the hydrology of rivers and water supply systems. It is fundamental to understanding and predicting the weather, ocean currents, and the analysis of pollution and is consequently a key ingredient in the science of climate change and the predictions of global warming.

I do not know if Brunel was familiar with the discovery of these equations that governed the motion of the ships he was designing, but he did have the insight and intuition to engage someone who probably was. That man was William Froude, who had studied mathematics at Oxford and had worked as a young engineer on the Great Western Railway some years earlier.

During the building of the *Great Eastern,* Brunel had asked Froude to investigate the problem of the rolling and stability of ships. This eventually led him to answer the critical question as to what the optimum shape of a hull should be that would minimize the viscous drag forces of water. The economic implications for shipping and global trade resulting from his work were huge. Thus was born the modern science of ship design. However, of even greater impact and longer-term importance was his introduction of the revolutionary concept of modeling a system in order to determine how the real system worked.

Although the Navier-Stokes equation describes fluid motion under essentially any conditions it is extremely difficult, and in almost all cases impossible, to solve exactly because it is inherently nonlinear. Roughly speaking, the nonlinearity arises from feedback mechanisms in which water interacts with itself. This is manifested in all kinds of fascinating behaviors and patterns such as those we see in the eddies and whirlpools of rivers and streams, or the wakes left by ships as they move through water, or the awesome specter of hurricanes and tornadoes and the beauty and infinite variation of ocean waves. All of these are manifestations of *turbulence* and are encapsulated in the hidden richness of the Navier-Stokes equation.

Indeed, studying turbulence gave us the first important mathematical insights into the concept of *complexity* and its relationship to nonlinearity. Complex systems often manifest chaotic behavior in which a small change or perturbation in one part of the system produces an exponentially enhanced response in some other part. As we discussed earlier, in traditional linear thinking a small perturbation produces a commensurately small response. The highly nonintuitive enhancement in nonlinear systems is popularly expressed as "the butterfly effect," in which the mythical flapping of a butterfly's wings in Brazil produces a hurricane in Florida. Despite 150 years of intense theoretical and experimental study, a general understanding of turbulence remains an unsolved problem in physics even though we have learned an enormous amount about it. Indeed the famous physicist Richard Feynman described turbulence as "the most important unsolved problem of classical physics."[15]

Froude may not have fully recognized just how big a challenge he was facing but he did perceive that for applications to shipbuilding a new strategy was needed. It was in this context that he invented the new methodology of *modeling* and, by extension, the concept of *scaling theory* for determining how quantitative results from small-scale investigations could be used to help predict how full-size ships would behave. In the spirit of Galileo, Froude realized that almost all scaling is nonlinear, so traditional models based on a faithful 1:1 representation were not useful for determining how the actual system works. His seminal contribution was to suggest a quantitative mathematical strategy for figuring out how to scale from the small-size model to the full-size object.

Like many new ideas that threaten to change the way we think about an old problem, Froude's endeavors were at first dismissed as irrelevant by the cognoscenti of the time. John Russell, who founded the Institution for Naval Architects in England in 1860 in order to encourage the formal education of ship designers, ridiculed Froude: "You will have on the small scale a series of beautiful, interesting little experiments which, I am sure, will afford Mr. Froude infinite pleasure in the making of them . . . and will afford you infinite pleasure in the hearing of them; but which are quite remote from any practical results upon the large scale."

Many of us recognize this kind of rhetoric, often aimed at scholarly or academic research with the implication that it is out of touch with the "real world." Well, no doubt, much of it is. But much of it isn't and, more to the point, it is very often difficult to perceive in the moment the potential impact of some piece of seemingly arcane research. Much of our entire technologically driven society and extraordinary quality of life that many of us are privileged to enjoy is the result of such research. There is a continuing tension in society between supporting what is perceived as pie-in-the-sky basic research with no obvious immediate benefit versus highly directed research focused on "useful, real world" problems.

In 1874 after Froude had revolutionized ship design, Russell to his credit relented and enthusiastically embraced Froude's methodology and ideas, lamely claiming, however, that he himself had already thought of all of this and done the experiments years earlier. Russell had, in fact, been Brunel's main partner in building the *Great Eastern* and it is true that he had actually tinkered with models, but he had unfortunately not appreciated either their significance or the underlying conceptual framework.

Froude made small models of ships, three to twelve feet long, that he dragged through water contained in long tanks and measured their resistance to flow as well as their stability characteristics. With his mathematical background he possessed the technical machinery for knowing how to scale up his findings to large ships.

He realized that the primary quantity that determined the character of their relative motion was something that later became known as the *Froude number*. It is defined as the square of the ship's velocity divided by its length multiplied by the acceleration due to gravity. This is a bit of a mouthful and may sound a little intimidating, but it's actually quite simple because "the acceleration due to gravity" that occurs in this expression is the same for all objects, regardless of their size, shape, or composition. This is just a restatement of Galileo's observations that falling objects of different weights reach the ground in the same time. So in terms of the quantities that actually vary, Froude's number is simply proportional to the velocity squared divided by the length. This ratio plays a central role in all problems involving motion, ranging from speeding bullets and running dinosaurs to flying airplanes and sailing ships.

The crucial point recognized by Froude was that because the underlying physics remains the same, objects of different sizes moving at different speeds behave in the same way *if their Froude numbers have the same value.* Thus, by making the length and speed of the model ship have the same Froude number as that of the full-size version, one can determine the dynamical behavior of the full-size ship before building it.

Let me give a simple illustration of this by asking how fast a 10-foot-long model ship would have to move to mimic the motion of the 700-foot-long *Great Eastern* moving at 20 knots (a little over 20 mph). If they are to have the same Froude number (that is, the same value of the square of their velocity divided by their length), then the velocity must scale as the square root of their lengths. Now, the ratio of the square root of their lengths is $\sqrt{(700 \text{ ft.}/10 \text{ ft.})}$, which is $\sqrt{70} = 8.4$. So, for the 10-foot model to mimic the *Great Eastern* it must move with a speed of approximately $20/8.4 = 2.5$ knots, which is just about walking speed. In other words, the dynamics of a 10-foot-long model ship moving at just 2.5 knots simulates that of the 700-foot-long *Great Eastern* moving at 20 knots.

I have oversimplified his methodology in that there are typically other numbers analogous to Froude's number that can enter the problem and which explicitly include other dynamical effects such as, for example, the viscosity of water. Nevertheless, the example illustrates the essence of Froude's method and provides the generic template for modeling and scaling theory. It represents the transition from a primitive trial-and-error, rule-of-thumb approach that served us well for thousands of years toward a more analytic, principled scientific strategy for solving problems and designing modern arti-facts ranging from computers and ships to airplanes, buildings, and even com-panies. Froude's water tank design is still used today to study ships, and its extension to wind tunnels, which strongly influenced the Wright brothers, plays an analogous role for airplanes and automobiles. Sophisticated computer analyses are now central in the design process, thereby simulating the princi-ples of scaling theory in order to optimize performance. The phrase "computer model" is an integral part of our vocabulary. Effectively, we are now able to "solve," or simulate solutions to, the Navier-Stokes equations, or their equiva-lent, leading to greater accuracy in predicting performance.

One of the curious unintended consequences of these advances is that al-most all automobiles, for example, now look alike because all manufacturers are solving the same equations to optimize similar performance parameters. Fifty years ago, before we had access to such high-powered computation and therefore less accuracy in predicting outcomes, and before we became so con-cerned about fuel performance and exhaust pollution, the diversity of car de-sign was much more varied and consequently much more interesting. Compare a 1957 Studebaker Hawk or a 1927 Rolls-Royce to a relatively boring-looking 2006 Honda Civic or a 2014 Tesla, even though these latter vehicles are far su-perior machines.

11. SIMILARITY AND SIMILITUDE: DIMENSIONLESS AND SCALE-INVARIANT NUMBERS

The scaling methodology introduced by Froude has now evolved to become a powerful and sophisticated component of the tool kit of science and engineer-ing and has been applied with great effect to a very broad and diverse range of problems. It was not formalized as a general technique until the beginning of the twentieth century, when the eminent mathematical physicist Lord Rayleigh wrote a provocative and highly influential paper in the journal *Na-ture* titled "The Principle of Similitude."[16] This was his term for what we have been calling scaling theory. His major emphasis was on the primary role played in *any* physical system by special quantities that have the property of being *dimensionless.* These are combinations of variables, like the Froude number, whose value is the same regardless of the system of units used to measure them. Let me elaborate.

Typical quantities that you are accustomed to measuring in your daily life such as lengths, times, and pressures depend upon the units used to measure them, such as feet, seconds, pounds per square inch, and so on. The same quantity can, however, be measured using different units: for instance, the distance from New York to Los Angeles is 3,210 miles, but it could just as well be expressed as 5,871 kilometers. Even though these are *different* numbers they represent the *same* thing. Similarly, the distance between London and Manchester can be expressed as either 278 miles or 456 kilometers. However, the ratio of the distance from New York to Los Angeles to that of London to Manchester (either 3,210 miles ÷ 278 miles, or 5,871 kilometers ÷ 456 kilometers) is the *same* (14.89) regardless of what units are used.

This is the simplest example of a dimensionless quantity: it is a "pure" number that does not change when a different system of units is used to measure it. This *scale invariance* expresses something absolute about the quantities they represent in that the dependence on the arbitrariness of the human choice of units and measurement has been removed. Specific units are convenient inventions of human beings for communicating measures in a standardized language, especially regarding construction, commerce, and the exchange of goods and services. Indeed, the introduction of standardized measures marked a critical stage in the evolution of civilization and the rise of cities, as they were crucial in developing a trustworthy political fabric subject to the rule of law.

Perhaps the most famous dimensionless number is pi (π), the ratio of the circumference of a circle to its diameter. This has no units, because it is the ratio of two lengths, and has the same value for all circles everywhere, at all times, no matter how small or how large they are. π therefore embodies the universal quality of "circleness."

This concept of "universality" is the reason why the acceleration due to gravity was included in the definition of the Froude number, even though it played no explicit role in how to scale from model ships to the real thing. It turns out that the ratio of the square of the velocity to the length is *not* dimensionless and so depends on the units used, whereas dividing by gravity renders it dimensionless and therefore scale invariant.

But why was gravity chosen and not some other acceleration? Because gravity ubiquitously constrains *all* motion on Earth. This is pretty evident in our own walking and running where we have to continually fight gravity to raise our legs with each step forward, especially when going uphill. Not quite so obvious is how it enters into the motion of ships, because the buoyancy of water balances gravity (remember Archimedes' principle). However, as a ship moves through water it continually creates wakes and surface waves whose behavior is constrained by the pull of gravity—in fact, the waves you are familiar with on oceans and lakes are technically called gravity waves. So indirectly gravity plays an important role in the motion of ships. Consequently, Froude's number embodies a "universal" quality associated with all motion on Earth transcending the specific details of the object that is moving. Its value is therefore a major determinant not only in the motion of ships, but also for cars, airplanes, and ourselves. Furthermore, it also tells us how these motions on another planet with a *different* gravitational strength would differ from the same motion on Earth.

Because the essence of any measurable quantity cannot depend on an arbitrary choice of units made by human beings, neither can the laws of physics. Consequently, all of these and indeed all of the laws of science must be expressible as relationships between scale-invariant dimensionless quantities, even though conventionally they are not typically written that way. This was the underlying message of Rayleigh's seminal paper.

His paper elegantly illustrates the technique with many well-chosen examples, including one that provides the scientific explanation for one of the great mysteries of life that all of us have pondered at some time, namely, *why is the sky blue?* Using an elegant argument based solely on relating purely dimensionless quantities, he shows that the intensity of light waves scattered by small particles must *decrease* with the fourth power of their wavelength. Thus, when sunlight, which is a combination of all of the colors of the rainbow, scatters from microscopic particles suspended in the atmosphere, the shortest wavelengths, corresponding to blue light, dominate.

Actually, Rayleigh had derived this stunning result much earlier in a tour

de force based on a masterful mathematical analysis of the problem that provided a detailed mechanistic explanation for the origin of the shift toward the
blue end of the spectrum. His point in presenting the simple derivation in his
similitude paper was to show that the same result could have been derived
"after a few minutes' consideration," as he put it, using a scaling argument in
the guise of the "great principle of similitude" without the detailed sophisticated mathematics. His scaling argument showed that the shift to shorter
wavelengths is an inevitable outcome of *any* analysis once you know what the
important variables are. What is missing from such a derivation is a deeper
understanding of the mechanism by which the result comes about. This is
characteristic of many scaling arguments: general results can be derived, but
details of their mechanistic origins sometimes remain hidden.

Rayleigh's mathematical analysis of the scattering of waves formed the
basis of what became known as "scattering theory." Its application to many
problems, from water waves to electromagnetic waves and especially radar,
and more recently IT communication, has been incredibly important, not least
its role in the development of quantum mechanics. It provided the basis for the
formalism developed for extracting discoveries from "scattering experiments"
performed at large particle accelerators such as CERN in Geneva, where the
famous Higgs particle was recently discovered.

If you look up his original paper, published in 1870 when he was only
twenty-eight years old, you will find that the author's name isn't Lord Rayleigh
but rather the more earthy John Strutt, sounding more like a Thomas Hardy
character than a distinguished professor of physics at Cambridge. This was
Rayleigh's name before he inherited his aristocratic title from his father in
1873, after which he identified himself as Lord Rayleigh. The name Strutt is
best known to the public via his younger brother Edward, who founded a famous real estate and property management firm, Strutt & Parker, which today
is one of the largest property partnerships in the United Kingdom. Look for
their signs on upscale properties in central London next time you're there.

Rayleigh was a great polymath. Among his many other great accomplishments, he developed the mathematical theory of sound and discovered argon
gas, for which he received one of the very first Nobel Prizes in 1904 (the fourth
ever awarded).

3

THE SIMPLICITY, UNITY,
AND COMPLEXITY OF LIFE

As emphasized in the opening chapter, living systems, from the smallest bacteria to the largest cities and ecosystems, are quintessential complex adaptive systems operating over an enormous range of multiple spatial, temporal, energy, and mass scales. In terms of mass alone, the overall scale of life covers more than thirty orders of magnitude (10^{30}) from the molecules that power metabolism and the genetic code up to ecosystems and cities. This range vastly exceeds that of the mass of the Earth relative to that of our entire galaxy, the Milky Way, which covers "only" eighteen orders of magnitude, and is comparable to the mass of an electron relative to that of a mouse.

Over this immense spectrum, life uses essentially the same basic building blocks and processes to create an amazing variety of forms, functions, and dynamical behaviors. This is a profound testament to the power of natural selection and evolutionary dynamics. All of life functions by transforming energy from physical or chemical sources into organic molecules that are metabolized to build, maintain, and reproduce complex, highly organized systems. This is accomplished by the operation of two distinct but closely interacting systems: the genetic code, which stores and processes the information and "instructions" to build and maintain the organism, and the metabolic system, which acquires, transforms, and allocates energy and materials

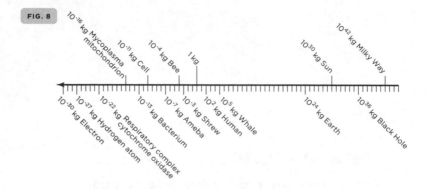

The extraordinary scale of life from complex molecules and microbes to whales and sequoias in relation to galactic and subatomic scales.

for maintenance, growth, and reproduction. Considerable progress has been made in elucidating both of these systems at levels from molecules to organisms, and later I will discuss how it can be extended to cities and companies. However, understanding how information processing ("genomics") integrates with the processing of energy and resources ("metabolics") to sustain life remains a major challenge. Finding the universal principles that underlie the structures, dynamics, and integration of these systems is fundamental to understanding life and to managing biological and socioeconomic systems in such diverse contexts as medicine, agriculture, and the environment.

A unified framework for understanding genetics has been developed that can account for phenomena from the replication, transcription, and translation of genes to the evolutionary origin of species. Slower to emerge has been a comparable unified theory of metabolism that links processes by which the energy and material transforming processes that are generated by biochemical reactions within cells are scaled up to sustain life, power biological activities, and set the timescales of vital processes at levels from organisms to ecosystems.

The search for fundamental principles that govern how the complexity of life emerges from its underlying simplicity is one of the grand challenges of twenty-first-century science. Although this has been, and will continue to be, primarily the purview of biologists and chemists, it is becoming an activity where other disciplines, and in particular physics and computer science, are playing an increasingly important role. Understanding more generally the emergence of complexity from simplicity, an essential characteristic of adaptive evolving systems, is one of the founding cornerstones of the new science of complexity.

The field of physics is concerned with fundamental principles and concepts at *all* levels of organization that are quantifiable and mathematizable (meaning amenable to computation), and can consequently lead to precise predictions that can be tested by experiment and observation. From this perspective, it is natural to ask if there are "universal laws of life" that are mathematizable so that biology could also be formulated as a predictive, quantitative science much like physics. Is it conceivable that there are yet-to-be-discovered "Newton's Laws of Biology" that would lead, at least in principle, to precise calculations of any biological process so that, for instance, one could accurately predict how long you and I would live?

This seems very unlikely. After all, life is the complex system par excellence, exhibiting many levels of emergent phenomena arising from multiple contingent histories. Nevertheless, it may not be unreasonable to conjecture that the generic coarse-grained behavior of living systems might obey quantifiable universal laws that capture their essential features. This more modest view presumes that at every organizational level average idealized biological systems can be constructed whose general properties are calculable. Thus we ought to be able to calculate the average and maximum life span of human beings even if we'll never be able to calculate our own. This provides a point of departure or baseline for quantitatively understanding actual biosystems, which can be viewed as variations or perturbations around idealized norms due to local environmental conditions or historical evolutionary divergence. I will elaborate on this perspective in much greater depth below, as it forms the conceptual strategy for attacking most of the questions posed in the opening chapter.

1. FROM QUARKS AND STRINGS
TO CELLS AND WHALES

Before launching into some of the big questions that have been raised, I want to make a short detour to describe the serendipitous journey that led me from investigating fundamental questions in physics to fundamental questions in biology and eventually to fundamental questions in the socioeconomic sciences that have a bearing on pivotal questions concerning global sustainability.

In October 1993 the U.S. Congress with the consent of President Bill Clinton officially canceled the largest scientific project ever conceived after having spent almost $3 billion on its construction. This extraordinary project was the mammoth Superconducting Super Collider (SSC), which, together with its detectors, was arguably the greatest engineering challenge ever attempted. The SSC was to be a giant microscope designed to probe distances down to a hundred trillionth of a micron with the aim of revealing the structure and dynamics of the fundamental constituents of matter. It would provide critical evidence for testing predictions derived from our theory of the elementary particles, potentially discover new phenomena, and lay the foundations for what was termed a "Grand Unified Theory" of all of the fundamental forces of nature. This grand vision would not only give us a deep understanding of what everything is made of but would also provide critical insights into the evolution of the universe from the Big Bang. In many ways it represented some of the highest ideals of mankind as the sole creature endowed with sufficient consciousness and intelligence to address the unending challenge of unraveling some of the deepest mysteries of the universe—perhaps even providing the very reason for our existence as the agents through which the universe would know itself.

The scale of the SSC was gigantic: it was to be more than fifty miles in circumference and would accelerate protons up to energies of 20 trillion electron volts at a cost of more than $10 billion. To give a sense of scale, an electron volt is a typical energy of the chemical reactions that form the basis of life. The energy of the protons in the SSC would have been eight times greater than that of the Large Hadron Collider now operating in Geneva that was recently in the limelight for discovering the Higgs particle.

The demise of the SSC was due to many, almost predictable, factors, including inevitable budget issues, the state of the economy, political resentment against Texas where the machine was being built, uninspired leadership, and so on. But one of the major reasons for its collapse was the rise of a climate of negativity toward traditional big science and toward physics in particular.[1] This took many forms, but one that many of us were subjected to was the oft-repeated pronouncement I have already quoted earlier that "while the nineteenth and twentieth centuries were the centuries of physics, the twenty-first century will be the century of biology."

Even the most arrogant hard-nosed physicist had a hard time disagreeing with the sentiment that biology would very likely eclipse physics as the dominant science of the twenty-first century. But what incensed many of us was the implication, oftentimes explicit, that there was no longer any need for further basic research in this kind of fundamental physics because we already knew all that was needed to be known. Sadly, the SSC was a victim of this misguided parochial thinking.

At that time I was overseeing the high energy physics program at the Los Alamos National Laboratory, where we had a significant involvement with one of the two major detectors being constructed at the SSC. For those not familiar with the terminology, "high energy physics" is the name of the subfield of physics concerned with fundamental questions about the elementary particles, their interactions and cosmological implications. I was a theoretical physicist (and still am) whose primary research interests at that time were in this area. My visceral reaction to the provocative statements concerning the diverging trajectories of physics and biology was that, yes, biology will almost certainly be the predominant science of the twenty-first century, but for it to become truly successful, it will need to embrace some of the quantitative, analytic, predictive culture that has made physics so successful. Biology will need to integrate into its traditional reliance on statistical, phenomenological, and qualitative arguments a more theoretical framework based on underlying mathematizable or computational principles. I am embarrassed to say that I knew very little about biology at that time, and this outburst came mostly from arrogance and ignorance.

Nevertheless, I decided to put my money where my mouth was and started to think about how the paradigm and culture of physics might help solve interesting challenges in biology. There have, of course, been several physicists who made extremely successful forays into biology, the most spectacular of which was probably Francis Crick, who with James Watson determined the structure of DNA, which revolutionized our understanding of the genome. Another is the great physicist Erwin Schrödinger, one of the founders of quantum mechanics, whose marvelous little book titled *What Is Life?*, published in 1944, had a huge influence on biology.[2] These examples were inspirational evidence that physics might have something of interest to say to biology, and had stimulated a small but growing stream of physicists crossing the divide, giving rise to the nascent field of biological physics.

I had reached my early fifties at the time of the demise of the SSC, and as I remarked at the opening of the book, I was becoming increasingly conscious of the inevitable encroachment of the aging process and the finiteness of life. Given the poor track record of males in my ancestry, it seemed natural to begin my thinking about biology by learning about aging and mortality. Because these are among the most ubiquitous and fundamental characteristics of life, I naively assumed that almost everything was known about them. But, to my great surprise, I learned that not only was there no accepted general theory of aging and death but the field, such as it was, was relatively small and something of a backwater. Furthermore, few of the questions that would be natural for a physicist to ask, such as those I posed in the opening chapter, seemed to have been addressed. In particular, where does the scale of one hundred years for human life span come from, and what would constitute a quantitative, predictive theory of aging?

Death is an essential feature of life. Indeed, implicitly it is an essential feature of the theory of evolution. A necessary component of the evolutionary process is that individuals eventually die so that their offspring can propagate new combinations of genes that eventually lead to adaptation by natural selection of new traits and new variations leading to the diversity of species. We must all die so that the new can blossom, explore, adapt, and evolve. Steve Jobs put it succinctly[3]:

No one wants to die. Even people who want to go to heaven don't want to die to get there. And yet death is the destination we all share. No one has ever escaped it, and that is how it should be, because death is very likely the single best invention of life. It's life's change agent. It clears out the old to make way for the new.

Given the critical importance of death and of its precursor, the aging process, I assumed that I would be able to pick up an introductory biology textbook and find an entire chapter devoted to it as part of its discussion of the basic features of life, comparable to discussions of birth, growth, reproduction, metabolism, and so on. I had expected a pedagogical summary of a mechanistic theory of aging that would include a simple calculation showing why we live for about a hundred years, as well as answering all of the questions I posed above. No such luck. Not even a mention of it, nor indeed was there any hint that these were questions of interest. This was quite a surprise, especially because, after birth, death is the most poignant biological event of a person's life. As a physicist I began to wonder to what extent biology was a "real" science (meaning, of course, that it was like physics!), and how it was going to dominate the twenty-first century if it wasn't concerned with these sorts of fundamental questions.

This apparent general lack of interest by the biological community in aging and mortality beyond a relatively small number of devoted researchers stimulated me to begin pondering these questions. As it appeared that almost no one was thinking about them in quantitative or analytic terms, there might be a possibility for a physics approach to lead to some small progress. Consequently, during interludes between grappling with quarks, gluons, dark matter, and string theory, I began to think about death.

As I embarked on this new direction, I received unexpected support for my ruminations about biology as a science and its relationship to mathematics from an unlikely source. I discovered that what I had presumed was subversive thinking had been expressed much more articulately and deeply almost one hundred years earlier by the eminent and somewhat eccentric biologist Sir D'Arcy Wentworth Thompson in his classic book *On Growth and Form,* published in 1917.[4] It's a wonderful book that has remained quietly revered not just in biology but in mathematics, art, and architecture, influencing thinkers and artists from Alan Turing and Julian Huxley to Jackson Pollock. A testament to its continuing popularity is that it still remains in print. The distinguished biologist Sir Peter Medawar, the father of organ transplants, who received the Nobel Prize for his work on graft rejection and acquired immune tolerance, called *On Growth and Form* "the finest work of literature in all the annals of science that have been recorded in the English tongue."

Thompson was one of the last "Renaissance men" and is representative of a breed of multi- and transdisciplinary scientist-scholars that barely exists today. Although his primary influence was in biology, he was a highly accomplished classicist and mathematician. He was elected president of the British Classical Association, president of the Royal Geographical Society, and was a good enough mathematician to be made an honorary member of the prestigious Edinburgh Mathematical Society. He came from an intellectual Scottish family and had a name, much like Isambard Kingdom Brunel, that one might associate with a minor fictional character in a Victorian novel.

Thompson begins his book with a quote from the famous German philosopher Immanuel Kant, who had remarked that the chemistry of his day was *"eine Wissenschaft, aber nicht Wissenschaft,"* which Thompson translates as: chemistry is "a science, but not Science," implying that "the criterion of true science lay in its relation to mathematics." Thompson goes on to discuss how there now existed a predictive "mathematical chemistry" based on underlying principles, thereby elevating chemistry from "science" with a small *s* to "Science" with a capital *S*. On the other hand, biology had remained qualitative, without mathematical foundations or principles, so that it was still just "a science" with a lowercase *s*. It would only graduate to becoming "Science" when it incorporated mathematizable physical principles. Despite the extraordinary progress that has been made in the intervening century, I began to discover that the spirit of Thompson's provocative characterization of biology still has some validity today.

Although he was awarded the prestigious Darwin Medal by the Royal Society in 1946, Thompson was critical of conventional Darwinian evolutionary theory because he felt that biologists overemphasized the role of natural selection and the "survival of the fittest" as the fundamental determinants of the form and structure of living organisms, rather than appreciating the importance of the role of physical laws and their mathematical expression in the evolutionary process. The basic question implicit in his challenge remains unanswered: are there "universal laws of life" that can be mathematized so that biology can be formulated as a predictive quantitative Science? He put it this way:

> It behoves us always to remember that in physics it has taken great men to discover simple things. . . . How far even then mathematics will suffice to describe, and physics to explain, the fabric of the body, no man can foresee. It may be that all the laws of energy, and all the properties of matter, and all the chemistry of all the colloids are as powerless to explain the body as they are impotent to comprehend the soul. For my part, I think it is not so. Of how it is that the soul informs the body, physical science teaches me nothing; and that living matter influences and is influenced by mind is a mystery without a clue. Consciousness is not explained to my comprehension by all the nerve-paths and neurons of the physiologist; nor do I ask of physics how goodness shines in one man's face, and evil betrays itself in another. But of the construction and growth and working of the body, as of all else that is of the earth earthy, physical science is, in my humble opinion, our only teacher and guide.

This pretty much expresses the credo of modern-day "complexity science," including even the implication that consciousness is an emergent systemic phenomenon and not a consequence of just the sum of all the "nerve-paths and neurons" in the brain. The book is written in a scholarly but eminently readable style with surprisingly little mathematics. There are no pronouncements of great principles other than the belief that the physical laws of nature, written in the language of mathematics, are the major determinant of biological growth, form, and evolution.

Although Thompson's book did not address aging or death, nor was it particularly helpful or sophisticated technically, its philosophy provided support and inspiration for contemplating and applying ideas and techniques from physics to all sorts of problems in biology. In my own thinking, this led me to perceive our bodies as metaphorical machines that need to be fed, maintained, and repaired but which eventually wear out and "die," much like cars and washing machines. However, to understand how something ages and dies, whether an animal, an automobile, a company, or a civilization, one first needs to understand what the processes and mechanisms are that are keeping it alive, and then discern how these become degraded with time. This naturally leads to considerations of the energy and resources that are required for sustenance and possible growth, and their allocation to maintenance and repair for combating the production of entropy arising from destructive forces associated with damage, disintegration, wear and tear, and so on. This line of thinking led me to focus initially on the central role of metabolism in keeping us alive before asking why it can't continue doing so forever.

2. METABOLIC RATE AND NATURAL SELECTION

Metabolism is *the fire of life* . . . and food, *the fuel of life*. Neither the neurons in your brain nor the molecules of your genes could function without being supplied by metabolic energy extracted from the food you eat. You could not walk, think, or even sleep without being supplied by metabolic energy. It supplies the power organisms need for maintenance, growth, and reproduction, and for specific processes such as circulation, muscle contraction, and nerve conduction.

Metabolic rate is *the* fundamental rate of biology, setting the pace of life for almost everything an organism does, from the biochemical reactions within its cells to the time it takes to reach maturity, and from the rate of uptake of carbon dioxide in a forest to the rate at which its litter breaks down. As was discussed in the opening chapter, the basal metabolic rate of the average human being is only about 90 watts, corresponding to a typical incandescent lightbulb and equivalent to the approximately 2,000 food calories you eat every day.

Like all of life, we evolved by a process of natural selection, interacting with and adapting to our fellow creatures, whether bacteria and viruses, ants and beetles, snakes and spiders, cats and dogs, or grass and trees, and everything else in a continuously challenging and evolving environment. We have been coevolving together in a never-ending multidimensional interplay of interaction, conflict, and adaptation. Each organism, each of its organs and subsystems, each cell type and genome, has therefore evolved following its own unique history in its own ever-changing environmental niche. The principle of natural selection, introduced independently by Charles Darwin and Alfred Russell Wallace, is key to the theory of evolution and the origin of species. Natural selection, or the "survival of the fittest," is the gradual process by which a successful variation in some inheritable trait or characteristic becomes fixed in a population through the differential reproductive success of organisms that have developed this trait by interacting with their environment. As Wallace expressed it, there is sufficiently broad variation that "there is always material for natural selection to act upon in *some* direction that may be advantageous," or as put more succinctly by Darwin: "each slight variation, if useful, is preserved."

Out of this melting pot, each species evolves with its suite of physiological traits and characteristics that reflect its unique path through evolutionary time, resulting in the extraordinary diversity and variation across the spectrum of life from bacteria to whales. So after many millions of years of evolutionary tinkering and adaptation, of playing the game of the survival of the fittest, we human beings have ended up walking on two legs, being around five to six feet high, living for up to one hundred years, having a heart that beats about sixty times a minute, with a systolic blood pressure of about 100 mm Hg, sleeping about eight hours a day, an aorta that's about eighteen inches long, having about five hundred mitochondria in each of our liver cells, and a metabolic rate of about 90 watts.

Is all of this solely arbitrary and capricious, the result of millions of tiny accidents and fluctuations in our long history that have been frozen in place by the process of natural selection, at least for the time being? Or is there some order here, some hidden pattern reflecting other mechanisms at work?

Indeed there is, and explaining it is the segue back to scaling.

3. SIMPLICITY UNDERLYING COMPLEXITY: KLEIBER'S LAW, SELF-SIMILARITY, AND ECONOMIES OF SCALE

We need about 2,000 food calories a day to live our lives. How much food and energy do other animals need? What about cats and dogs, mice and elephants? Or, for that matter, fish, birds, insects, and trees? These questions were posed at the opening of the book where I emphasized that despite the naive expectation from natural selection, and despite the extraordinary complexity and diversity of life, and despite the fact that metabolism is perhaps the most complex physical-chemical process in the universe, metabolic rate exhibits an extraordinarily systematic regularity across all organisms. As was shown in Figure 1, metabolic rate scales with body size in the simplest possible manner one could imagine when plotted logarithmically against mass, namely, as a straight line indicative of a simple power law scaling relationship.

The scaling of metabolic rate has been known for more than eighty years. Although primitive versions of it were known before the end of the nineteenth century, its modern incarnation is credited to the distinguished physiologist Max Kleiber, who formalized it in a seminal paper published in an obscure Danish journal in 1932.[5] I was quite excited when I first came across Kleiber's law because I had presumed that the randomness and unique historical path dependency implicit in how each species had evolved would have resulted in a huge uncorrelated variability among them. Even among mammals, after all, whales, giraffes, humans, and mice don't look very much alike except for some very general features, and each operates in a vastly different environment facing very different challenges and opportunities.

In his pioneering work, Kleiber surveyed the metabolic rates for a spectrum of animals ranging from a small dove weighing about 150 grams to a large steer weighing almost 1,000 kilograms. Over the ensuing years his analysis has been extended by many researchers to include the entire spectrum of mammals ranging from the smallest, the shrew, to the largest, the blue whale, thereby covering more than eight orders of magnitude in mass. Remarkably, and of equal importance, the same scaling has been shown to be valid across all multicellular taxonomic groups including fish, birds, insects, crustacea, and plants, and even to extend down to bacteria and other unicellular organisms.[6] Overall, it encompasses an astonishing twenty-seven orders of magnitude, perhaps the most persistent and systematic scaling law in the universe.

Because the range of animals in Figure 1 spans well over five orders of magnitude (a factor of more than 100,000), from a little mouse weighing only 20 grams (0.02 kg) to a huge elephant weighing almost 10,000 kilograms, we are forced to plot the data *logarithmically*, meaning that the scales on both axes increase by successive factors of ten. For instance, mass increases along the horizontal axis from 0.001 to 0.01 to 0.1 to 10 to 100 kilograms, and so on, rather than *linearly* from 1 to 2 to 3 to 4 kilograms, et cetera. Had we tried to plot this on a standard-size piece of paper using a conventional linear scale, *all* of the data points except the elephant would pile up in the bottom left-hand corner of the graph because even the next lightest animals after the elephant, the bull and the horse, are more than ten times lighter. To be able to distinguish all of them with any reasonable resolution would require a ridiculously large piece of paper more than a kilometer wide. And to resolve the eight orders of magnitude between the shrew and the blue whale it would have to be more than 100 kilometers wide.

So as we saw when discussing the Richter scale for earthquakes in the previous chapter, there are very practical reasons for using logarithmic coordinates for representing data such as this which span many orders of magnitude. But there are also deep conceptual reasons for doing so related to the idea that the structures and dynamics being investigated have *self-similar* properties, which are represented mathematically by simple *power laws*, as I will now explain.

We've seen that a straight line on a logarithmic plot represents a power law whose exponent is its slope (⅔ in the case of the scaling of strength, shown in Figure 7). In Figure 1 you can readily see that for every *four* orders of magnitude increase in mass (along the horizontal axis), metabolic rate increases by only *three* orders of magnitude (along the vertical axis), so the slope of the straight line is ¾, the famous exponent in Kleiber's law. To illustrate more specifically what this implies, consider the example of a cat weighing 3 kilograms that is 100 times heavier than a mouse weighing 30 grams. Kleiber's law can straightforwardly be used to calculate their metabolic rates, leading to around 32 watts for the cat and about 1 watt for the mouse. Thus, even though the cat is 100 times heavier than the mouse, its metabolic rate is only about 32 times greater, an explicit example of economy of scale.

If we now consider a cow that is 100 times heavier than the cat, then Kleiber's law predicts that its metabolic rate is likewise 32 times greater than the cat's, and if we extend this to a whale that is 100 times heavier than the cow, its metabolic rate would be 32 times greater than the cow's. This repetitive behavior, the recurrence in this case of the same factor 32 as we move up in mass by the same repetitive factor of 100, is an example of the general *self-similar* feature of power laws. More generally: if the mass is increased by *any* arbitrary factor at *any* scale (100, in the example), then the metabolic rate increases by the same factor (32, in the example) no matter what the value of the initial mass is, that is, whether it's that of a mouse, cat, cow, or whale. This remarkably systematic repetitive behavior is called *scale invariance* or *self-similarity* and is a property inherent to power laws. It is closely related to the concept of a *fractal,* which will be discussed in detail in the following chapter. To varying degrees, fractality, scale invariance, and self-similarity are ubiquitous across nature from galaxies and clouds to your cells, your brain, the Internet, companies, and cities.

We just saw that a cat that is 100 times heavier than a mouse requires only about 32 times as much energy to sustain it even though it has approximately 100 times as many cells—a classic example of an *economy of scale* resulting from the essential nonlinear nature of Kleiber's law. Naive linear reasoning would have predicted the cat's metabolic rate to have been 100 times larger, rather than only 32 times. Similarly, if the size of an animal is doubled it doesn't need 100 percent more energy to sustain it; it needs only about 75 percent more—thereby saving approximately 25 percent with each doubling. Thus, in a systematically predictable and quantitative way, the larger the organism the less energy has to be produced per cell per second to sustain a gram of tissue. Your cells work less hard than your dog's, but your horse's work even less hard. Elephants are roughly 10,000 times heavier than rats but their metabolic rates are only 1,000 times larger, despite having roughly 10,000 times as many cells to support. Thus, an elephant's cells operate at about a tenth the rate of a rat's, resulting in a corresponding decrease in the rates of cellular damage, and consequently to a greater longevity for the elephant, as will be explained in greater detail in chapter 4. This is an example of how the systematic economy of scale has profound consequences that reverberate across life from birth and growth to death.

4. UNIVERSALITY AND THE MAGIC NUMBER FOUR THAT CONTROLS LIFE

The systematic regularity of Kleiber's law is pretty amazing, but equally surprising is that similar systematic scaling laws hold for almost any physiological trait or life-history event across the entire range of life from cells to whales to ecosystems. In addition to metabolic rates, these include quantities such as growth rates, genome lengths, lengths of aortas, tree heights, the amount of cerebral gray matter in the brain, evolutionary rates, and life spans; a sampling of these is illustrated in Figures 9-12. There are probably well over fifty such scaling laws and—another big surprise—their corresponding exponents (the analog of the ¾ in Kleiber's law) are invariably very close to simple multiples of ¼.

For example, the exponent for growth rates is very close to ¾, for lengths of aortas and genomes it's ¼, for heights of trees ¼, for cross-sectional areas of *both* aortas *and* tree trunks ¾, for brain sizes ¾, for cerebral white and gray matter 5/4, for heart rates *minus* ¼, for mitochondrial densities in cells *minus* ¼, for rates of evolution *minus* ¼, for diffusion rates across membranes *minus* ¼, for life spans ¼ . . . and many, many more. The "minus" here simply indicates that the corresponding quantity *decreases* with size rather than increases, so, for instance, heart rates *decrease* with increasing body size following the ¼ power law, as shown in Figure 10. I can't resist drawing your attention to the intriguing fact that aortas and tree trunks scale in the same way.

Particularly fascinating is the emergence of the number *four* in the guise of the ¼ powers that appear in all of these exponents. It occurs ubiquitously across the entire panoply of life and seems to play a special, fundamental role in determining many of the measurable characteristics of organisms regardless of their evolved design. Viewed through the lens of scaling, a remarkably general *universal* pattern emerges, strongly suggesting that evolution has been constrained by other general physical principles beyond natural selection.

These systematic scaling relationships are highly counterintuitive. They show that almost all the physiological characteristics and life-history events of any organism are primarily determined simply by its size. For example, the pace of biological life decreases systematically and predictably with increasing size: large mammals live longer, take longer to mature, have slower heart rates, and have cells that work less hard than those of small mammals, all to the same *predictable* degree. Doubling the mass of a mammal increases all of its timescales such as its life span and time to maturity by about 25 percent on average and, concomitantly, decreases all rates, such as its heart rate, by the same amount.

Whales live in the ocean, elephants have trunks, giraffes have long necks, we walk on two legs, and dormice scurry around, yet despite these obvious differences, we are all, to a large degree, nonlinearly scaled versions of one another. If you tell me the size of a mammal, I can use the scaling laws to tell you almost everything about the average values of its measurable characteristics: how much food it needs to eat each day, what its heart rate is, how long it will take to mature, the length and radius of its aorta, its life span, how many offspring it will have, and so on. Given the extraordinary complexity and diversity of life, this is pretty amazing.

Opposite and following pages: A small sampling of the many examples of scaling showing their remarkable universality and diversity. (9) Biomass production of both individual insects and insect colonies showing how they both scale with mass with an exponent of ¾, just like metabolic rate of animals shown in Figure 1. (10) Heart rates of mammals scale with mass with an exponent of -¼. (11) The volume of white matter in mammalian brains scales with the volume of gray matter with an exponent of ¾. (12) The scaling of metabolic rate of single cells and bacteria with their mass following the classic ¾ exponent of Kleiber's law for multicellular animals.

FIG. 9

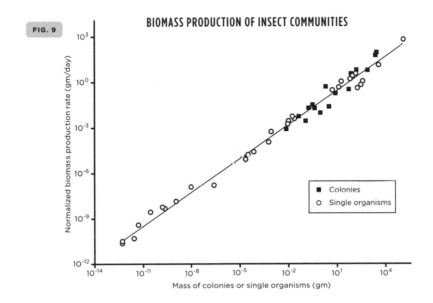

BIOMASS PRODUCTION OF INSECT COMMUNITIES

FIG. 10

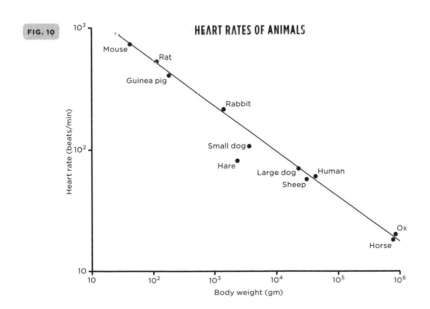

HEART RATES OF ANIMALS

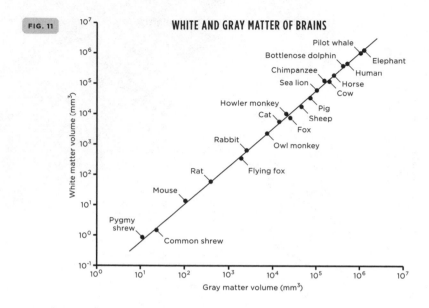

FIG. 11

WHITE AND GRAY MATTER OF BRAINS

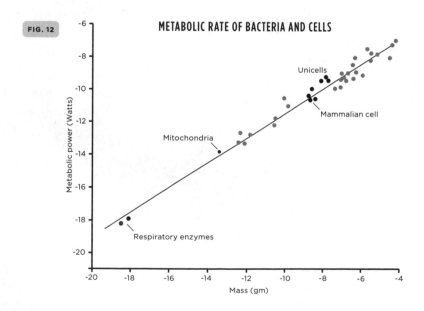

FIG. 12

METABOLIC RATE OF BACTERIA AND CELLS

I was very excited when I realized that my quest to learn about some of the mysteries of death had unexpectedly led me to learn about some of the more surprising and intriguing mysteries of life. For here was an area of biology that was explicitly quantitative, expressible in mathematical terms, and, at the same time, manifested a spirit of "universality" beloved of physicists. In addition to the surprise that these "universal" laws seemed to be at odds with a naive interpretation of natural selection, it was equally surprising that they seemed not to have been fully appreciated by most biologists, even though many were aware of them. Furthermore, there was no general explanation for their origin. Here was something ripe for a physicist to get his teeth into.

Actually, it isn't quite true that scaling laws had been entirely unappreciated by biologists. Scaling laws had certainly maintained an ongoing presence in ecology and, until the advent of the molecular and genomics revolution in biology in the 1950s, had attracted the attention of many eminent biologists, including Julian Huxley, J. B. S. Haldane, and D'Arcy Thompson.[7] Indeed, Huxley coined the term *allometric* to describe how physiological and morphological characteristics of organisms scale with body size, though his focus was primarily on how that occurred during growth. *Allometric* was introduced as a generalization of the Galilean concept of *isometric* scaling, discussed in the previous chapter, where body shape and geometry do not change as size increases, so all lengths associated with an organism increase in the same proportion; *iso* is Greek for "the same," and *metric* is derived from *metrikos*, meaning "measure." *Allometric*, on the other hand, is derived from *allo*, meaning "different," and refers to the typically more general situation where shapes and morphology change as body size increases and different dimensions scale differently. For example, the radii and lengths of tree trunks, or for that matter the limbs of animals, scale differently from one another as size increases: radii scale as the ⅜ power of mass, whereas lengths scale more slowly with ¼ power (that is, as the ⅛ power). Consequently, trunks and limbs become more thickset and stockier as the size of a tree or animal increases; just think of an elephant's legs relative to those of a mouse. This is a generalization of Galileo's original argument regarding the scaling of strength. Had it been isometric, then radii and lengths would have scaled in the same way and the shape of trunks and limbs would have remained unchanged, making the support of the animal or tree unstable as it increases in size. An elephant whose legs have the same spindly shape as a mouse would collapse under its own weight.

Huxley's term *allometric* was extended from its more restrictive geometric, morphological, and ontogenetic origins to describe the kinds of scaling laws that I discussed above, which include more dynamical phenomena such as how flows of energy and resources scale with body size, with metabolic rate being the prime example. All of these are now commonly referred to as *allometric scaling laws*.

Julian Huxley, himself a very distinguished biologist, was the grandson of the famous Thomas Huxley, the biologist who championed Charles Darwin and the theory of evolution by natural selection, and the brother of the novelist and futurist Aldous Huxley. In addition to the word *allometric*, Julian Huxley brought several other new words and concepts into biology, including replacing the much-maligned term *race* with the phrase *ethnic group*.

In the 1980s several excellent books were written by mainstream biologists summarizing the extensive literature on allometry.[8] Data across all scales and all forms of life were compiled and analyzed and it was unanimously concluded that quarter-power scaling was a pervasive feature of biology. However, there was surprisingly little theoretical or conceptual discussion, and no general explanation was given for why there should be such systematic laws, where they came from, or how they related to Darwinian natural selection.

As a physicist, it seemed to me that these "universal" quarter-power scaling laws were telling us something fundamental about the dynamics, structure, and organization of life. Their existence strongly suggested that generic underlying dynamical processes that transcend individual species were at work constraining evolution. This therefore opened a possible window onto underlying emergent laws of biology and led to the conjecture that the generic coarse-grained behavior of living systems obeys *quantifiable* laws that capture their essential features.

It would seem impossible, almost diabolical, that these scaling laws could be just a coincidence, each an independent phenomenon, a "special" case reflecting its own unique dynamics and organization, a wicked series of accidents of evolutionary dynamics, so that the scaling of heart rates is unrelated to the scaling of metabolic rates and the heights of trees. Of course, every individual organism, biological species, and ecological assemblage is unique, reflecting differences in genetic makeup, ontogenetic pathways, environmental conditions, and evolutionary history. So in the absence of any additional physical constraints, one might have expected that different organisms, or at least each group of related organisms inhabiting similar environments, might exhibit different size-related patterns of variation in structure and function. The fact that they don't—that the data almost always closely approximate a simple power law across a broad range of size and diversity—raises some very challenging questions. The fact that the exponents of these power laws are nearly always simple multiples of ¼ poses an even greater challenge.

The question as to what the underlying mechanism for their origin could be seemed a wonderful conundrum to think about, especially given my morbid interest in aging and death and the fact that even life spans seem to scale allometrically with ¼ power (albeit with large variance).

5. ENERGY, EMERGENT LAWS, AND THE HIERARCHY OF LIFE

As I have emphasized, no aspect of life can function without energy. Just as every muscle contraction or any activity requires metabolic energy, so does every random thought in your brain, every twitch of your body even while you sleep, and even the replication of your DNA in your cells. At the most fundamental biochemical level metabolic energy is created in semiautonomous molecular units within cells called *respiratory complexes*. The critical molecule that plays the central role in metabolism goes by the slightly forbidding name of *adenosine triphosphate*, usually referred to as ATP. The detailed biochemistry of metabolism is extremely complicated but in essence it involves the breaking down of ATP, which is relatively unstable in the cellular environment, from adenosine *tri*phosphate (with three phosphates) into ADP, adenosine *di*phosphate (with just two phosphates), thereby releasing the energy stored in the binding of the additional phosphate. The energy derived from breaking this phosphate bond is the source of your metabolic energy and therefore what is keeping you alive. The reverse process converts ADP back into ATP using energy from food via oxidative respiration in mammals such as ourselves (that's why we have to breathe in oxygen), or via photosynthesis in plants. The cycle of releasing energy from the breakup of ATP into ADP and its recycling back from ADP to store energy in ATP forms a continuous loop process much like the charging and recharging of a battery. A cartoon of this process is shown on page 101. Unfortunately it doesn't do justice to the beauty and elegance of this extraordinary mechanism that fuels most of life.

Given its central role, it's not surprising that the flux of ATP is often referred to as the currency of metabolic energy for almost all of life. At any one time our bodies contain only about half a pound (about 250 g) of ATP, but here's something truly extraordinary that you should know about yourself: *every day* you typically make about 2×10^{26} ATP molecules—that's two hundred trillion trillion molecules—corresponding to a mass of about 80 kilograms (about 175 lbs.). In other words, each day you produce and recycle the equivalent of your own body weight of ATP! Taken together, all of these ATPs add up to meet our total metabolic needs at the rate of the approximately 90 watts we require to stay alive and power our bodies.

These little energy generators, the respiratory complexes, are situated on crinkly membranes inside *mitochondria*, which are potato-shaped objects floating around inside cells. Each mitochondrion contains about five hundred to one thousand of these respiratory complexes . . . and there about five hundred to one thousand of these mitochondria inside each of your cells, depending on the cell type and its energy needs. Because muscles require greater access to energy, their cells are densely packed with mitochondria, whereas fat cells have many fewer. So on average each cell in your body may have up to a million of these little engines distributed among its mitochondria working away night and day, collectively manufacturing the astronomical number of ATPs needed to keep you viable, healthy, and strong. The rate at which the total number of these ATPs is produced is a measure of your metabolic rate.

Your body is composed of about a hundred trillion (10^{14}) cells. Even though they represent a broad range of diverse functionalities from neuronal and muscular to protective (skin) and storage (fat), they all share the same basic features. They all process energy in a similar way via the hierarchy of respiratory complexes and mitochondria. Which raises a huge challenge: the five

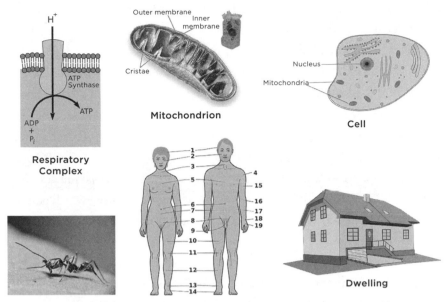

Social Organizations

The energy flow hierarchy of life beginning with respiratory complexes (top left) that produce our energy up through mitochondria and cells (middle and top right) to multicellular organisms and community structures. From this perspective, cities are ultimately powered and sustained by the ATP produced in our respiratory complexes. Although each of these looks quite different with very different engineered structures, energy is distributed through each of them by space-filling hierarchical networks having similar properties.

hundred or so respiratory complexes inside your mitochondria cannot behave as independent entities but have to act collectively in an integrated coherent fashion in order to ensure that mitochondria function efficiently and deliver energy in an appropriately ordered fashion to cells. Similarly, the five hundred or so mitochondria inside each of your cells do not act independently but, like respiratory complexes, have to interact in an integrated coherent fashion to ensure that the 10^{14} cells that constitute your body are supplied with the energy they need to function efficiently and appropriately. Furthermore, these hundred trillion cells have to be organized into a multitude of subsystems such as your various organs, whose energy needs vary significantly depending on demand and function, thereby ensuring that you can do all of the various activities that constitute living, from thinking and dancing to having sex and repairing your DNA. And this entire interconnected multilevel dynamic structure has to be sufficiently robust and resilient to continue functioning for up to one hundred years!

It's natural to generalize this hierarchy of life beyond individual organisms and extend it to community structures. Earlier I talked about how ants collectively cooperate to create fascinating social communities that build remarkable structures by following emergent rules arising from their integrated interactions. Many other organisms, such as bees and plants, form similar integrated communities that take on a collective identity.

The most extreme and astonishing version of this is us. In a very short period of time we have evolved from living in small, somewhat primitive bands of relatively few individuals to dominating the planet with our mammoth cities and social structures encompassing many millions of individuals. Just as organisms are constrained by the integration of the emergent laws operating at the cellular, mitochondrion, and respiratory complex levels, so cities have emerged from, and are constrained by, the underlying emergent dynamics of social interactions. Such laws are not "accidents" but the result of evolutionary processes acting across multiple integrated levels of structure.

This hugely multifaceted, multidimensional process that constitutes life is manifested and replicated in myriad forms across an enormous scale ranging over more than twenty orders of magnitude in mass. A huge number of dynamical agents span and interconnect the vast hierarchy ranging from respiratory complexes and mitochondria to cells and multicellular organisms and up to community structures. The fact that this has persisted and remained so robust, resilient, and sustainable for more than a billion years suggests that *effec-*

tive laws that govern their behavior must have emerged at all scales. Revealing, articulating, and understanding these emergent laws that transcend all of life is the great challenge.

It is within this context that we should view allometric scaling laws: their systematic regularity and universality provides a window onto these emergent laws and underlying principles. As external environments change, all of these various systems must be *scalable* in order to meet the continuing challenges of adaptability, evolvability, and growth. The same generic underlying dynamical and organizational principles must operate across multiple spatial and temporal scales. The scalability of living systems underlies their extraordinary resilience and sustainability both at the individual level and for life itself.

6. NETWORKS AND THE ORIGINS OF QUARTER-POWER ALLOMETRIC SCALING

As I began to ponder what the origins of these surprising scaling laws might be, it became clear that whatever was at play had to be independent of the evolved design of any specific type of organism, because the same laws are manifested by mammals, birds, plants, fish, crustacea, cells, and so on. All of these organisms ranging from the smallest, simplest bacterium to the largest plants and animals depend for their maintenance and reproduction on the close integration of numerous subunits—molecules, organelles, and cells—and these microscopic components need to be serviced in a relatively "democratic" and efficient fashion in order to supply metabolic substrates, remove waste products, and regulate activity.

Natural selection has solved this challenge in perhaps the simplest possible way by evolving hierarchical branching *networks* that distribute energy and materials between macroscopic reservoirs and microscopic sites. Functionally, biological systems are ultimately constrained by the rates at which energy, metabolites, and information can be supplied through these networks. Examples include animal circulatory, respiratory, renal, and neural systems, plant vascular systems, intracellular networks, and the systems that supply food, water, power, and information to human societies. In fact, when you think about it,

Examples of biological networks, counter-clockwise from the top left-hand corner: the circulatory system of the brain; microtubial and mitochondrial networks inside cells; the white and gray matter of the brain; a parasite that lives inside elephants; a tree; and our cardiovascular system.

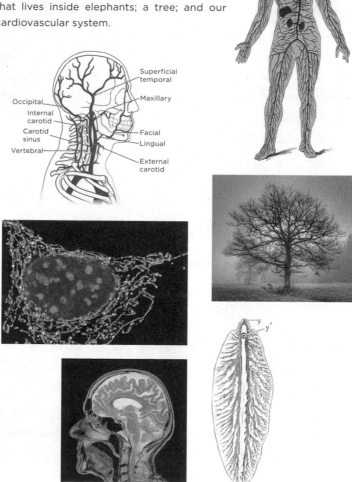

you realize that underneath your smooth skin you are effectively an integrated series of such networks, each busily transporting metabolic energy, materials, and information across all scales. Some of these are above.

Because life is sustained at all scales by such hierarchical networks, it was natural to conjecture that the key to the origin of quarter-power allometric scaling laws, and consequently to the generic coarse-grained behavior of biological systems, lay in the universal physical and mathematical properties of these networks. In other words, despite the great diversity in their evolved structure—some are constructed of tubes like the plumbing in your house, some are bundles of fibers like electrical cables, and some are just diffusive pathways—they are all presumed to be constrained by the same physical and mathematical principles.

7. PHYSICS MEETS BIOLOGY: ON THE NATURE OF THEORIES, MODELS, AND EXPLANATIONS

As I was struggling to develop the network-based theory for the origin of quarter-power scaling, a wonderful synchronicity occurred: I was serendipitously introduced to James Brown and his then student Brian Enquist. They, too, had been thinking about this problem and had also been speculating that network transportation was a key ingredient. Jim is a distinguished ecologist (he was president of the Ecological Society of America when we met) and is well known, among many other things, for his seminal role in inventing an increasingly important subfield of ecology called macroecology.[9] As its name suggests, this takes a large-scale, top-down systemic approach to understanding ecosystems, having much in common with the philosophy inherent in complexity science, including an appreciation of using a coarse-grained description of the system. Macroecology has whimsically been referred to as "seeing the forest for the trees." As we become more concerned about global environmental issues and the urgent need for a deeper understanding of their origins, dynamics, and mitigation, Jim's big-picture vision, articulated in the ideas of macroecology, is becoming increasingly important and appreciated.

When we first met, Jim had only recently moved to the University of New Mexico (UNM), where he is a Distinguished Regents Professor. He had concomitantly become associated with the Santa Fe Institute (SFI), and it was through SFI that the connection was made. Thus began "a beautiful relationship" with Jim, SFI, and Brian and, by extension, with the ensuing cadre of wonderful postdocs and students, as well as other senior researchers who worked with us. Over the ensuing years, the collaboration between Jim, Brian, and me, begun in 1995, was enormously productive, extraordinarily exciting, and tremendous fun. It certainly changed my life, and I venture to say that it did likewise for Brian and Jim, and possibly even for some of the others. But like all excellent, fulfilling, and meaningful relationships, it has also occasionally been frustrating and challenging.

Jim, Brian, and I met every Friday beginning around nine-thirty in the morning and finishing in mid-afternoon by around three with only short breaks for necessities (neither Jim nor I eat lunch). This was a huge commitment, as both of us ran sizable groups elsewhere—Jim had a large ecology group at UNM and I was still running high energy physics at Los Alamos. Jim and Brian very generously drove up most weeks from Albuquerque to Santa Fe, which is about an hour's drive, whereas I did the reverse trip only every few months or so. Once the ice was broken and some of the cultural and language barriers that inevitably arise between fields were crossed, we created a refreshingly open atmosphere where all questions and comments, no matter how "elementary," speculative, or "stupid," were encouraged, welcomed, and treated with respect. There were lots of arguments, speculations, and explanations, struggles with big questions and small details, lots of blind alleys and an occasional aha moment, all against a backdrop of a board covered with equations and hand-drawn graphs and illustrations. Jim and Brian patiently acted as my biology tutors, exposing me to the conceptual world of natural selection, evolution and adaptation, fitness, physiology, and anatomy, all of which were embarrassingly foreign to me. Like many physicists, I was horrified to learn that there were serious scientists who put Darwin on a pedestal above Newton and Einstein. Given the primacy of mathematics and quantitative analysis in my own thinking, I could hardly believe it. However, since I became seriously engaged with biology my appreciation for Darwin's monumental achievements has grown enormously, though I must admit that it's still difficult for me to see how anyone could rank him above the even more monumental achievements of Newton and Einstein.

For my part, I tried to reduce complicated nonlinear mathematical equations and technical physics arguments to relatively simple, intuitive calculations and explanations. Regardless of the outcome, the entire process was a wonderful and fulfilling experience. I particularly enjoyed being reminded of the primal excitement of why I loved being a scientist: the challenge of learning and developing concepts, figuring out what the important questions were, and occasionally being able to suggest insights and answers. In high energy physics, where we struggle to unravel the basic laws of nature at the most microscopic level, we mostly know what the questions are and most of one's effort goes into trying to be clever enough to carry out the highly technical calculations. In biology I found it to be mostly the other way around: months were spent trying to figure out what the problem actually was that we were trying to solve, the questions we should be asking, and the various relevant quantities that were needed to be calculated, but once that was accomplished, the actual technical mathematics was relatively straightforward.

In addition to a strong commitment to solving a fundamental long-standing problem that clearly needed close collaboration between physicists and biologists, a crucial ingredient of our success was that Jim and Brian, as well as being outstanding biologists, thought a lot like physicists and were appreciative of the importance of a mathematical framework grounded in underlying principles for addressing problems. Of equal importance was their appreciation that, to varying degrees, all theories and models are approximate. It is often difficult to see that there are boundaries and limitations to theories, no matter how successful they might have been. This does not mean that they are wrong, but simply that there is a finite range of their applicability. The classic case of Newton's laws is a standard example. Only when it was possible to probe very small distances on the atomic scale or very large velocities on the scale of the speed of light did serious deviations from the predictions from Newton's laws become apparent. And these led to the revolutionary discovery of quantum mechanics to describe the microscopic, and to the theory of relativity to describe ultrahigh speeds comparable to the speed of light. Newton's laws are still applicable and correct outside of these two extreme domains. And here's something of great importance: modifying and extending Newton's laws to these wider domains led to a deep and profound shift in our philosophical conceptual understanding of how everything works. Revolutionary ideas like the realization that the nature of matter itself is fundamentally probabilistic, as embodied in Heisenberg's uncertainty principle, and that space and time are not fixed and absolute, arose out of addressing the limitations of classical Newtonian thinking.

Lest you think that these revolutions in our understanding of fundamental problems in physics are just arcane academic issues, I want to remind you that they have had profound consequences for the daily life of everyone on the planet. Quantum mechanics is the foundational theoretical framework for understanding materials and plays a seminal role in much of the high-tech machinery and equipment that we use. In particular, it stimulated the invention of the laser, whose many applications have changed our lives. Among them are bar code scanners, optical disk drives, laser printers, fiber-optic communications, laser surgery, and much more. Similarly, relativity together with quantum mechanics spawned atomic and nuclear bombs, which changed the entire dynamic of international politics and continue to hang over all of us as a constant, though often suppressed and sometimes unacknowledged, threat to our very existence.

To varying degrees, *all* theories and models are incomplete. They need to be continually tested and challenged by increasingly accurate experiments and observational data over wider and wider domains and the theory modified or extended accordingly. This is an essential ingredient in the scientific method. Indeed, understanding the boundaries of their applicability, the limits to their predictive power, and the ongoing search for exceptions, violations, and failures has provoked even deeper questions and challenges, stimulating the continued progress of science and the unfolding of new ideas, techniques, and concepts.

A major challenge in constructing theories and models is to identify the important quantities that capture the essential dynamics at each organizational level of a system. For instance, in thinking about the solar system, the masses of the planets and the sun are clearly of central importance in determining the motion of the planets, but their color (Mars red, the Earth mottled blue, Venus white, etc.) is irrelevant: the color of the planets is irrelevant for calculating the details of their motion. Similarly, we don't need to know the color of the satellites that allow us to communicate on our cell phones when calculating their detailed motion.

However, this is clearly a scale-dependent statement in that if we look at the Earth from a very close distance of, say, just a few miles above its surface rather than from millions of miles away in space, then what was perceived as its color is now revealed as a manifestation of the huge diversity of the Earth's surface phenomena, which include everything from mountains and rivers to lions, oceans, cities, forests, and us. So what was irrelevant at one scale can become dominant at another. The challenge at every level of observation is to abstract the important variables that determine the dominant behavior of the system.

Physicists have coined a concept to help formalize a first step in this approach, which they call a "toy model." The strategy is to simplify a complicated system by abstracting its essential components, represented by a small number of dominant variables, from which its leading behavior can be determined. A classic example is the idea first proposed in the nineteenth century that gases are composed of molecules, viewed as hard little billiard balls, that are rapidly moving and colliding with one another and whose collisions with the surface of a container are the origin of what we identify as pressure. What we call temperature is similarly identified as the average kinetic energy of the molecules. This was a highly simplified model which in detail is not strictly correct, though it captured and explained for the first time the essential macroscopic coarse-grained features of gases, such as their pressure, temperature, heat conductivity, and viscosity. As such, it provided the point of departure for developing our modern, significantly more detailed and precise understanding not only of gases, but of liquids and materials, by refining the basic model and ultimately incorporating the sophistication of quantum mechanics. This simplified toy model, which played a seminal role in the development of modern physics, is called the "kinetic theory of gases" and was first proposed independently by two of the greatest physicists of all time: James Clerk Maxwell, who unified electricity and magnetism into electromagnetism, thereby revolutionizing the world with his prediction of electromagnetic waves, and Ludwig Boltzmann, who brought us statistical physics and the microscopic understanding of entropy.

A concept related to the idea of a toy model is that of a "zeroth order" approximation of a theory, in which simplifying assumptions are similarly made in order to give a rough approximation of the exact result. It is usually employed in a quantitative context as, for example, in the statement that "a zeroth order estimate for the population of the Chicago metropolitan area in 2013 is 10 million people." Upon learning a little more about Chicago, one might make what could be called a "first order" estimate of its population of 9.5 million, which is more precise and closer to the actual number (whose precise value from census data is 9,537,289). One could imagine that after more detailed investigation, an even better estimate would yield 9.54 million, which would be called a "second order" estimate. You get the idea: each succeeding "order" represents a refinement, an improved approximation, or a finer resolution that converges to the exact result based on more detailed investigation and analysis. In what follows, I shall be using the terms "coarse-grained" and "zeroth order" interchangeably.

This was the philosophical framework that Jim, Brian, and I were exploring when we embarked on our collaboration. Could we first construct a coarse-grained zeroth order theory for understanding the plethora of quarter-power allometric scaling relations based on generic underlying principles that would capture the essential features of organisms? And could we then use it as a point of departure for quantitatively deriving more refined predictions, the higher order corrections, for understanding the dominant behavior of real biological systems?

I later learned that compared with the majority of biologists, Jim and Brian were the exception rather than the rule in appreciating this approach. Despite some of the seminal contributions that physics and physicists have made to biology, a prime example being the unraveling of the structure of DNA, many biologists appear to retain a general suspicion and lack of appreciation of theory and mathematical reasoning.

Physics has benefited enormously from a continuous interplay between the development of theory and the testing of its predictions and implications by performing dedicated experiments. A great example is the recent discovery of the Higgs particle at the Large Hadron Collider at CERN in Geneva. This had been predicted many years earlier by several theoretical physicists as a necessary and critical component of our understanding of the basic laws of physics, but it took almost fifty years for the technical machinery to be developed and the large experimental team assembled to mount a successful search for it. Physicists take for granted the concept of the "theorist" who "only" does theory, whereas by and large biologists do not. A "real" biologist has to have a "lab" or a field site with equipment, assistants, and technicians who observe, measure, and analyze data. Doing biology with just pen, paper, and laptop, in the way many of us do physics, is considered a bit dilettantish and simply doesn't cut it. There are, of course, important areas of biology, such as biomechanics, genetics, and evolutionary biology, where this is not the case. I suspect that this situation will change as big data and intense computation increasingly encroach on all of science and we aggressively attack some of the big questions such as understanding the brain and consciousness, environmental sustainability, and cancer. However, I agree with Sydney Brenner, the distinguished biologist who received the Nobel Prize for his work on the genetic code and who provocatively remarked that "technology gives us the tools to analyze organisms at all scales, but we are drowning in a sea of data and thirsting for some theoretical framework with which to understand it. . . . We need theory and a firm grasp on the nature of the objects we study to predict the rest." His article begins, by the way, with the astonishing pronouncement that "biological research is in crisis."[10]

Many recognize the cultural divide between biology and physics.[11] Nevertheless, we are witnessing an enormously exciting period as the two fields become more closely integrated, leading to new interdisciplinary subfields such as biological physics and systems biology. The time seems right for revisiting D'Arcy Thompson's challenge: "How far even then mathematics will suffice to describe, and physics to explain, the fabric of the body, no man can foresee. It may be that all the laws of energy, and all the properties of matter, all . . . chemistry . . . are as powerless to explain the body as they are impotent to comprehend the soul. For my part, I think it is not so." Many would agree with the spirit of this remark, though new tools and concepts, including closer collaboration, may well be needed to accomplish his lofty goal. I would like to think that the marvelously enjoyable collaboration between Jim, Brian, and me, and all of our colleagues, postdocs, and students has contributed just a little bit to this vision.

8. NETWORK PRINCIPLES AND THE ORIGINS OF ALLOMETRIC SCALING

Prior to this digression into the interrelationship between the cultures of biology and physics, I argued that the mechanistic origins of scaling laws in biology were rooted in the universal mathematical, dynamical, and organizational properties of the multiple networks that distribute energy, materials, and information to local microscopic sites that permeate organisms, such as cells and mitochondria in animals. I also argued that because the structures of biological networks are so varied and stand in marked contrast to the uniformity of the scaling laws, their generic properties must be independent of their specific evolved design. In other words, there must be a common set of network properties that transcends whether they are constructed of tubes as in mammalian circulatory systems, fibers as in plants and trees, or diffusive pathways as in cells.

Formulating a set of general network principles and distilling out the essential features that transcend the huge diversity of biological networks proved to be a major challenge that took many months to resolve. As is often the case when moving into uncharted territory and trying to develop new ideas and ways of looking at a problem, the final product seems so obvious once the discovery or breakthrough has been made. It's hard to believe that it took so long, and one wonders why it couldn't have been done in just a few days. The frustrations and inefficiencies, the blind alleys, and the occasional eureka moments are all part and parcel of the creative process. There seems to be a natural gestation period, and this is simply the nature of the beast. However, once the problem comes into focus and it's been solved it is extremely satisfying and enormously exciting.

This was our collective experience in deriving our explanation for the origin of allometric scaling laws. Once the dust had settled, we proposed the following set of generic network properties that are presumed to have emerged as a result of the process of natural selection and which give rise to quarter-power scaling laws when translated into mathematics. In thinking about them it might be useful to reflect on their possible analogs in cities, economies, companies, and corporations, to which we will turn in some detail in later chapters.

I. Space Filling

The idea behind the concept of space filling is simple and intuitive. Roughly speaking, it means that the tentacles of the network have to extend everywhere throughout the system that it is serving, as is illustrated in the networks on page 104. More specifically: whatever the geometry and topology of the network is, it must service all local biologically active subunits of the organism or subsystem. A familiar example will make it clear: Our circulatory system is a classic hierarchical branching network in which the heart pumps blood through the many levels of the network beginning with the main arteries, passing through vessels of regularly decreasing size, ending with the capillaries, the smallest ones, before looping back to the heart through the venal network system. Space filling is simply the statement that the capillaries, which are the terminal units or last branch of the network, have to service every cell in our body so as to efficiently supply each of them with sufficient blood and oxygen. Actually, all that is required is for capillaries to be close enough to cells for sufficient oxygen to diffuse efficiently across capillary walls and thence through the outer membranes of the cells.

Quite analogously, many of the infrastructural networks in cities are also space filling: for example, the terminal units or end points of the utility networks—gas, water, and electricity—have to end up supplying all of the various buildings that constitute a city. The pipe that connects your house to the water line in the street and the electrical line that connects it to the main cable are analogs of capillaries, while your house can be thought of as an analog to cells. Similarly, all employees of a company, viewed as terminal units, have to be supplied by resources (wages, for example) and information through multiple networks connecting them with the CEO and the management.

II. The Invariance of Terminal Units

This simply means that the terminal units of a given network design, such as the capillaries of the circulatory system that we just discussed, all have approximately the same size and characteristics regardless of the size of the organism. Terminal units are critical elements of the network because they are points of delivery and transmission where energy and resources are exchanged. Other examples are mitochondria within cells, cells within bodies, and petioles (the last branch) of plants and trees. As individuals grow from newborn to adult, or as new species of varying sizes evolve, terminal units do not get reinvented nor are they significantly reconfigured or rescaled. For example, the capillaries of all mammals, whether children, adults, mice, elephants, or whales, are essentially all the same despite the enormous range and variation of body sizes.

This invariance of terminal units can be understood in the context of the parsimonious nature of natural selection. Capillaries, mitochondria, cells, et cetera, act as "ready-made" basic building blocks of the corresponding networks for new species, which are rescaled accordingly. The invariant properties of the terminal units within a specific design characterize the taxonomic class. For instance, all mammals share the same capillaries. Different species within that class such as elephants, humans, and mice are distinguished from one another by having larger or smaller, but closely related, network configurations. From this perspective, the difference between taxonomic groups, that is, between mammals, plants, and fish, for example, is characterized by different properties of the terminal units of their various corresponding networks. Thus while all mammals share similar capillaries and mitochondria, as do all fish, the mammalian ones differ from those of fish in their size and overall characteristics.

Analogously, the terminal units of networks that service and sustain buildings in a city, such as electrical outlets or water faucets, are also approximately invariant. For example, the electrical outlets in your house are essentially identical to those of almost any building anywhere in the world, no matter how big or small it is. There may be small local variations in detailed design but they are all pretty much the same size. Even though the Empire State Building in New York City and many other similar buildings in Dubai, Shanghai, or São Paulo may be more than fifty times taller than your house, all of them, including your house, share outlets and faucets that are very similar. If outlet size was naively scaled isometrically with the height of buildings, then a typical electrical outlet in the Empire State Building would have to be more than fifty times larger than the ones in your house, which means it would be more than ten feet tall and three feet wide rather than just a few inches. And as in biology, basic terminal units, such as faucets and electrical outlets, are not reinvented every time we design a new building regardless of where or how big it is.

III. Optimization

The final postulate states that the continuous multiple feedback and fine-tuning mechanisms implicit in the ongoing processes of natural selection and which have been playing out over enormous periods of time have led to the network performance being "optimized." So, for example, the energy used by the heart of any mammal, including us, to pump blood through the circulatory system is on average minimized. That is, it is the smallest it could possibly be given its design and the various network constraints. To put it slightly differently: of the infinite number of possibilities for the architecture and dynamics of circulatory systems that *could* have evolved, and that are space filling with invariant terminal units, the ones that actually did evolve and are shared by all mammals minimize cardiac output. Networks have evolved so that the energy needed to sustain an average individual's life and perform the mundane tasks of living is *minimized* in order to *maximize* the amount of energy available for sex, reproduction, and the raising of offspring. This maximization of offspring is an expression of what is referred to as *Darwinian fitness,* which is the genetic contribution of an average individual to the next generation's gene pool.

This naturally raises the question as to whether the dynamics and structure of cities and companies are the result of analogous optimization principles. What, if anything, is optimized in their multiple network systems? Are cities organized to maximize social interactions, or to optimize transport by minimizing mobility times, or are they ultimately driven by the ambition of each citizen and company to maximize their assets, profits, and wealth? I will return to these issues in chapters 8, 9, and 10.

Optimization principles lie at the very heart of all of the fundamental laws of nature, whether Newton's laws, Maxwell's electromagnetic theory, quantum mechanics, Einstein's theory of relativity, or the grand unified theories of the elementary particles. Their modern formulation is a general mathematical framework in which a quantity called the *action,* which is loosely related to energy, is minimized. All the laws of physics can be derived from the *principle of least action* which, roughly speaking, states that, of all the possible configurations that a system can have or that it can follow as it evolves in time, the one that is physically realized is the one that minimizes its action. Consequently, the dynamics, structure, and time evolution of the universe since the Big Bang, everything from black holes and the satellites transmitting your cell phone messages to the cell phones and messages themselves, all electrons, photons, Higgs particles, and pretty much everything else that is physical, are determined from such an optimization principle. So why not life?

This question returns us to our earlier discussion concerning the differences between *simplicity* and *complexity.* You may recall that almost all the laws of physics come under the umbrella of simplicity, primarily because they can be expressed in a parsimonious way in terms of just a few compact mathematical equations such as Newton's laws, Maxwell's equations, Einstein's theory of relativity, and so on, all of which can be formulated and elegantly derived from the principle of least action. This is one of the crowning achievements of science and has contributed enormously to our understanding of the world around us and to the remarkable development of our modern technological society. Is it conceivable that the coarse-grained dynamics and structure of *complex adaptive systems,* whether organisms, cities, or companies, could be analogously formulated and derived from such a principle?

It is important to recognize that the three postulates enunciated above are to be understood in a coarse-grained average sense. Let me explain. It may have occurred to you that there must be variation among the almost trillion capillaries in any individual human body, as there must be across all the species of a given taxonomic group, so strictly speaking capillaries cannot be invariant. However, this variation has to be viewed in a relative scale-dependent way. The point is that any variation among capillaries is extremely small compared with the many orders of magnitude variation in body size. For instance, even if the length of mammalian capillaries varied by a factor of two, this is still tiny compared with the factor of 100 million in the variation of their body masses. Similarly, there is relatively little variation in petioles, the last branch of a tree prior to the leaf, or even in the size of leaves themselves, during the growth of a tree from a tiny sapling to a mature tree that might be a hundred or more feet high. This is also true across species of trees: leaves do vary in size but by a relatively small factor, despite huge factors in the variation of their heights and masses. A tree that is just twenty times taller than another does not have leaves whose diameter is twenty times larger. Consequently, the variation among terminal units within a given design is a relatively small secondary effect. The same goes for possible variations in the other postulates: networks may not be precisely space filling or precisely optimized. Corrections due to such deviations and variations are considered to be "higher order" effects in the sense we discussed earlier.

These postulates underlie the zeroth order, coarse-grained theory for the structure, organization, and dynamics of biological networks, and allow us to calculate many of the essential features of what I referred to as the *average idealized organism* of a given size. In order to carry out this strategy and calculate quantities such as metabolic rates, growth rates, the heights of trees, or the number of mitochondria in a cell, these postulates have to be translated into mathematics. The goal is to determine the consequences, ramifications, and predictions of the theory and confront them with data and observations. The details of the mathematics depend on the specific kind of network being considered. As discussed earlier, our circulatory system is a network of pipes driven by a beating heart, whereas plants and trees are networks of bundles of thin fibers driven by a steady nonpulsatile hydrostatic pressure. Fundamental to the conceptual framework of the theory is that, despite these completely different physical designs, both kinds of networks are constrained by the same three postulates: *they are space filling, have invariant terminal units, and minimize the energy needed to pump fluid through the system.*

Carrying out this strategy proved to be quite a challenge, both conceptually and technically. It took almost a year to iron out all of the details, but ultimately we showed how Kleiber's law for metabolic rates and, indeed, quarter-power scaling in general arises from the dynamics and geometry of optimized space-filling branching networks. Perhaps most satisfying was to show how the magic number four arises and where it comes from.[12]

In the following subsections I am going to translate the mathematics of how all of this comes about into English to give you an insight into some surprising ways that our bodies work and how we are intimately related not only to all of life but to the entire physical world around us. This was an extraordinary experience that I hope you will find as fascinating and exciting as I did. Equally satisfying was to extend this framework to address all sorts of other problems such as forests, sleep, rates of evolution, and aging and mortality, some of which I will turn to in the following chapter.

9. METABOLIC RATE AND CIRCULATORY SYSTEMS IN MAMMALS, PLANTS, AND TREES

As was explained earlier, oxygen is crucial for maintaining the continuous supply of ATP molecules that are the basic currency of the metabolic energy that keeps us alive—that's why we have to be continuously breathing. Inhaled oxygen is transported across the surface membranes of our lungs, which are suffused with capillaries, where it is absorbed by our blood and pumped through the cardiovascular system to be delivered to our cells. Oxygen molecules bind to the iron-rich hemoglobin in blood cells, which act as the carriers of oxygen. It is this oxidation process that is responsible for our blood being red in much the same way that iron turns red when it oxidizes to rust in the atmosphere. After the blood has delivered its oxygen to the cells, it loses its red color and turns bluish, which is why veins, which are the vessels that return blood back to the heart and lungs, look blue.

The rate at which oxygen is delivered to cells and likewise the rate at which blood is pumped through our circulatory system are therefore measures of our metabolic rate. Similarly, the rate at which oxygen is inhaled through our mouths and into the respiratory system is also a measure of metabolic rate. These two systems are tightly coupled together so blood flow rates, respiratory rates, and metabolic rates are all proportional to one another and related by simple linear relationships. Thus, hearts beat approximately four times for each breath that is inhaled, regardless of the size of the mammal. This tight coupling of the oxygen delivery systems is why the properties of the cardiovascular and respiratory networks play such an important role in determining and constraining metabolic rate.

The rate at which you use energy to pump blood through the vasculature of your circulatory system is called your *cardiac power output*. This energy expenditure is used to overcome the viscous drag, or friction, on blood as it flows through increasingly narrower and narrower vessels in its journey through the aorta, which is the first artery leaving your heart, down through multiple levels of the network to the tiny capillaries that feed cells. A human aorta is an approximately cylindrical pipe that is about 18 inches long (about 45 cm) and about an inch (about 2.5 cm) in diameter, whereas our capillaries are only about 5 micrometers wide (about a hundredth of an inch), which is somewhat smaller than a hairbreadth.[13] Although a blue whale's aorta is almost a foot in diameter (30 cm), its capillaries are still pretty much the same size as yours and mine. This is an explicit example of the invariance of the terminal units in these networks.

It's much harder to push fluid through a narrow tube than a wider one, so almost all of the energy that your heart expends is used to push blood through the tiniest vessels at the end of the network. It's a bit like having to push juice through a sieve, in this case one that is made up of about 10 billion little holes. On the other hand, you use relatively little energy pumping blood through your arteries or indeed through any of the other larger tubes in the network, even though that's where most of your blood resides.

One of the basic postulates of our theory is that the network configuration has evolved to minimize cardiac output, that is, the energy needed to pump blood through the system. For an arbitrary network where the flow is driven by a pulsatile pump such as our hearts there is another potential source of energy loss in addition to that associated with the viscous drag of blood flowing through capillaries and smaller vessels. This is a subtle effect arising from its pulsatile nature and nicely illustrates the beauty of the design of our cardiovascular system that has resulted from optimizing its performance.

When blood leaves the heart, it travels down through the aorta in a wave motion that is generated by the beating of the heart. The frequency of this wave is synchronous with your heart rate, which is about sixty beats a minute. The aorta branches into two arteries, and when blood reaches this first branch point some of it flows down one tube and some down the other, both in a wavelike motion. A generic feature of waves is that they suffer reflections when they meet a barrier, a mirror being the most obvious example. Light is an electromagnetic wave, so the image that you see is just the reflection from the surface of the mirror of the light waves that originate from your body. Other familiar examples are the reflection of water waves from a barrier, or an echo in which sound waves are reflected from a hard surface.

In a similar fashion, the blood wave traveling along the aorta is partially reflected back when it meets the branch point, the remainder being transmitted down through the daughter arteries. These reflections have potentially very bad consequences because they mean that your heart is effectively pumping against itself. Furthermore, this effect gets hugely enhanced as blood flows down through the hierarchy of vessels because the same phenomenon occurs at each ensuing branch point in the network, resulting in a large amount of energy being expended by your heart just in overcoming these multiple reflections. This would be an extremely inefficient design, resulting in a huge burden on the heart and a huge waste of energy.

To avoid this potential problem and minimize the work our hearts have to do, the geometry of our circulatory systems has evolved so that there are *no* reflections at any branch point throughout the network. The mathematics and physics of how this is accomplished is a little bit complicated, but the result is simple and elegant: the theory predicts that there will be no reflections at any branch point if *the sum of the cross-sectional areas of the daughter tubes leaving the branch point is the same as the cross-sectional area of the parent tube coming into it.*

As an example, consider the simple case where the two daughter tubes are identical and therefore have the same cross-sectional areas (which is approximately correct in real circulatory systems). Suppose that the cross-sectional area of the parent tube is 2 square inches; then, in order to ensure that there are no reflections, the cross-sectional area of each daughter has to be 1 square inch. Because the cross-sectional area of any vessel is proportional to the square of its radius, another way of expressing this result is to say that the square of the radius of the parent tube has to be just twice the square of the radius of each of the daughters. So to ensure that there is no energy loss via reflections as one progresses down the network, the radii of successive vessels must scale in a regular *self-similar* fashion, *decreasing by a constant factor of the square root of two ($\sqrt{2}$) with each successive branching.*

This so-called *area-preserving branching* is, indeed, how our circulatory system is constructed, as has been confirmed by detailed measurements across many mammals—and many plants and trees. This at first seems quite surprising given that plants and trees do not have beating hearts—the flow through their vasculature is steady and nonpulsatile, yet their vasculature scales just like the pulsatile circulatory system. However, if you think of a tree as a bundle of fibers all tightly tied together beginning in the trunk and then sequentially spraying out up through its branches, then it's clear that the cross-sectional area must be preserved all the way up through the hierarchy. This is illustrated below, where this fiber bundle structure is compared with the pipe structure of mammals. An interesting consequence of area-preserving branching is that the cross-sectional area of the trunk is the same as the sum of the cross-sectional areas of all the tiny branches at the end of the network (the petioles). Amazingly, this was known to Leonardo da Vinci. I have reproduced the requisite page of his notebook where he demonstrates this fact.

Although this simple geometric picture demonstrates why trees obey area-

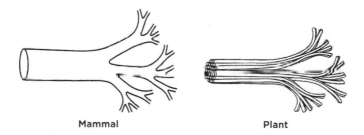

Mammal   Plant

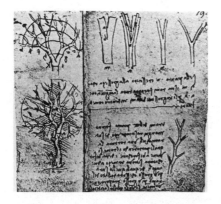

(Above) A schematic of the hierarchical branching pipe structure of mammals (left) and the fiber bundle structure of the vasculature of plants and trees (right); the sequential "unraveling" of fibers forms their physical branch structure. In both cases cutting across any level of branching and adding up the cross-sectional areas results in the same value throughout the network. (Left) A page from da Vinci's notebooks showing that he understood the area-preserving branching of trees.

preserving branching, it is in actuality an oversimplification. However, area preserving can be derived from a much more realistic model for trees using the general network principles of space filling and optimization enunciated above, supplemented with biomechanical constraints that require branches to be resilient against perturbations from wind by bending without breaking. Such an analysis shows that in almost all respects plants and trees scale just like mammals, both within individuals as well as across species, including the ¾ power law for their metabolic rates, even though their physical design is quite different.[14]

10. DIGRESSION ON NIKOLA TESLA, IMPEDANCE MATCHING, AND AC/DC

It's a lovely thought that the optimum design of our circulatory system obeys the same simple area-preserving branching rules that trees and plants do. It's equally satisfying that the condition of nonreflectivity of waves at branch points in pulsatile networks is essentially identical to how national power grids are designed for the efficient transmission of electricity over long distances.

This condition of nonreflectivity is called *impedance matching*. It has multiple applications not only in the working of your body but across a very broad spectrum of technologies that play an important part in your daily life. For example, telephone network systems use matched impedances to minimize echoes on long-distance lines; most loudspeaker systems and musical instruments contain impedance matching mechanisms; and the bones in the middle ear provide impedance matching between the eardrum and the inner ear. If you have ever witnessed or been subject to an ultrasound examination you will be familiar with the nurse or technician smearing a gooey gel over your skin before sliding the probe over it. You probably thought that this was for lubrication purposes but in fact it's actually for matching impedances. Without the gel, the impedance mismatch in ultrasound detection would result in almost all of the energy being reflected back from the skin, leaving very little to go into the body to be reflected back from the organ or fetus under investigation.

The term *impedance matching* can be a very useful metaphor for connoting important aspects of social interactions. For example, the smooth and efficient functioning of social networks, whether in a society, a company, a group activity, and especially in relationships such as marriages and friendships, requires good communication in which information is faithfully transmitted between groups and individuals. When information is dissipated or "reflected," such as when one side is not listening, it cannot be faithfully or efficiently processed, inevitably leading to misinterpretation, a process analogous to the loss of energy when impedances are not matched.

As we became more and more reliant on electricity as a major source of power, the necessity for transmitting it across long distances became a matter of some urgency as the nineteenth century progressed. Not surprisingly, Thomas Edison was a major player in thinking about how this could be accomplished. He subsequently became the great proponent of *direct current* (DC) transmission. You are probably familiar with the idea that electricity comes in two major varieties: direct current (DC), beloved of Edison, in which electricity flows in a continuous fashion like a river, and alternating current (AC), in which it flows in a pulsatile wave motion much like ocean waves or the blood in your arteries. Until the 1880s all commercial electrical current was DC, partly because AC electrical motors had not yet been invented and partly because most transmission was over relatively short distances. However, there were good scientific reasons for favoring AC transmission, especially for long distances, not least of which is that one can take advantage of its pulsatile nature and match impedances at branch nodes in the power grid so as to minimize power loss, just as we do in our circulatory system.

The invention of the AC induction motor and transformer in 1886 by the brilliant charismatic inventor and futurist Nikola Tesla marked a turning point and signaled the beginning of the "war of currents." In the United States this turned into a battle royal between the Thomas Edison Company (later General Electric) and the George Westinghouse Company. Ironically, Tesla had come to the United States from his native Serbia to work for Edison to perfect DC transmission. Despite his success in this endeavor, he moved on to develop the superior AC system, ultimately selling his patents to Westinghouse. Although AC eventually won out and now dominates electrical transmission globally, DC persisted well into the twentieth century. I grew up in houses in England with DC electricity and well remember when our neighborhood was converted to AC and we joined the twentieth century.

You have no doubt heard of Nikola Tesla primarily because his name was co-opted by the much-publicized automobile company that produces sleek upscale electric cars. Until recently he had been all but forgotten except by physicists and electrical engineers. He was famous in his lifetime not only for his major achievements in electrical engineering technology but for his somewhat wild ideas and outrageous showmanship, so much so that he made it onto the cover of *Time* magazine. His research and speculations on lightning, death rays, and improving intelligence via electrical impulses, as well as his photographic memory, his apparent lack of the need for sleep or close human relationship, and his Central European accent led him to become the prototype of the "mad scientist." Although his patents earned him a considerable fortune, which he used to fund his own research, he died destitute in New York in 1943. Over the last twenty years his name has experienced a major resurgence in the popular culture, culminating fittingly in its use for the automobile company.

11. BACK TO METABOLIC RATE, BEATING HEARTS, AND CIRCULATORY SYSTEMS[15]

The theoretical framework discussed in the previous sections explains how cardiovascular systems scale *across* species from the shrew to the blue whale. Equally important, it also explains how they scale *within* an average individual from the aorta to the capillaries. So if for some perverse reason you wanted to know the radius, length, blood flow rate, pulse rate, velocity, pressure, et cetera, in the fourteenth branch of the circulatory system of the average hippopotamus, the theory will provide you with the answer. In fact, the theory will tell you the answer for *any* of these quantities for *any* branch of the network in *any* animal.

As blood flows through smaller and smaller vessels on its way down through the network, viscous drag forces become increasingly important, leading to the dissipation of more and more energy. The effect of this energy loss is to progressively dampen the wave on its way down through the network hierarchy until it eventually loses its pulsatile character and turns into a steady flow. In other words, the nature of the flow makes a transition from being pulsatile in the larger vessels to being steady in the smaller ones. That's why you feel a pulse only in your main arteries—there's almost no vestige of it in your smaller vessels. In the language of electrical transmission, the nature of the blood flow changes from being AC to DC as it progresses down through the network.

Thus, by the time blood reaches the capillaries its viscosity ensures that it is no longer pulsatile and that it is moving extremely slowly. It slows down to a speed of only about 1 millimeter per second, which is tiny compared with its speed of 40 centimeters per second when it leaves the heart. This is extremely important because this leisurely speed ensures that oxygen carried by the blood has sufficient time to diffuse efficiently across the walls of the capillaries and be rapidly delivered to feed cells. Interestingly, the theory predicts that these velocities at the two extremities of the network, the capillaries and the aorta, are the same for all mammals, as observed. You are very likely aware of this huge difference in speeds between capillaries and the aorta. If you prick your skin, blood oozes out very slowly from the capillaries with scant resulting damage, whereas if you cut a major artery such as your aorta, carotid, or femoral, blood gushes out and you can die in just a matter of minutes.

But what's really surprising is that blood pressures are also predicted to be the same across all mammals, regardless of their size. Thus, despite the shrew's heart weighing only about 12 milligrams, the equivalent of about 25 grains of salt, and its aorta having a radius of only about 0.1 millimeter and consequently barely visible, whereas a whale's heart weighs about a ton, almost the weight of a Mini Cooper, and its aorta has a radius of about 30 centimeters, their blood pressures are approximately the same. This is pretty amazing—just think of the enormous stresses on the walls of the shrew's tiny aorta and arteries compared with the pressures on yours or mine, let alone those on a whale's. No wonder the poor creature dies after only a year or two.

The first person to study the physics of blood flow was the polymath Thomas Young. In 1808 he derived the formula for how its velocity depends on the density and elasticity of the arterial walls. His seminal results were of great importance in paving the way for understanding how the cardiovascular system works and for using measurements on pulse waves and blood velocities to probe and diagnose cardiovascular disease. For example, as we age, our arteries harden, leading to significant changes in their density and elasticity, and therefore to predictable changes in the flow and pulse velocity of the blood.

In addition to his work on the cardiovascular system, Young was famous for several other quite diverse and profound discoveries. He is perhaps best known for establishing the wave theory of light, in which each color is associated with a particular wavelength. But he also contributed to early work in linguistics and Egyptian hieroglyphics, including being the first person to decipher the famous Rosetta Stone now sitting in the British Museum in London. As a fitting tribute to this remarkable man, Andrew Robinson wrote a spirited biography of Young titled *The Last Man Who Knew Everything: Thomas Young, the Anonymous Polymath Who Proved Newton Wrong, Explained How We See, Cured the Sick, and Deciphered the Rosetta Stone, Among Other Feats of Genius.* I have a certain soft spot for Young because he was born in Milverton, in the county of Somerset in the West of England, just a few short miles from Taunton, where I was born.

12. SELF-SIMILARITY AND THE ORIGIN OF THE MAGIC NUMBER FOUR

Most biological networks like the circulatory system exhibit the intriguing geometric property of being a *fractal*. You're probably familiar with the idea. Simply put, fractals are objects that look approximately the same at all scales or at any level of magnification. A classic example is a cauliflower or a head of broccoli shown opposite. Fractals are ubiquitous throughout nature, appearing everywhere from lungs and ecosystems to cities, companies, clouds, and rivers. I want to spend this section elaborating on what they are, what they mean, how they are related to power law scaling, and how they are manifested in the circulatory system that we have been discussing.

If broccoli is broken into smaller pieces, each piece looks like a reduced-size version of the original. When scaled back up to the size of the whole, each piece appears indistinguishable from the original. If each of these smaller pieces is likewise broken into even smaller pieces, then these too look like reduced-size versions of the original broccoli. You can imagine repeating this process over and over again with basically the same result, namely that each subunit looks like a scaled-down version of the original whole. To put it slightly differently: if you took a photograph of any of the pieces of broccoli, whatever their size, and blew it up to the size of the original head, you would have a difficult time telling the difference between the blown-up version and the original.

This is in marked contrast to what we normally see when, for example, we use a microscope to zoom in on an object using a higher and higher resolution

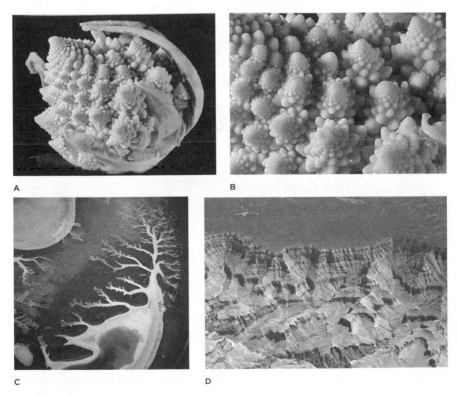

A

B

C

D

Examples of classic fractals and scale invariance; in all cases it's not straightforward to discern the absolute scale. (A) and (B): Romanesco cauliflower at two different resolutions showing its self-similarity. (C): A dried-up riverbed in California. The similarity with a tree in winter, a dried leaf, or our circulatory system is obvious. (D): The Grand Canyon. It could just as well be erosion along the dirt roadway to my house after the runoff following a big storm.

in order to reveal greater detail and new structure that is qualitatively different from those of the whole. Obvious examples are cells in tissue, molecules in materials, or protons in atoms. If, on the other hand, the object is a fractal, no new pattern or detail arises when the resolution is increased: the same pattern repeats itself over and over again. In reality, this is an idealized description for, of course, the images at various levels of resolution differ very slightly from one another and eventually the recursive repetition ceases and new patterns of structural design appear. If you continue breaking down broccoli to increasingly smaller pieces, these eventually lose the geometric characteristics of broccoli and eventually reveal the structure of its tissue, its cells, and its molecules.

This repetitive phenomenon is called *self-similarity* and is a generic characteristic of fractals. Analogous to the repetitive scaling exhibited by broccoli are the infinite reflections in parallel mirrors, or the nesting of Russian dolls (*matryoshka*) of regularly decreasing sizes inside one another. Long before the concept was invented, self-similarity was poetically expressed by the Irish satirist Jonathan Swift, the author of *Gulliver's Travels*, in this whimsical quatrain:

> So, naturalists observe, a flea
> Hath smaller fleas that on him prey;
> And these have smaller still to bite 'em;
> And so proceed ad infinitum.

So it is with the hierarchical networks we have been discussing. If you cut a piece out of such a network and appropriately scale it up, then the resulting network looks just like the original. Locally, each level of the network essentially replicates a scaled version of the levels adjacent to it. We saw an explicit example of this when discussing the consequences of impedance matching in the pulsatile regime of the circulatory system where area-preserving branching resulted in the radii of successive vessels decreasing by a constant factor ($\sqrt{2} = 1.41\ldots$) with each successive branching. So, for example, if we compare the radii of vessels separated by 10 such branchings, then they are related by a scale factor of $(\sqrt{2})^{10} = 32$. Because our aorta has a radius of about 1.5 centimeters, this means that the radii of vessels at the tenth branching level are only about half a millimeter.

Because blood flow changes from pulsatile to nonpulsatile as one progresses down the network, our circulatory system is actually *not* continuously self-similar nor therefore a precise fractal. In the nonpulsatile domain where the flow is dominated by viscous forces, minimizing the amount of power being dissipated leads to a self-similarity in which the radii of successive vessels decrease by a constant factor of the cube root of two $\sqrt[3]{2}$ (= 1.26...), rather than the square root $\sqrt{2}$ (= 1.41...) as in the pulsatile region. Thus the fractal nature of the circulatory system subtly changes from the aorta to the capillaries, reflecting the change in the nature of the flow from pulsatile to nonpulsatile. Trees, on the other hand, maintain approximately the same self-similarity from the trunk to their leaves, with radii successively decreasing by the area-preserving ratio of $\sqrt{2}$.

The space-filling requirement that the network must service the entire volume of the organism at all scales also requires it to be self-similar in terms of the lengths of the vessels. To fill the three-dimensional space, lengths of successive vessels have to decrease by a constant factor of $\sqrt[3]{2}$ with each successive branching and, in contrast to radii, this remains valid down through the entire network, including both the pulsatile and nonpulsatile domains.

Having determined how networks scale *within* individuals following these simple rules, the last piece of the derivation is to determine how this connects *across* species of different weights. This is accomplished from a further consequence of the energy minimization principle: namely, that the total volume of the network—that is, the total volume of blood in the body—must be directly proportional to the volume of the body itself, and therefore proportional to its weight, as observed. In other words, the volume of blood is a constant proportion of the volume of the body, regardless of size. For a tree this is obvious because the network of its vessels constitutes the entire tree—there is no analog of flesh in between all of its branches, so the volume of the network is the volume of the tree.[16]

Now, the volume of the network is just the sum of the volumes of all of its vessels or branches, and these can be straightforwardly calculated from knowing how their lengths and radii scale, thereby connecting the self-similarity of the *internal* network to body size. It is the mathematical interplay between the *cube root* scaling law for lengths and the *square root* scaling law for radii, constrained by the linear scaling of blood volume and the invariance of the terminal units, that leads to quarter-power allometric exponents *across* organisms.

The resulting magic number *four* emerges as an effective extension of the usual *three* dimensions of the volume serviced by the network by an additional dimension resulting from the fractal nature of the network. I shall go into this in more detail in the following chapter, where I discuss the general concept of *fractal dimension*, but suffice it to say here that natural selection has taken advantage of the mathematical marvels of fractal networks to optimize their distribution of energy so that *organisms operate as if they were in four dimensions*, rather than the canonical three. In this sense the ubiquitous number four is actually 3 + 1. More generally, it is the dimension of the space being serviced plus one. So had we lived in a universe of eleven dimensions, as some of my string theory friends believe, the magic number would have been 11 + 1 = 12, and we would have been talking about the universality of $\frac{1}{12}$ power scaling laws rather than $\frac{1}{4}$ power ones.

13. FRACTALS: THE MYSTERIOUS CASE OF THE LENGTHENING BORDERS

Mathematicians had recognized for a long time that there were geometries that lay outside of the canonical boundaries of the classical Euclidean geometry that has formed the basis for mathematics and physics since ancient times. The traditional framework that many of us have been painfully and joyfully exposed to implicitly assumes that all lines and surfaces are smooth and continuous. Novel ideas that evoked concepts of discontinuities and crinkliness, which are implicit in the modern concept of fractals, were viewed as fascinating formal extensions of academic mathematics but were not generally perceived as playing any significant role in the real world. It fell to the French mathematician Benoit Mandelbrot to make the crucial insight that, quite to the contrary, crinkliness, discontinuity, roughness, and self-similarity—in a word, fractality—are, in fact, ubiquitous features of the complex world we live in.[17]

In retrospect it is quite astonishing that this insight had eluded the greatest mathematicians, physicists, and philosophers for more than two thousand years. Like many great leaps forward, Mandelbrot's insight now seems almost "obvious," and it beggars belief that his observation hadn't been made hundreds of years earlier. After all, "natural philosophy" has been one of the major categories of human intellectual endeavor for a very long time, and almost everyone is familiar with cauliflowers, vascular networks, streams, rivers, and mountain ranges, all of which are now perceived as being fractal. However, almost no one had conceived of their structural and organizational regularities in general terms, nor the mathematical language used to describe them. Perhaps, like the erroneous Aristotelian assumption that heavier things "obviously" fall faster, the Platonic ideal of smoothness embodied in classical Euclidean geometry was so firmly ingrained in our psyches that it had to wait a very long time for someone to actually check that it was valid with real-life examples. That person was an unusual British polymath named Lewis Fry Richardson, who almost accidentally laid the foundation for inspiring Mandelbrot's invention of fractals. The tale of how Richardson came to this is an unusually interesting one, which I shall recount very briefly.

Mandelbrot's insights imply that when viewed through a coarse-grained lens of varying resolution, a hidden simplicity and regularity is revealed underlying the extraordinary complexity and diversity in much of the world around us. Furthermore, the mathematics that describes self-similarity and its implicit recursive rescaling is identical to the power law scaling discussed in previous chapters. In other words, power law scaling is the mathematical expression of self-similarity and fractality. Consequently, because animals obey power law scaling both *within* individuals, in terms of the geometry and dynamics of their internal network structures, as well as *across* species, they, and therefore all of us, are living manifestations of self-similar fractals.

Lewis Fry Richardson was a mathematician, physicist, and meteorologist who at the age of forty-six also earned a degree in psychology. He was born in 1881 and early in his career made seminal contributions to our modern methodology of weather forecasting. He pioneered the idea of computationally modeling the weather using the fundamental equations of hydrodynamics (the Navier-Stokes equations introduced earlier when discussing the modeling of ships), augmented and updated with continuous feedback from real-time weather data, such as changes in air pressure, temperature, density, humidity, and wind velocity. He conceived this strategy early in the twentieth century, well before the development of modern high-speed computers, so his computations had to be carried out painfully slowly by hand, resulting in very limited predictive power. Nevertheless, this strategy and the general mathematical techniques he developed provided the foundation for science-based forecasts and pretty much form the template now used to give us relatively accurate weather forecasts for up to several weeks into the future. The advent of high-speed computers, coupled with almost minute-by-minute updating from huge amounts of local data gathered across the globe, has enormously improved our ability to forecast weather.

Both Richardson and Mandelbrot came from relatively unusual backgrounds. Although both were trained in mathematics, neither followed a standard academic career path. Richardson, who was a Quaker, had been a conscientious objector during the First World War and was consequently prevented from having any subsequent university academic position, a rule that might strike us today as particularly vindictive. And Mandelbrot did not get his first tenured professorial appointment until he was seventy-five years old, thereby becoming the oldest professor in Yale's history to receive tenure. Perhaps it requires outliers and mavericks like Richardson and Mandelbrot working outside mainstream research to revolutionize our way of seeing the world.

Richardson had worked for the British Meteorological Office before the war and rejoined it after the war ended only to resign his post a couple of years later, again on conscientious grounds, when the office became part of the Air Ministry in charge of the Royal Air Force. It is curiously fitting that his deeply felt pacifism and consequent fringe connection to the world of mainstream academic research led to his most interesting and seminal observation, namely that measuring lengths isn't as simple as it might appear, thereby bringing to consciousness the role of fractals in our everyday world. To appreciate how he came to this I need to make a small detour into his other accomplishments.

Stimulated by his passionate pacifism, Richardson embarked on an ambitious program to develop a quantitative theory for understanding the origins of war and international conflict in order to devise a strategy for their ultimate prevention. His aim was nothing less than to develop a science of war. His main thesis was that the dynamics of conflict are primarily governed by the rates at which nations build up their armaments and that their continued accumulation is the major cause of war. He viewed the accumulation of weapons as a proxy for the collective psychosocial forces that reflect, but transcend, history, politics, economics, and culture and whose dynamics inevitably lead to conflict and instability. Richardson used the mathematics developed for understanding chemical reaction dynamics and the spread of communicable diseases to model the ever-increasing escalation of arms races in which the arsenal of each country increases in response to the increase in armaments of every other country.

His theory did not attempt to explain the fundamental origins of war, that is, why we collectively resort to force and violence to settle our conflicts, but rather to show how the dynamics of arms races escalate, resulting in catastrophic conflict. Although his theory is highly oversimplified, Richardson had some success in comparing his analyses with data, but more important, he provided an alternative framework for quantitatively understanding the origins of war that could be confronted with data. Furthermore, it had the virtue of showing what parameters were important, especially in providing scenarios under which a potentially peaceful situation could be achieved and sustained. In contrast to conventional, more qualitative theories of conflict, the roles of leadership, cultural and historical animosity, and specific events and personalities play no explicit role in his theory.[18]

In his desire to create a testable scientific framework, Richardson collected an enormous amount of historical data on wars and conflicts. In order to quantify them he introduced a general concept, which he called the *deadly quarrel*, defined as any violent conflict between human beings resulting in death. War is then viewed as a particular case of a deadly quarrel, but so is an individual murder. He quantified their magnitudes by the subsequent number of deaths: for an individual murder the size of the deadly quarrel is therefore just one, whereas for the Second World War it is more than 50 million, the exact number depending on how civilian casualties are counted. He then took the bold leap of asking whether there was a continuum of deadly quarrels beginning with the individual and progressing up through gang violence, civil unrest, small conflicts, and ending up with the two major world wars, thereby covering a range of almost eight orders of magnitude. Trying to plot these on a single axis leads to the same challenge we faced earlier when trying to accommodate all earthquakes or all mammalian metabolic rates on a simple linear scale. Practically, it simply isn't possible, and one has to resort to using a logarithmic scale to see the entire spectrum of deadly quarrels.

Thus, by analogy with the Richter scale, the Richardson scale begins with zero for a single individual murder and ends with a magnitude of almost eight for the two world wars (eight orders of magnitude would represent a hundred million deaths). In between, a small riot with ten victims would have magnitude one, a skirmish in which one hundred combatants were killed would be two, and so on. Obviously there are very few wars of magnitude seven but an enormous number of conflicts with magnitude zero or one. When he plotted the number of deadly quarrels of a given size versus their magnitude on a logarithmic scale, he found an approximately straight line just like the straight lines we saw when physiological quantities like metabolic rate were plotted in this way versus animal size (see Figure 1).

Consequently, the frequency distribution of wars follows simple power law scaling indicating that conflicts are approximately self-similar.[19] This remarkable result leads to the surprising conclusion that, in a coarse-grained sense, a large war is just a scaled-up version of a small conflict, analogous to the way that elephants are approximately scaled-up mice. Thus underlying the extraordinary complexity of wars and conflicts seems to be a common dynamic operating across all scales. Recent work has confirmed such findings for recent wars, terrorist attacks, and even cyberattacks.[20] No general theory has yet been advanced for understanding these regularities, though they very likely reflect the fractal-like network characteristics of national economies, social behavior, and competitive forces. In any case, any ultimate theory of war needs to account for them.

This, at last, leads to the main point of telling the story of Lewis Richardson. He had viewed the power law scaling of conflicts as just one of potentially other systematic regularities concerning war from which he hoped to discover the general laws governing human violence. In trying to develop a theory, he hypothesized that the probability of war between neighboring states was proportional to the length of their common border. Driven by his passion to test his theory, he turned his attention to figuring out how the lengths of borders are measured . . . and in so doing inadvertently discovered fractals.

To test his idea, he set about collecting data on lengths of borders and was surprised to discover that there was considerable variation in the published data. For example, he learned that the length of the border between Spain and Portugal was sometimes quoted as 987 kilometers but other times as 1,214 kilometers, and similarly that the border between the Netherlands and Belgium was sometimes 380 kilometers and at other times 449 kilometers. It was hard to believe that such large discrepancies were errors in measurement. By that time surveying was already a highly developed, well-established, and accurate science. For instance, the height of Mount Everest was known to within a few feet by the end of the nineteenth century. So discrepancies of hundreds of kilometers in the lengths of borders were totally weird. Clearly something else was going on.

Until Richardson's investigation, the methodology of measuring lengths was taken completely for granted. The idea seems so simple it's hard to see what could go wrong. Let's then analyze the process of how we measure lengths. Suppose you want to make a rough estimate of the length of your living room. This can be straightforwardly accomplished by laying down a meter stick end to end (in a straight line) and counting how many times it fits in between the walls. You discover that it takes just over 6 times and so conclude that the room is roughly 6 meters long. Sometime later you find that you need a more accurate estimate and so use the finer-grained resolution of a 10-centimeter ruler to make the estimate. Carefully placing it end to end you find that it takes just under 63 times to fit it across the room, leading to a more accurate approximation for its length of 63 × 10 centimeters, which is 630 centimeters, or 6.3 meters. Obviously you can repeat this process over and over again with finer and finer resolutions depending on how accurately you want to know the answer. If you were to measure the room to an accuracy of millimeters, you might find that its length is 6.289 meters.

In actuality, we don't usually lay down rulers end to end but, for convenience, employ appropriately long continuous tape measures or other measuring devices to relieve us of this tedious process. But the principle remains exactly the same: a tape measure or any other measuring device is simply a sequence of shorter rulers of a given standard length, such as a meter or 10 centimeters, sewed together end to end.

Implicit in our measurement process, whatever it is, is the assumption that with increasing resolution the result converges to an increasingly accurate fixed number, which we call *the* length of the room, a presumably objective property of your living room. In the example, its length converged from 6 to 6.3 to 6.289 meters as the resolution increased. This convergence to a well-defined length seems completely obvious and, indeed, was not questioned for several thousand years until 1950, when Richardson stumbled upon the surprising mystery of the lengthening borders and coastlines.

Let's now imagine measuring the length of the border between two neighboring countries, or the length of a country's coastline, following the standard procedure outlined above. To get a very rough estimate we might start by using 100-mile segments laid end to end to cover its entire length. Suppose we find that with this resolution the border is approximated by just over 12 such segments so that its length is roughly a little over 1,200 miles. To get a more accurate measurement we might then use 10-mile segments to estimate the length. According to the usual "rules of measurement" articulated with the living room example, we might find something like 124 segments, leading to a better estimate of 1,240 miles. Greater accuracy could then be obtained by increasing the resolution to one mile, in which case we might expect to find 1,243 segments, say, leading to a value of 1,243 miles. This can be continued using progressively finer and finer resolutions to obtain as accurate a number as needed.

However, to his great surprise, Richardson found that when he carried out this standard iterative procedure using calipers on detailed maps, this simply wasn't the case. In fact, he discovered that the finer the resolution, and therefore the greater the expected accuracy, the longer the border got, rather than converging to some specific value! Unlike lengths of living rooms, the lengths of borders and coastlines continually *get longer* rather than *converging* to some fixed number, violating the basic laws of measurement that had implicitly been presumed for several thousand years. Equally surprising, Richardson discovered that this increase in length progressed in a systematic fashion. When he plotted the length of various borders and coastlines versus the resolution used to make the measurements on a logarithmic scale, it revealed a straight line indicative of the power law scaling we've seen in many other places (see Figure 14). This was extremely strange, as it indicated that contrary to conventional belief, these lengths seem to depend on the scale of the units used to make the measurement and, in this sense, are *not* an objective property of what is being measured.[21]

So what's going on here? A moment's reflection and you will quickly realize what it is. Unlike your living room, most borders and coastlines are not straight lines. Rather, they are squiggly meandering lines either following local geography or having "arbitrarily" been determined via politics, culture, or history. If you lay a straight ruler of length 100 miles between two points on a coastline or border, as is effectively done when surveying, then you will obviously miss all of the many meanderings and wiggles in between (see Figure 13). If, however, you were to instead use a 10-mile-long ruler then you become sensitive to all of those meanderings and wiggles that you missed whose scale is bigger than 10 miles. This finer resolution can see these finer details and follow the wiggles, thereby leading to an estimate that is necessarily larger than that obtained with the coarser 100-mile scale. Likewise, the 10-mile scale will be blind to similar meanderings and wiggles whose scale is smaller than 10 miles, but which would be included if we increased the resolution to one mile, leading to a further increase in the length. Thus, for lines like the borders and coastlines that Richardson studied with many wiggles and squiggles, we can readily understand how their measured lengths continuously increase with resolution.

Because the increase follows simple power law behavior, these borders are in fact self-similar fractals. In other words, the wiggles and squiggles at one scale are, on average, scaled versions of the wiggles and squiggles at another. So when you've marveled at how the erosion on the bank of a stream looks just like a scaled-down version of the erosion you've seen on a large river, or that it even looks like a mini-version of the Grand Canyon, you weren't fantasizing, it actually is (see page 127).

This is wonderful stuff. Once again we see that underlying the daunting complexity of the natural world lies a surprising simplicity, regularity, and unity when viewed through the coarse-grained lens of scale. Although Richardson discovered this strange, revolutionary, nonintuitive behavior in his investigations of borders and coastlines and understood its origins, he didn't

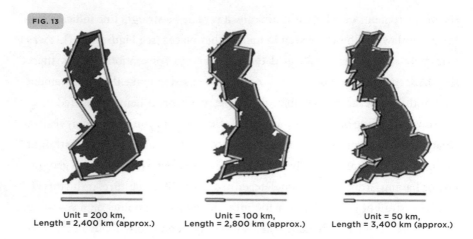

FIG. 13

| Unit = 200 km, Length = 2,400 km (approx.) | Unit = 100 km, Length = 2,800 km (approx.) | Unit = 50 km, Length = 3,400 km (approx.) |

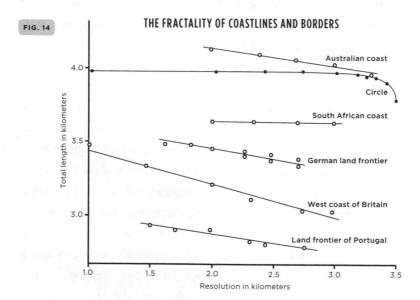

FIG. 14

THE FRACTALITY OF COASTLINES AND BORDERS

Total length in kilometers (y-axis)
Resolution in kilometers (x-axis)

Australian coast
Circle
South African coast
German land frontier
West coast of Britain
Land frontier of Portugal

Measuring the lengths of coastline using different resolutions (Britain in the example). (13) The lengths increase systematically with resolution following a power law as indicated by the examples in the graph. (14) The slope gives the fractal dimension for the coastline: the more squiggly it is, the steeper the slope.

fully appreciate its extraordinary generality and far-reaching implications. This bigger insight fell to Benoit Mandelbrot.

Richardson's discovery was almost entirely ignored by the scientific community. This is not too surprising because it was published in a relatively obscure journal and, in addition, it was buried in the middle of his investigations into the origins of war. His paper, published in 1961, carries the marvelously obscure title "The Problem of Contiguity: An Appendix to Statistics of Deadly Quarrels," barely revealing, even to the cognoscenti, what the content might be. Who was to know that this was to herald a paradigm shift of major significance?

Well, Benoit Mandelbrot did. He deserves great credit not only for resurrecting Richardson's work but for recognizing its deeper significance. In 1967 he published a paper in the high-profile journal *Science* with the more transparent title "How Long Is the Coast of Britain? Statistical Self-Similarity and Fractional Dimension."[22] This brought Richardson's work to light by expanding on his findings and generalizing the idea. Crinkliness, later to become known as fractality, is quantified by how steep the slopes of the corresponding straight lines are on Richardson's logarithmic plots: the steeper the slope, the more crinkly the curve. These slopes are just the exponents of the power laws relating length to resolution and are the analog of the ¾ exponent relating metabolic rate to mass for organisms. For very smooth traditional curves, like circles, the slope or exponent is zero because its length does not change with increasing resolution but converges to a definite value, as in the living room example. However, for rugged, crinkly coastlines the slope is nonzero. For example, for the west coast of Britain, it's 0.25. For more crinkly coastlines like those of Norway with its magnificent fjords and multiple levels of bays and inlets that successively branch into increasingly smaller bays and inlets, the slope has the whopping value of 0.52. On the other hand, Richardson found that the South African coast is unlike almost any other coastline, with a slope of only 0.02, closely approximating a smooth curve. As for the frontier between Spain and Portugal, whose "discrepancies" had originally piqued his interest in this problem, he found a slope of 0.18; see Figure 14.

To appreciate what these numbers mean in English, imagine increasing the resolution of the measurement by a factor of two; then, for instance, the measured length of the west coast of Britain would increase by about 25 percent and that of Norway by over 50 percent. This is an enormous effect, which had been completely overlooked until Richardson stumbled across it just seventy years ago. So for the process of measurement to be meaningful, knowing the resolution is crucial and integral to the entire process.

The take-home message is clear. *In general, it is meaningless to quote the value of a measured length without stating the scale of the resolution used to make it.* In principle, it is as meaningless as saying that a length is 543, 27, or 1.289176 without giving the units it's measured in. Just as we need to know if it is in miles, centimeters, or angstroms, we also need to know the resolution that was used.

Mandelbrot introduced the concept of a *fractal dimension*, defined by adding 1 to the exponent of the power law (the value of the slopes). Thus the fractal dimension of the South African coast is 1.02, Norway 1.52, and so on. The point of adding the 1 was to connect the idea of fractals to the conventional concept of ordinary dimensions discussed in chapter 2. Recall that a smooth line has dimension 1, a smooth surface dimension 2, and a volume dimension 3. Thus the South African coast is very close to being a smooth line because its fractal dimension is 1.02, which is very close to 1, whereas Norway is far from it because its fractal dimension of 1.52 is so much greater than 1.

You could imagine an extreme case of this in which the line is so crinkly and convoluted that it effectively fills an entire area. Consequently, even though it's still a line with "ordinary" dimensions 1, it behaves as if it were an *area* in terms of its scaling properties, therefore having a fractal dimension of 2. This curious gain of an effective additional dimension is a general feature of space-filling curves, to which I will return in the next chapter.

In the natural world almost nothing is smooth—most things are crinkly, irregular, and crenulated, very often in a self-similar way. Just think of forests, mountain ranges, vegetables, clouds, and the surfaces of oceans. Consequently, most physical objects have no absolute objective length, and it is crucial to quote the resolution when stating the measurement. So why did it take more than two thousand years for people to recognize something so basic and which now seems almost obvious? Very likely this has its origins in the duality that emerged as we gradually separated from a close connection to the natural world and became more and more distant from the forces of nature that have determined our biology. Once we invented language, learned how to take advantage of economies of scale, formed communities, and began making artifacts, we effectively changed the geometry of our daily world and its immediate surroundings. In designing and manufacturing human-engineered artifacts, whether primitive pots and tools or modern sophisticated automobiles, computers, and skyscrapers, we employed and aspired to the simplicity of straight lines, smooth curves, and smooth surfaces. This was brilliantly formalized and reflected in the development of quantified measurement and the invention of mathematics, manifested, in particular, in the idealized paradigm of Euclidian geometry. This is the mathematics appropriate to the world of artifacts we created around us as we evolved from being a mammal like any other to become social *Homo sapiens*.

In this new world of artifacts we inevitably became conditioned to seeing it through the lens of Euclidian geometry—straight lines, smooth curves, and smooth surfaces—blinding ourselves, at least as scientists and technologists, to the seemingly messy, complex, convoluted world of the environment from which we had emerged. This was mostly left to the imagination of artists and writers. Although measurement plays a central role in this new, more regular artificial world, it has the elegant simplicity of Euclid, so there is no need to be concerned with awkward questions like that of resolution. In this new world, a length is a length is a length, and that's it. Not so, however, in the immediate "natural" world around us, which is highly complex and dominated by crinkles, wrinkles, and crenulations. As Mandelbrot succinctly put it: "Smooth shapes are very rare in the wild but extremely important in the ivory tower and the factory."

From the beginning of the nineteenth century mathematicians had already contemplated curves and surfaces that were not smooth but they had not been motivated by the pervasiveness of such geometries in the natural world. Their motivation was simply to explore new ideas and concepts that were primarily of academic interest, such as whether it was possible to formulate consistent geometries that violate the sacred tenets of Euclid.

To which the answer was yes, and Mandelbrot was well positioned to take advantage of this. In contrast to Richardson, Mandelbrot had been educated in the more formal tradition of classical French mathematics and was familiar with the strange world of abstract, densely crinkled, non-Euclidean curves and surfaces. His great contribution was to see that what Richardson had discovered could be put on a firm mathematical basis and that the weird geometries that academic mathematicians had been playing with and which seemed to have nothing to do with "reality" had, in fact, everything to do with reality—and, in some respects, possibly even more so than Euclidean geometry.

Perhaps of greater importance is that he realized that these ideas are generalizable far beyond considerations of borders and coastlines to almost anything that can be measured, even including times and frequencies. Examples include our brains, balls of crumpled paper, lightning, river networks, and time series like electrocardiograms (EKGs) and the stock market. For instance, it turns out that the pattern of fluctuations in financial markets during an hour of trading is, on average, the same as that for a day, a month, a year, or a decade. They are simply nonlinearly scaled versions of one another. Thus if you are shown a typical plot of the Dow Jones average over some period of time, you can't tell if it's for the last hour or for the last five years—the distributions of dips, bumps, and spikes is pretty much the same, regardless of the time frame. In other words, the behavior of the stock market is a self-similar fractal pattern that repeats itself across all timescales following a power law that can be quantified by its exponent or, equivalently, its fractal dimension.

You might think that with this knowledge you might soon become rich. Although this certainly gives new insight into hidden regularities in stock markets, unfortunately it has predictive power only in an average coarse-grained sense and does not give specific information about the behavior of individual stocks. Nevertheless, it's an important ingredient for understanding the dynamics of the market over different timescales. This has stimulated the development of a new transdisciplinary subfield of finance called *econophysics* and motivated investment companies to hire physicists, mathematicians, and computer scientists to use these sorts of ideas to develop novel investment strategies.[23] Many have done very well, though it is unclear just how big a role their physics and mathematics actually played in their success.

Likewise, the self-similarity observed in EKGs is a potentially important gauge of the condition of our hearts. You might have thought that the healthier the heart the smoother and more regular would be the EKG, that is, that a healthy heart would have a low fractal dimension compared with a more diseased one. Quite the contrary. Healthy hearts have relatively high fractal dimensions, reflecting more spiky and ragged EKGs, whereas diseased hearts have low values with relatively smooth EKGs. In fact, those that are most seriously at risk have fractal dimensions close to one with an uncharacteristically smooth EKG. Thus the fractal dimension of the EKG provides a potentially powerful complementary diagnostic tool for quantifying heart disease and health.[24]

The reason that being healthy and robust equates with greater variance and larger fluctuations, and therefore a larger fractal dimension as in an EKG, is closely related to the resilience of such systems. Being overly rigid and constrained means that there isn't sufficient flexibility for the necessary adjustments needed to withstand the inevitable small shocks and perturbations to which any system is subjected. Think of the stresses and strains your heart is exposed to every day, many of which are unexpected. Being able to accommodate and naturally adapt to these is critical for your long-term survival. These continuous changes and impingements require all of your organs, including your brain as well as its psyche, to be both flexible and elastic and therefore to have a significant fractal dimension.

This can be extended, at least metaphorically, beyond individuals to companies, cities, states, and even life itself. Being diverse and having many interchangeable, adaptable components is another manifestation of this paradigm. Natural selection thrives on and consequently manufactures greater diversity. Resilient ecosystems have greater diversity of species. It is no accident that successful cities are those that offer a greater spectrum of job opportunities and businesses, and that successful companies have a diversity of products and people with the flexibility to change, adapt, and reinvent in response to changing markets. I shall discuss this further in chapters 8 and 9 when I turn to cities and companies.

In 1982 Mandelbrot published a highly influential and very readable semipopular book titled *The Fractal Geometry of Nature*.[25] This inspired tremendous interest in fractals by showing their ubiquity across both science and the natural world. It stimulated a mini industry searching for fractals, finding them everywhere, measuring their dimensions, and showing how their magical properties result in wonderfully exotic geometric figures.

Mandelbrot had shown how relatively simple algorithmic rules based on fractal mathematics can produce surprisingly complex patterns. He, and later many others, produced amazingly realistic simulations of mountain ranges and landscapes, as well as intriguing psychedelic patterns. This was enthusiastically embraced by the film and media industries, so much so that a great deal of what you now see on the screen and in advertisements, whether "realistic" battle scenes, glorious landscapes, or futuristic fantasy, is based on fractal paradigms. *The Lord of the Rings, Jurassic Park,* and *Game of Thrones* would be drab versions of realistic fantasies without the early work and insights on fractals.

Fractals even showed up in music, painting, and architecture. It is claimed that the fractal dimensions of musical scores can be used to quantify the signature nature and characteristics of different composers, such as between Beethoven, Bach, and Mozart, while the fractal dimensions of Jackson Pollock's paintings were used to distinguish fakes from the real thing.[26]

Although there is a mathematical framework for describing and quantifying fractals, no fundamental theory based on underlying physical principles for mechanistically understanding why they arise in general, or for calculating their dimensions, has been developed. Why are coastlines and borders fractal, what were the dynamics that gave rise to their surprising regularity and determined that South Africa should have a relatively smooth coastline, whereas Norway a rugged one? And what are the common principles and dynamics that link these disparate phenomena to the behavior of stock markets, cities, vascular systems, and EKGs?

A fractal dimension is just one single metric out of many that characterize such systems. It is amazing how much store we put into single metrics such as these. For example, the Dow Jones Industrial Average is almost religiously perceived as *the* indicator of the overall state of the U.S. economy, just as body temperature is typically used as an indicator of our overall health. Better is to have a suite of such metrics such as you get from an annual physical examination, or what economists generate in order to get a broader picture of the state of the economy. But much better still is to have a general quantitative theory and conceptual framework supplemented by dynamical models for mechanistically understanding why the various metrics are the sizes they are and to be able to predict how they will evolve.

In this context, just knowing Kleiber's law for how metabolic rates scale, or even knowing all of the other allometric scaling laws obeyed by organisms, does *not* constitute a theory. Rather, these phenomenological laws are a sophisticated summary of enormous amounts of data that reveal and encapsulate the systematic, generic features of life. Being able to derive them analytically from a parsimonious set of general underlying principles such as the geometry and dynamics of networks at increasingly finer levels of granularity provides a deep understanding of their origins, leading to the possibility of addressing and predicting other and new phenomena. In the following chapter I will show how the network theory provides such a framework and present a few chosen examples to illustrate the point.

One final note: Mandelbrot showed surprisingly little interest in understanding the mechanistic origins of fractals. Having revealed to the world their extraordinary universality, his passion remained more with their mathematical description than with their physical origins. His attitude seemed to be that they were a fascinating property of nature and we should delight in their ubiquity, simplicity, complexity, and beauty. We should develop a mathematics to describe and use them, but we should not be too concerned about the underlying principles for how they are generated. In a word, he approached them more as a mathematician than as a physicist. This may have been one of the reasons that his great discovery did not receive quite the appreciation in the physics community and scientific establishment that it perhaps deserved and, as a result, he did not receive the Nobel Prize, despite broad recognition in many quarters and a litany of prestigious awards and prizes.

THE FOURTH DIMENSION OF LIFE

Growth, Aging, and Death

Almost all of the networks that sustain life are approximately self-similar fractals. In the previous chapter, I explained how the nature and origin of these fractal structures are a consequence of generic geometric, mathematical, and physical principles such as optimization and space filling, thereby leading to the derivation for how networks scale both within an average individual as well as across species.

Most of my discussion concentrated on circulatory systems, but the same principles apply to respiratory systems, plants, trees, insects, and cells. Indeed, a major success of the theory is that the same set of network principles leads to similar scaling laws in systems with very different evolved designs. Not only do they explain the origin of ubiquitous quarter-power scaling across different taxonomic groups but they also show, for instance, why aortas scale in the same way as tree trunks. Many such quantities can be calculated using the theory, and the accompanying tables, reproduced from our original papers published in *Science* and *Nature*, offer a sampling to illustrate the extent of its predictive power. Shown are predictions for a plethora of measured quantities associated with the circulatory, respiratory, plant, and forest community systems compared with observation. As you can readily see the agreement is generally excellent.

TABLE 1 **Cardiovascular**

QUANTITY	PREDICTED	OBSERVED
Aorta radius	⅜ = 0.375	0.36
Aorta pressure	0 = 0.00	0.032
Aorta blood velocity	0 = 0.00	0.07
Blood volume	1 = 1.00	1.00
Circulation time	¼ = 0.25	0.25
Circulation distance	¼ = 0.25	No data
Cardiac stroke volume	1 = 1.00	1.03
Cardiac frequency	-¼ = -0.25	-0.25
Cardiac output	¾ = 0.75	0.74
Number of capillaries	¾ = 0.75	No data
Service volume radius	No data	No data
Womersley number	¼ = 0.25	0.25
Density of capillaries	-1/12 = -0.083	-0.095
O_2 affinity of blood	-1/12 = -0.083	-0.089
Total resistance	-¾ = -0.75	-0.76
Metabolic rate	¾ = 0.75	0.75

TABLE 2 Respiratory

QUANTITY	PREDICTED	OBSERVED
Tracheal radius	⅜ = 0.375	0.39
Interpleural pressure	0 = 0.00	0.004
Air velocity in trachea	0 = 0.00	0.02
Lung volume	1 = 1.00	1.05
Volume flow to lung	¾ = 0.75	0.80
Volume of alveolus	¼ = 0.25	No data
Tidal volume	1 = 1.00	1.041
Respiratory frequency	-¼ = -0.25	-0.26
Power dissipated	¾ = 0.75	0.78
Number of alveoli	¾ = 0.75	No data
Radius of alveolus	¹⁄₁₂ = 0.083	0.13
Area of alveolus	⅙ = 0.167	No data
Area of lung	¹¹⁄₁₂ = 0.92	0.95
O_2 diffusing capacity	1 = 1.00	0.99
Total resistance	-¾ = -0.75	-0.70
O_2 consumption rate	¾ = 0.75	0.76

TABLE 3 Predicted Values of Scaling Exponents for Physiological and Anatomical Variables of Plant Vascular Systems

QUANTITY	AS A FUNCTION OF PLANT MASS	AS A FUNCTION OF TRUNK OR STEM RADIUS	
	EXPONENT	EXPONENT	
	PREDICTED	PREDICTED	OBSERVED
Number of leaves	¾ (0.75)	2 (2.00)	2.007
Number of branches	¾ (0.75)	-2 (-2.00)	-2.00
Number of tubes	¾ (0.75)	2 (2.00)	No data
Trunk length	¼ (0.25)	⅔ (0.67)	0.652
Trunk radius	⅜ (0.375)		
Area of conductive tissue	⅞ (0.0625)	7/3 (2.33)	2.13
Tube radius	1/16 (0.0625)	⅙ (0.167)	No data
Conductivity	1 (1.00)	8/3 (2.67)	2.63
Leaf-specific conductivity	¼ (0.25)	⅔ (0.67)	0.727
Fluid flow rate		2 (2.00)	No data
Metabolic rate	¾ (0.75)		
Pressure gradient	-¼ (-0.25)	-⅔ (-0.67)	No data
Fluid velocity	-⅛ (-0.125)	-⅓ (-0.33)	No data
Branch resistance	-¾ (-0.75)	-⅓ (-0.33)	No data

Despite being based on the same set of principles, the actual mathematics and dynamics are typically quite different in each case, reflecting the different physical structures of the networks. I will not belabor how these various systems exploit the same set of principles, but in all cases the results are very similar and quarter-power scaling emerges.

This is all very satisfying, but a lingering question remains: why does the same ¼ power exponent emerge from each of these different networks rather than, say, a ⅙ power emerging in one, a ⅛ power in another, and so on? In other words, what necessitates that this same set of principles should lead to the same scaling exponents when applied to different network systems with a variety of structures and dynamics? Are there additional design principles that transcend the dynamics that ensure that the ¼ emerges in virtually all organismic groups? This is an important conceptual question, especially for understanding why this universal behavior extends even to systems such as bacteria, where an explicit hierarchical branching network structure is much less obvious.

1. THE FOURTH DIMENSION OF LIFE

A general argument to address this can be made by recognizing that, in addition to *minimizing* energy loss, natural selection has also led to a *maximization* of metabolic capacity because metabolism produces the energy and materials required to sustain and reproduce life.[1] This has been achieved by maximizing surface areas across which resources and energy are transported. These surfaces are in actuality the total surface areas of all the terminal units of the network. For instance, all of our metabolic energy is transmitted across the total surface area of all of our capillaries to fuel our cells, just as the metabolism of a tree is governed by the transmission of energy gathered from sunlight through all of its leaves to fuel photosynthesis and of water from soil through all of the terminal fibers of its root system. Terminal units therefore play a critical role not only because they are invariant but also because they are the interface with the resource environment, whether internal as in the case of capillaries or external as in the case of leaves. As we shall see later, this central role as the gateway for the exchange of energy is critical to many aspects of life, from determining how long you sleep to how long you live.

Natural selection has taken advantage of the fractal nature of space-filling networks to maximize the total effective surface area of these terminal units and thereby maximize metabolic output. Geometrically, the nested levels of continuous branching and crenulations inherent in fractal-like structures optimize the transport of information, energy, and resources by maximizing the surface areas across which these essential features of life flow. Because of their fractal nature, these effective surface areas are very much larger than their apparent physical size. Let me give you some remarkable examples from your own body to illustrate the point.

Even though your lungs are only about the size of a football with a volume of about 5 to 6 liters (about one and a half gallons), the total surface area of the alveoli, which are the terminal units of the respiratory system where oxygen and carbon dioxide are exchanged with the blood, is almost the size of a tennis court and the total length of all the airways is about 2,500 kilometers, almost the distance from Los Angeles to Chicago, or London to Moscow. Even more striking is that if all the arteries, veins, and capillaries of your circulatory system were laid end to end, their total length would be about 100,000 kilometers, or nearly two and a half times around the Earth or over a third of the distance to the moon . . . and all of this neatly fits inside your five-to-six-foot-tall body. It's quite fantastic and yet another amazing feature of your body where natural selection has exploited the wonders of physics, chemistry, and mathematics.

This remarkable phenomenon is an extreme case of what Richardson discovered and Mandelbrot articulated about coastlines and frontiers, namely that lengths and areas are not always what they seem to be. As explained in the previous chapter, a crinkly enough line that is space filling can scale as if it's an area. Its fractality effectively endows it with an additional dimension. Its conventional Euclidean dimension, discussed in chapter 2, still has the value 1, indicating that it's a line, but its fractal dimension is 2, indicating that it's maximally fractal and scaling as if it were an area. In a similar fashion an area, if crinkly enough, can behave as if it's a volume, thereby gaining an effective extra dimension: its Euclidean dimension is 2, indicating that it's an area, but its fractal dimension is 3.

A familiar example will make this clear. Think of washing sheets. Being sensitive to conserving energy and at the same time wanting to save yourself money and time, you wait several weeks until you have more than a sufficient number of dirty ones to fill the entire tub of your washing machine. So when the time comes you stuff in as much and as many as you possibly can to fill the entire volume of the tub. Now, recall that ordinary volumes scale faster than areas, so if you were to double the size of your washing machine by doubling all of its lengths while keeping its shape the same, its volume would increase by a factor of eight (2^3) whereas all of its surface areas would increase by a factor of four (2^2). Naively, you might therefore conclude that because sheets are essentially all area and consequently two-dimensional (their thickness being negligible), you could accommodate four times as many sheets by doubling the size of your washing machine. However, if we stuff all of the sheets into the tub so that they completely fill its entire volume and because this volume has increased by a factor of eight, then it's clear that you can actually accommodate eight times as many sheets, rather than just four times. In other words, the total *effective* area of two-dimensional sheets filling three-dimensional washing machines scales like a volume rather than an area, so in this sense, we have turned an area into a volume.

The reason for this is that we have taken smooth Euclidean surfaces, the sheets, and crumpled them up to create a huge number of crinkles and wrinkles, thereby turning them into fractals. Indeed, the distribution of the sizes of the wrinkles follows a classic power law: there are very few long creases but lots of very small ones, their numbers following a power law distribution. This has actually been tested and verified in experiments on crumpled balls of paper.[2] In reality, you can't completely crumple all of the sheets in your washing machine, or balls of paper for that matter, so that they are entirely space filling, but you can come pretty close; and this is reflected in their measured fractal dimensions actually being a little less than 2. Nor would you want to completely crumple them up as it's likely that the machine wouldn't do a very good job of cleaning them if they were so tightly compressed.

However, driven by the forces of natural selection to maximize exchange surfaces, biological networks do achieve maximal space filling and consequently scale like three-dimensional volumes rather than two-dimensional Euclidean surfaces. This additional dimension, which arises from optimizing network performance, leads to organisms' functioning as if they are operating in four dimensions. This is the geometric origin of the quarter power. Thus, instead of scaling with classic ⅓ exponents, as would be the case if they were smooth nonfractal Euclidean objects, they scale with ¼ exponents. Although living things occupy a three-dimensional space, their internal physiology and anatomy operate as if they were four-dimensional.

It is no accident, therefore, that many biological networks exhibit area-preserving branching, even though different anatomical designs exploit different dynamical scenarios. Unlike the genetic code, which has evolved only once in the history of life, fractal-like distribution networks that confer an additional effective fourth dimension have originated many times. Examples include surface areas of leaves, gills, lungs, guts, kidneys, mitochondria, and the branching architectures of diverse respiratory and circulatory systems from trees to sponges. It is not surprising, therefore, that even unicellular organisms, such as bacteria, have taken advantage of this and exhibit quarter-power scaling.

Quarter-power scaling laws are perhaps as universal and as uniquely biological as the biochemical pathways of metabolism, the structure and function of the genetic code, and the process of natural selection. The vast majority of organisms exhibit scaling exponents very close to ¾ for metabolic rate and ¼ for internal times and distances. These are the maximal and minimal values, respectively, for the effective surface area and linear dimensions of a volume-filling fractal-like network. This is testimony to the power of natural selection to have exploited variations on this fractal theme to produce an incredible variety of biological forms and functions. But it is also testimony to the severe geometric and physical constraints on metabolic processes which have dictated that all of these organisms obey a common set of quarter-power scaling laws. Fractal geometry has literally given life an added dimension.

In marked contrast to this, almost none of our man-made engineered artifacts and systems, whether automobiles, houses, washing machines, or television sets, invoke the power of fractals to optimize performance. To a very limited extent, electronic equipment such as computers and smart phones does, but compared with how you work they are extraordinarily primitive. On the other hand, human-engineered systems that have grown organically such as cities, and to a limited extent corporations, have unconsciously evolved self-similar fractal structures which have tended to optimize their performance. More on this in chapters 8 and 9.

2. WHY AREN'T THERE MAMMALS
THE SIZE OF TINY ANTS?

Idealized mathematical fractals continue "forever." The repetitive self-similarity persists ad infinitum unbounded from the infinitesimally small to the infinitely large. But in real life there are clear limits. You can break down broccoli only so far before it eventually loses its self-similar characteristics and reveals the underlying structure and geometry of its tissues, cells, and ultimately its molecular constituents. A related question is how far you can scale down, or for that matter scale up, a mammal before it is no longer a mammal. In other words, what determines the maximum and minimum size of mammals? Or maybe there aren't any limits, in which case one might still ask why there's no mammal smaller than a shrew, which weighs just a few grams, or larger than a blue whale, which weighs more than a hundred million grams.

The answer lies in the subtleties of networks and their interplay with physiological limits in the spirit of Galileo's original argument that there are limits to the maximum size of structures. Unlike most biological networks, mammalian circulatory systems are not single self-similar fractals but an admixture of two different ones, reflecting the change in flow from predominantly pulsatile AC to predominantly nonpulsatile DC as blood flows from the aorta to the capillaries. Most of the blood resides in the larger vessels of the upper part of the network where AC dominates, leading to the ¾ power scaling law for metabolic rates.

Although the branching changes continuously from one mode to the other, the region over which it changes is relatively narrow and its location (as measured by the number of branchings up from the capillaries) is independent of body size and therefore the same for all mammals. In other words, all mammals have roughly the same number of branching levels, about fifteen, where the flow is predominantly steady nonpulsatile DC. The distinction among mammals as their size increases is the increasing number of levels where the flow is pulsatile AC. For example, we have about seven to eight, the whale has about sixteen to seventeen, and the shrew just one or two. Impedance matching in these vessels ensures that relatively little energy is required to pump

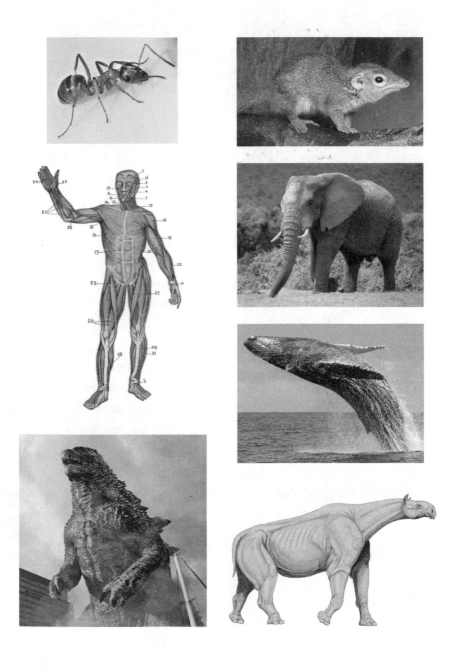

Mammals range in size from the 2 gm shrew (top right) to the 20,000 kg blue whale. Why couldn't we be the size of a 2 mg ant or a 2 million kg Godzilla, shown bottom left? The animal in the bottom right is the 2,000 kg *Paraceratherium*, the largest land mammal ever.

blood through them, so the more of them you have the better. Almost all of your cardiac output goes into pumping blood through the much smaller vessels of the nonpulsatile regime, whose number of levels is approximately the same for all mammals. Relatively speaking, then, the proportion of the network where the heart expends most of its energy systematically decreases as the size of a mammal increases, illustrating again that larger mammals are more efficient than smaller ones: the whale needs only one hundredth the amount of energy needed by a shrew to supply blood to one of its cells.

Now imagine continuously decreasing the size of the animal. Concomitantly, the number of area-preserving branchings where vessels are large enough to support pulsatile waves decreases until a tipping point is reached where the network can support only nonpulsatile DC flow. At that stage even the major arteries become so small and constricted that they are unable to support pulsatile waves. In such vessels, waves become so overdamped due to the viscosity of blood that they can no longer propagate and the flow shifts to becoming entirely steady DC, just like the flow of water in the pipes of your house: pulsatile waves generated by the beating heart are immediately damped as they enter the aorta.

This is really weird. Such an animal would have a beating heart but no pulse! This is not only weird, but more important, it would be an extremely inefficient design because it would have entirely lost the advantages of impedance matching and would lead to significant energy being dissipated in *all* of the vessels throughout its circulatory system. This loss of performance efficiency is reflected in how metabolic rates would scale. Instead of following the classic sublinear ¾ power scaling law, a calculation shows that it would now scale linearly—that is, directly proportional to body mass—and thereby lose the advantages of economies of scale. In this purely DC case the power needed to support a gram of tissue would now be the same regardless of size instead of systematically decreasing with size following quarter-power scaling. Consequently, no evolutionary advantage would be conferred by increasing size.

This argument shows that only mammals that are large enough for their circulatory systems to support pulsatile waves through at the very least the first couple of branching levels would have evolved, thereby providing a fundamental reason why there is a minimum size.[3] The theory can be used to derive a formula for when this tipping point occurs. Its actual value depends upon generic quantities such as the density and viscosity of blood and the elasticity of arterial walls. The calculation leads to an approximate value for the size of the smallest mammal of just a couple of grams, comparable to the mass of the Etruscan shrew, which is the smallest known mammal. It is only about 4 centimeters long, easily sitting on the palm of your hand. Its tiny heart beats at more than a thousand times a minute—about twenty times a second—as *it pumps blood with the same pressure and speed as you do*, and even more astounding, as does a blue whale. And all of this through its minuscule aorta, which is only a couple of millimeters long and an astonishingly couple of tenths of a millimeter wide, not much thicker than a hairbreadth. As I have remarked earlier, no wonder the poor creature doesn't live very long.

3. AND WHY AREN'T THERE ENORMOUS MAMMALS THE SIZE OF GODZILLA?

This is one of those intriguing seminal questions raised by Galileo—though he didn't, of course, raise the specter of Godzilla. Recall from chapter 2 that his argument was crafted on the deceptively simple idea that the weight of an animal increases faster than the ability of its limbs to support it so that if the design, shape, and materials remain unchanged, it will collapse under its own weight as its size increases. This gave an elegant demonstration that there are limits to the sizes of animals, plants, and buildings and provides a template for considerations of limits to growth and sustainability.

However, to actually implement this argument and make quantitative estimates of maximum sizes of animals requires a detailed analysis of their biomechanics beyond just the static situation envisioned by Galileo. The greatest mechanical stresses occur during movement and especially during running, which is an essential feature of survival for many animals. The largest land mammals that have ever existed are the *Paraceratheria*, a sort of precursor of the modern rhinoceros, which was almost 10 meters (more than 30 ft.) long and weighed up to 20 tons (20,000 kg)—pictured on page 156. It is very likely that the maximum size of land animals was achieved by the largest dinosaurs, who reached more than 25 meters in length and weighed in excess of 50 tons. There is some evidence of larger ones, but these come from bone fragments requiring extensive extrapolations about their design and anatomy. It has even been conjectured that some dinosaurs were so large that they had to be semi-aquatic in order to support their enormous weight, though there is no substantial evidence to support this. Whether this is true or not, it does provide a natural segue to the conjecture that to extend the boundaries of size, animals needed to relieve themselves of the burden of gravity and return to the sea.

Without having to combat the force of gravity Galileo's argument becomes moot, so it is not surprising that the largest animals that have ever existed are alive today and until modern humans came on the scene prospered in the vast oceans of the planet. The very largest of these is the magnificent blue whale, which is a mammal that can be up to 30 meters long (about 100 ft.) and weigh almost 200 tons, more than twenty times heavier than the infamous *Tyrannosaurus rex*. Could there conceivably be even larger mammals that are yet to evolve? Well, there are certainly biomechanical and ecological constraints acting on ocean animals just as there are for land animals. Whales need to swim sufficiently fast over long distances to be able to supply themselves with the enormous amount of food they need to support their huge metabolic rate, which is equivalent to almost a million food calories a day or about four hundred times more than you eat. Putting these constraints into mathematics and physics in order to quantitatively determine a maximum size for aquatic organisms is even more challenging than it is for land animals, and no credible estimates have been made.

However, as I will now show, there are additional constraints on maximum body size that transcend ecological biomechanics and that arise from the fundamental need to ensure that all cells are supplied with sufficient oxygen. This therefore involves the geometry and dynamics of network supply systems. Let me sketch a simplified version of the argument to show how this leads to a rough estimate of maximum body size.

One of the more arcane results of the network theory is that the average distance between terminal units such as capillaries scales with body mass as a power law with an exponent of $1/12$ (= 0.0833 . . .). This is an unusually small exponent that represents an exceedingly slow variation with body size, leading to the network very gradually opening up and becoming sparser as size increases, as observed. For instance, canopies of larger trees are typically more spread out than smaller ones, with the average distance between leaves following this very slow increase with size. Likewise, although a blue whale is a hundred million times (10^8) heavier than a shrew, the average distance between its capillaries is only about $(10^8)^{1/12} = 4.6$ times larger.

Capillaries service cells, so the opening up of the network means that there is increasingly more tissue that needs to be serviced situated between adjacent capillaries as size increases. So on average, each capillary systematically has to service more cells, another reflection of the increasing economy of scale discussed earlier. However, there is a limit to how far this can be pushed. Each capillary, being an invariant unit, can deliver only so much oxygen to the tissue. So if the group of cells needing to be supplied by a single capillary becomes too large, some of them will inevitably become oxygen deprived, a situation technically referred to as *hypoxia*.

The physics of how oxygen diffuses across capillary walls and through tissue to supply cells was first quantitatively addressed more than a hundred years ago by the Danish physiologist August Krogh, who received a Nobel Prize for his work. He recognized that there is a limit to how far oxygen can diffuse before there isn't sufficient left to sustain the cells that are too far away. This distance is known as the maximal Krogh radius, which is the radius of an imaginary cylinder surrounding the length of a capillary, like a sheath, and which contains all of the cells that can be sustained (just to remind you: a capillary is about half a millimeter long and about five times longer than its diameter). Based on this, one can calculate how large an animal could be before the separation distance between its capillaries gets so large that significant hypoxia develops. This leads to an estimate of about 100 kilograms for the maximum size, roughly equivalent to the largest blue whales, suggesting that they represent the end of the line for the mammalian family.

Before exploring other critical implications of the subtleties at the interface between capillaries and cells, such as how they influence your growth, aging, and subsequent death, I want to briefly return to the issue of Godzilla. It's clear from what has already been discussed that if Godzilla is anything like the rest of us in the biosphere, he will remain a mythical figure. Even if he didn't collapse under his own weight à la Galileo, he wouldn't be able to supply most of his cells with oxygen and so would not be viable. Of course, like Superman, he could be made of very different materials that can sustain the huge stresses involved in his support and mobility and have properties that allow his internal networks to deliver sufficient nutrients to his presumed cells so that he can function in the way depicted in the movies.

In principle it's possible using the ideas that I've discussed to estimate the values of the various properties of the materials he would have to be made of for him to function like us. One could estimate, for instance, how large the compression strength of his limbs, the viscosity of his "blood," the elasticity of his tissue, and so on would have to be for him to function. I'm not entirely sure how useful such an exercise would be because, being a complex adaptive system, any meddling with parameters and design could potentially have huge unintended consequences and therefore not be terribly meaningful. One would have to think quite carefully and extensively about all of the myriad interconnectivities and detailed consequences of making such changes before believing that such a beast could conceivably exist. This is the challenge that is usually and probably necessarily ignored when arbitrarily fantasizing about alternative designs and scenarios that dominate science fiction. Nevertheless, such fantasies can be a wonderful exercise in imagination, unbounded by some of the facts and constraints of science, and can conceivably stimulate innovative and wild ways of thinking about some of our big issues. So it's not that one should not fantasize, but rather that one should remain conscious of science fact before jumping to too many conclusions and acting out the fantasy, whatever it is.

When I was asked by journalists about Godzilla's various characteristics, such as how much he weighed, how long he slept, how fast he walked, et cetera, I immediately responded in my straitlaced professorial role and told them that, as any scientist knows, Godzilla couldn't exist and that's the end of the story. However, not wanting to be a complete nerdy party pooper I said that I was willing to ignore fundamental science and calculate what his various physiological and life history characteristics would be if we naively follow the allometric scaling laws and assume Godzilla to be just another animal. Although this is fundamentally inconsistent, it turned out to be an amusing exercise. So here are the "facts" about Godzilla.

In his latest incarnation Godzilla is 350 feet long, which translates into a weight of about 20,000 tons, about 100 times heavier than the biggest blue whales. To support this gargantuan amount of tissue Godzilla would have to eat about 25 tons of food a day, corresponding to a metabolic rate of about 20 million food calories a day, the food requirements of a small town of 10,000 people. His heart, which would weigh about 100 tons and have a diameter of about 50 feet, would have to pump almost 2 million liters of blood around his body. However, to counterbalance that, it would have to beat only just over a couple of times a minute and sustain a blood pressure similar to ours. Note, by the way, that his heart alone is comparable in size to an entire blue whale. His aorta through which this enormous amount of blood flows would be about 10 feet across, easily big enough for us to walk through quite comfortably. Godzilla might live for up to two thousand years and would need to sleep less than an hour a day. Relatively speaking, he would have a tiny brain representing less than 0.01 percent of his body weight, compared with the approximately 2 percent of ours. This doesn't mean that he would be stupid, but that's all he would need to carry out all of his neurological and physiological functions. As to the possibly less savory parts of his life, he would need to pee about 20,000 liters of urine a day, comparable to the size of a small swimming pool, and poop about 3 tons of feces, a good-size truckload. I shall leave speculations about his sex life to your imagination.

Estimating his walking and running speeds is even more speculative because of the biomechanical inconsistency inherent in such an animal. However, blindly extrapolating from other animals leads to an estimate of a modest 15 miles per hour for his walking speed, so the average person would have some difficulty escaping his clutches should he be aggressive. But this brings up the catch in all of this: the diameter of each of his legs would have to be about 60 feet and his thighs probably much bigger, possibly close to 100 feet. In other words, he would have to be almost all leg in order to avoid collapse and be mobile, so the design is no longer feasible. As stressed earlier, to evolve something this big requires new materials and probably new design principles.

One can speculate that natural selection has already begun to embark on this grand evolutionary process by first selecting human beings to be sufficiently intelligent to design such enormous "organisms." After all, on our planet there are now "trees," "birds," and "whales" that are significantly larger than their "natural" counterparts—we call them skyscrapers, airplanes, and ships, though we have yet to evolve mobile land "animals" larger than dinosaurs. On the other hand, we have made "organisms" that move and calculate faster and memorize more facts than any "natural" organism, including us. So much so that many believe we are on the way to creating cyborgs who will supersede anything we mere mortals can do. Despite these marvelous achievements, all of them thus far are at best pale imitations of their natural forebears, and most people would question designating them as "organisms" at all, even if they do share many characteristics in common with traditional life.

There is, however, one human invention that has evolved via this process which is comparable to what traditional natural selection has thus far produced, and that is the city. Cities clearly have an organic nature and share much in common with traditional organisms. They metabolize, they grow, they evolve, they sleep, they age, they contract disease, suffer damage and repair themselves, and so on. On the other hand, they rarely reproduce nor do they easily die. Furthermore, their size is enormously greater than even a mythical Godzilla. Godzilla extends only a few hundred feet and only metabolizes at a rate of 20 million food calories a day or about a million watts, whereas New York is more than fifteen miles across and metabolizes at well over 10 billion watts. In this sense, cities are perhaps the most astonishing "organisms" that have ever evolved. Chapters 7 and 8 are devoted to trying to understand some of their characteristics, including how they differ from "naturally" evolved organisms. What are the new materials and design principles that have brought this about?

4. GROWTH

All of us are familiar with growth. We've all experienced it at a very personal level and recognize it as an essential and ubiquitous feature of nature. Perhaps less familiar is thinking about it as a quintessential scaling phenomenon. As remarked earlier, the word *allometric* that we have been using to describe the scaling of organismic characteristics *across* species was originally coined by

Julian Huxley to describe how such characteristics change during growth *within* species. Biologists use the term *ontogenesis* to describe the developmental process that occurs within an individual during growth, beginning with the fertilization of an egg through birth and up to maturity. The *onto* part of this word is derived from the Greek word for "being," while *genesis* means "origins," so ontogenesis, or *ontogeny*, connotes the idea of the study of how we came to be.

Growth cannot happen without a continuous supply of energy and resources. You eat, you metabolize, you transport metabolic energy through networks to your cells, where some is allocated to the repair and maintenance of existing cells, some to replace those that have died, and some to create new ones that add to your overall biomass. This sequence of events is the template for how all growth occurs, whether for an organism, a city, a company, or even an economy, as symbolized in the figure below. Roughly speaking, incoming energy and resources are apportioned to general maintenance and repair, on the one hand, and on the other to the creation of new entities, whether cells, people, or infrastructure. This is none other than a statement of the conservation of energy: whatever goes in must be accounted for in terms of how it is

Incoming Metabolized Energy

↓

Maintenance (repair and replacement)

+

New Growth

The symbolic equation representing the energy budget of the growth process in which metabolic energy is allocated between general maintenance and new growth.

allocated between the various categories of what it acts upon and produces. There are subcategories of activities such as reproduction, movement, and the production of waste products that can either be explicitly incorporated or subsumed under one of the two major categories or, if more appropriate, can be treated separately.

One of the curiosities of our growth pattern that might well have intrigued you at various times in your life is why it is that we eventually *stop* growing, even though we continue to eat and metabolize throughout life. Why is it that we reach a relatively stable size and don't continue to grow by adding more and more tissue the way some organisms do? There are, of course, less dramatic, much smaller changes in size and shape that occur with age or with changes in diet and lifestyle and with which many of us become neurotically obsessed, such as putting on weight or growing a paunch, but these are secondary to the basic issue of *ontogenetic growth,* which starts at birth and finishes at maturity. I won't consider any of these secondary and much smaller changes here, though the framework I'll discuss can in principle be adapted to accommodate them.

Instead, I want to focus on ontogenetic growth and show how the network theory naturally leads to a quantitative derivation of how the weight of an organism changes with its age and, in particular, explains why we stop growing.[4] All mammals and many other animals share the same kind of growth trajectory that we follow, called *determinate growth* by biologists to distinguish it from *indeterminate growth,* typically observed in fish, plants, and trees, where growth continues indefinitely until death. Because the theory I'll present is based on general principles, it provides a unifying framework that accounts for both kinds of growth. In what follows I shall concentrate primarily on determinate growth, but suffice it to say that the data and the analyses support the idea that indeterminate growers die before they reach a stable size.

Because the supply of metabolic energy is apportioned between the maintenance of existing cells and the creation of new ones, the rate at which energy is used to create new tissue is just the difference between metabolic rate and what is needed for maintenance of existing cells. This latter term is directly proportional to the number of existing cells and therefore increases linearly with the mass of the organism, whereas metabolic rate increases sublinearly with a ¾ power exponent. This difference in the way these two contributions scale with increasing size plays a central role in growth, so to ensure that you understand what it implies here's a simple example to illustrate the point. Suppose the size of the organism doubles; then the number of cells also doubles, so the amount of energy needed for their maintenance increases by a factor of 2. However, metabolic rate (the supply of energy) increases by only a factor of $2^{¾} = 1.682$ . . . which is less than 2. So the rate at which energy is needed for maintenance increases faster than the rate at which metabolic energy can be supplied, forcing the amount of energy available for growth to systematically decrease and eventually go to zero, resulting in the cessation of growth. In other words, you stop growing because of the mismatch between the way maintenance and supply scale as size increases.

Let me deconstruct this a little further to give a more mechanistic understanding of what's going on. Recall that the reason metabolic rate scales with a sublinear ¾ power exponent lies in the hegemony of the network. Furthermore, because the entire flow through the network ends up going through all of the capillaries, and because they are invariant across species as well as during ontogeny (capillaries are approximately the same for mice, elephants, their babies and children, as well as for us), their number likewise scales with a ¾ power exponent. So as the organism grows and size increases, each capillary systematically has to service more cells following ¼ power scaling. It is this mismatch at the critical interface between capillaries and cells that controls growth and ultimately leads to its cessation: the increase in the number of supply units (the capillaries) cannot keep up with the demands from the increase in the number of customers (the cells).

All of this can be expressed in mathematical equations that can be solved analytically to give a compact formula predicting how size changes with age. This quantitatively explains why we start out growing quickly at birth, gradually slow down, and eventually stop. One of the great attributes of the growth equation is that it depends on only a very small number of "universal" parameters that transcend species such as the mass of the average cell, how much energy is needed to create a cell, and the overall scale of metabolism. These determine the growth curve for any animal. Shown in Figures 15–18 is a sampling of such predictions illustrating how the same equation with similar parameters leads to predictions for growth curves for a wide variety of animals (in this case, two mammals, a bird, and a fish) in good agreement with data.

The universality of growth can be elegantly exhibited by expressing the results in terms of *dimensionless* quantities introduced in chapter 2. Recall that these are scale-invariant combinations of variables that do not depend upon the units used to measure them. A trivial example is the ratio of two masses, because it has the same value whether they are measured in pounds or kilograms. I emphasized that *all* laws of science can be expressed as relationships between such quantities. So instead of simply plotting mass versus age as in Figures 15–18, where the quantities depend on units (kilograms and days in that case), plotting a dimensionless mass variable versus an appropriately defined dimensionless time variable yields a scale-invariant curve valid across all animals. The actual mathematical combination of variables defining these dimensionless quantities is determined by the theory and can be found in the original papers.

Thus, in terms of these dimensionless combinations the growth curves for all animals collapse to a single universal curve. When viewed through this lens all animals follow the same growth trajectory, as illustrated in Figures 19–21. The theory tells us how to rescale the spatial and temporal dimensions of animals so that they all appear to grow in the same way and at the same rate. Growth trajectories of a limited sampling of diverse mammals, birds, fish, and crustacea with very different body designs and life spans all collapse onto a single curve whose mathematical form is predicted by the theory. As you can see, this is beautifully supported by the data and graphically reveals the hidden commonality and unity underlying the ontogeny of all animals. And the theory tells us why: growth is primarily determined by how energy is delivered to cells, and this is constrained by universal properties of networks that transcend design. Among the many other aspects of growth that can be derived from the theory, it predicts how the allocation of metabolic energy between maintenance and growth changes with age. At birth almost all of it is devoted to growth and relatively little to maintenance, whereas beyond maturity all of it is devoted to maintenance, repair, and replacement.

The theory has been extended to understand the growth of tumors, plants, insects, and both forest[5] and social insect communities[6] such as ants and bees.

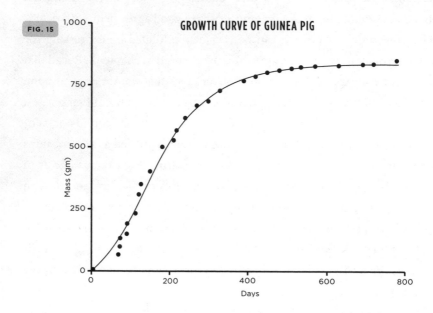

FIG. 15

GROWTH CURVE OF GUINEA PIG

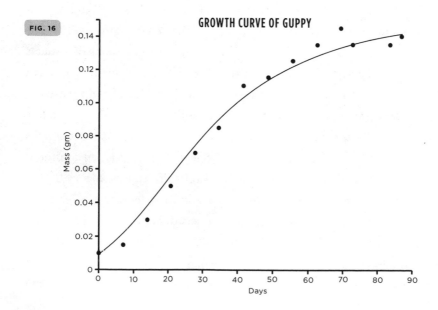

FIG. 16

GROWTH CURVE OF GUPPY

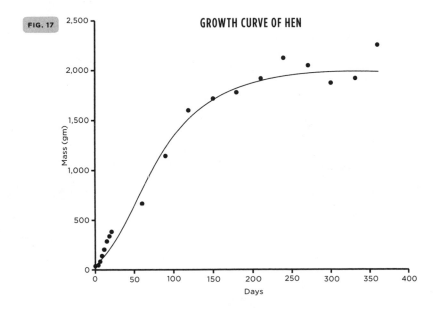

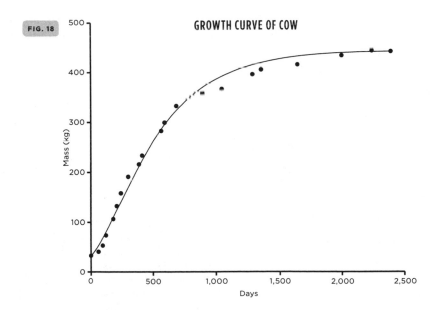

Growth curves for a sampling of animals showing how their mass increases with age, eventually ceasing at maturity. The solid lines are predictions from the general theory as explained in the text.

FIG. 19

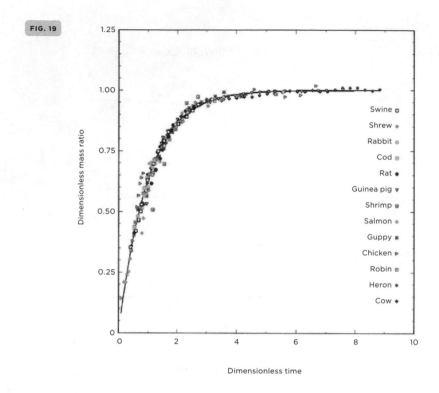

FIG. 20

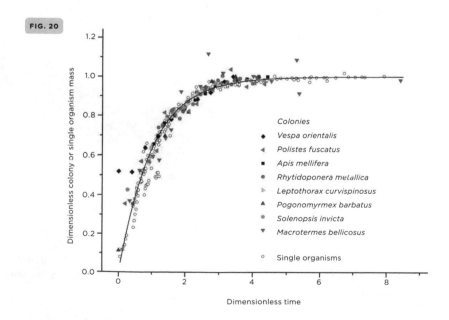

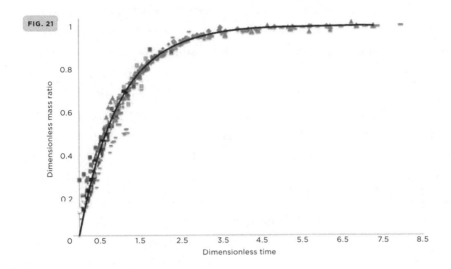

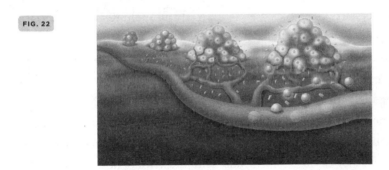

Graphs showing that when appropriately rescaled everything grows in the same way at the same rate. The solid lines as well as the scales on the axes, which are predictions from the theory, are identical in all three graphs: (19) A sampling of birds, fish, and mammals. (20) A sampling of insects and social insect communities. (21) The same data as in (19) but including a sampling of tumors. (22) A schematic showing how tumor networks feed off host networks.

These latter applications are forerunners of how we might start thinking about the growth of human organizations such as cities and companies, which I'll be turning to in chapters 8 and 9. Each of these very different systems represents a variation on the general thematic structure of the growth equation. For instance, tumors are parasitic and use metabolic energy derived from their host to grow, so their vasculature and metabolic rates depend not only on their own size but also on the size of the host.[7] Understanding this provides insight into how to scale up basic properties of tumors as well as potential therapeutic strategies from observations on mice to humans.[8] Trees, on the other hand, present the challenge that as they grow more and more of their physical structure becomes deadwood, which does not participate in their metabolic energy budget but plays an important role in their mechanical stability.[9] Figures 19–21 shows how all of these conform, to varying degrees, to the universal growth equation.

The overall agreement with theory is extremely satisfying. But much more than that, I find the extraordinary unity and interconnectivity of life that is revealed through this lens to be spiritually elevating in the pantheistic spirit articulated by the philosopher Baruch Spinoza. As Einstein wrote,[10] "We followers of Spinoza see our God in the wonderful order and lawfulness of all that exists and in its soul as it reveals itself in man and animal." Regardless of one's belief system, there is something supremely grand and reassuring when one perceives even a tiny piece of the mystifyingly chaotic world around us conforming to regularities and principles that transcend its awesome complexity and seeming meaninglessness.

As I argued earlier, analytic models such as the growth theory are deliberate oversimplifications of a more complex reality. Their utility depends on the extent to which they capture some fundamental essence of how nature works, the extent to which their assumptions are reasonable, their logic sound, and their simplicity or explanatory power and internal consistency in agreement with observations. Because the theory is deliberately simplified, it is inevitable that measurements of real organisms will, to varying degrees, deviate from model predictions. As can be seen in Figure 19 the agreement is surprisingly good with relatively few major outliers that deviate significantly from the idealized growth curve. We as primates are one of those. For instance, we take longer to mature than we "should" given our body weight. This is the result of our rapid evolution from being purely biological to becoming sophisticated socioeconomic creatures. Our effective metabolic rate is now one hundred times greater than what it was when we were truly "biological" animals, and this has had huge consequences for our recent life history. We take longer to mature, we have fewer offspring, and we live longer, all in qualitative agreement with having an effectively larger metabolic rate arising from socioeconomic activity. I shall return to this fascinating development in our history when discussing how these ideas apply to cities.

The critical take-home message from this section is that *sublinear scaling and the associated economies of scale arising from optimizing network performance lead to bounded growth and the systematic slowing of the pace of life.* This is the dynamic that dominates biology. How this transforms into open-ended growth and an increasing pace of life and how it's related to the huge enhancement in our "social" metabolic rate will be the focus of chapters 8 and 9.

5. GLOBAL WARMING, THE EXPONENTIAL SCALING OF TEMPERATURE, AND THE METABOLIC THEORY OF ECOLOGY

Because we are homeotherms, meaning that our body temperature remains approximately constant, we tend to forget that temperature plays a huge role across all of life. We are the exception. Perhaps only now with the advent of global warming are we beginning to appreciate how sensitive the natural world and the environment are to small changes in temperature and the threat that this imposes. What is shocking is how few people, even many scientists, appreciate that this sensitivity to temperature is *exponential.* The reason for this sensitivity is that all chemical reaction rates depend exponentially on temperature. In the previous chapter I showed how metabolism originates in the production of ATP molecules in cells. Consequently, metabolic rate scales exponentially with temperature rather than as a power law as it does with mass. Because metabolic rate—the rate at which energy is supplied to cells—is the fundamental driver of all biological rates and times, *all of the central features of life from gestation and growth to mortality are exponentially sensitive to temperature.*

Since the production of ATP is common to almost all animals, this exponential dependence is universal, much as quarter-power scaling with mass is.

Its overall scale is governed by just a single "universal" parameter: the average activation energy needed to produce an ATP molecule via the oxidative chemical process I discussed in the previous chapter. This is approximately 0.65 eV (electron volts, introduced in chapter 2), which is typical of chemical reactions and represents an average over many subprocesses. This leads to the fascinating conclusion that across the spectrum of life all biological rates and times such as those associated with growth, embryonic development, longevity, and evolutionary processes are determined by a joint universal scaling law in terms of just two parameters: the number ¼, arising from the network constraints that control the dependence on mass, and 0.65 eV, originating in the chemical reaction dynamics of ATP production. This result can be restated in a slightly different way: when adjusted for size and temperature, as determined by just these two numbers, all organisms run to a good approximation by the same universal clock with similar metabolic, growth, and evolutionary rates.

This parsimonious formulation of the coarse-grained mass and temperature dependence was introduced as a compact summary of the scaling work in a paper titled "Toward a Metabolic Theory of Ecology" published in the journal *Ecology* in 2004 and coauthored by Jim Brown and three of our then postdocs—Van Savage, Jamie Gillooly, and Drew Allen—together with me. Jim had deservedly been honored by the Ecological Society of America with its most prestigious prize, the Robert H. MacArthur Award for "meritorious contributions to ecology." In his acceptance speech at its annual meeting he chose to talk about our scaling work and this formed the basis for our joint paper. Although it summarizes only a subset of the scaling work, the term *metabolic theory of ecology* (MTE) has since taken on a life of its own.

In addition to the purely allometric quarter-power mass dependence that I've already discussed, the metabolic theory has been tested across a diverse range of organisms, including plants, bacteria, fish, reptiles, and amphibians. Shown, for instance, in Figure 23 is a plot of embryonic development times for eggs of birds and aquatic ectotherms (fish, amphibians, zooplankton, and aquatic insects) versus temperature plotted semilogarithmically so that an exponential appears as a straight line. Because these times depend on *both* temperature and mass, their mass dependence has been removed by rescaling the data according to the ¼ power scaling law in order to expose the purely temperature dependence. As can readily be seen, when this is done the data agree very well with the prediction of a straight line, confirming the prediction of an exponential dependence on temperature. Figure 24 shows a similar mass-adjusted plot for the life spans of a series of invertebrates as a function of *inverse absolute* temperature. The data are plotted in this slightly byzantine fashion for technical reasons: strictly speaking, fundamental chemical reaction theory tells us that reaction rates actually scale exponentially with the *inverse* of the absolute temperature (sometimes called the Kelvin scale), in which 0° corresponds to -273°C. It turns out that to a very good approximation, the prediction of an exponential dependence on temperature when expressed in the usual centigrade units is valid provided the range of variation is relatively small, as it is here in Figure 23.

I want to emphasize just how remarkable this is. The two most important events in an organism's life, its birth and its death, which are usually thought of as being independent, are intimately related to each other: the slopes of these two graphs are determined by exactly the same parameter, the 0.65 eV, representing the average energy needed to produce an ATP molecule. Below I will explore this further when I discuss how a more fundamental theory of aging based on network dynamics explains the mechanistic origins of this temperature dependence.

The important take-home message is that these quite disparate fundamental life-history events scale as predicted with both temperature and mass and, equally important, the parameters that govern the corresponding exponentials are the same. *Thus at a deep level, birth, growth, and death are all governed by the same underlying dynamics driven by metabolic rate and encapsulated in the dynamics and structure of networks.*

The exponential dependence of ATP production, which is governed by the 0.65 eV activation energy, can be translated into the simple statement that for every 10°C rise in temperature the production rate doubles. Consequently, a relatively small increase of only 10°C leads to a doubling of metabolic rate and therefore to a doubling of the rate of living. By the way, this is why you don't see many insects in the morning when it's cool—they have to wait till it warms up to increase their metabolism.

More pertinent, a modest 2°C change in ambient temperature leads to a

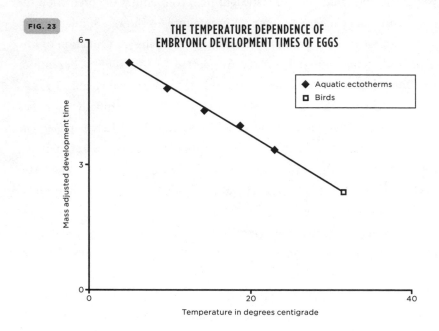

FIG. 23

THE TEMPERATURE DEPENDENCE OF EMBRYONIC DEVELOPMENT TIMES OF EGGS

Mass adjusted development time

Temperature in degrees centigrade

◆ Aquatic ectotherms
☐ Birds

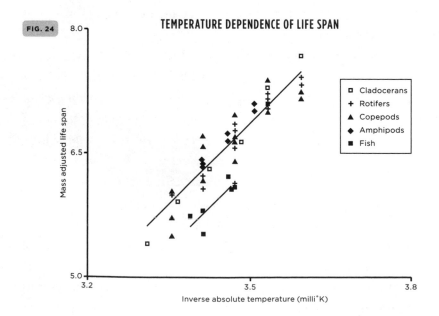

FIG. 24

TEMPERATURE DEPENDENCE OF LIFE SPAN

Mass adjusted life span

Inverse absolute temperature (milli°K)

☐ Cladocerans
+ Rotifers
▲ Copepods
◆ Amphipods
■ Fish

20 percent to 30 percent change in growth and mortality rates.[11] This is huge and therein lies our problem. If global warming induces a temperature increase of around 2°C, which it is on track to do, then the pace of almost all biological life across all scales will increase by a whopping 20 percent to 30 percent. This is highly nontrivial and will potentially wreak havoc with the ecosystem. It's analogous to the huge leap that Brunel attempted to make when building his mammoth ship the *Great Eastern,* which ended in disaster primarily because the science of shipbuilding had not yet been sufficiently well developed. Ships are extremely simple compared with the profound complexity of ecosystems and societies. Without a comprehensive systemic scientific framework for understanding the bigger picture, we are in an analogous position to Brunel when it comes to being able to confidently predict the detailed consequences of such a huge climate change and especially its effects on agricultural production, let alone the entire ecology of the planet. Developing a metabolic theory of ecology is a tiny step in this direction.

One last note: the underlying physics and chemistry of reaction theory have been known for a very long time, having been developed by the Swedish physicist-turned-chemist Svante August Arrhenius, who won the Nobel Prize in Chemistry in 1903. He holds the distinction of being the first Swede to become a Nobel laureate. Arrhenius was a man of very broad interests whose many novel ideas and contributions to science have been very influential.

He was one of the first people to seriously suggest that life on Earth might have originated from the transport of spores having been transported from another planet, a rather speculative theory with a surprisingly large following that now goes by the name of *panspermia.* Of greater significance is that he

Opposite page: (23) The exponential scaling of embryonic development times with temperature for eggs of birds and aquatic ectotherms (measured in centigrade), rescaled according to the ¼ power scaling law to remove their mass dependence (see text). These "mass adjusted" times are plotted logarithmically on the vertical axis against the temperature plotted linearly on the horizontal one. On such a semilogarithmic plot exponentials appear as straight lines, as observed. (24) Similar "mass adjusted" plot showing the exponential dependence of life spans of various invertebrates on temperature. Notice that for technical reasons explained in the text the data are plotted versus *inverse absolute* temperature (milli degrees Kelvin) so that temperature actually *decreases* as one moves to the right.

was the first scientist to calculate how changes in the levels of carbon dioxide in the atmosphere could alter the surface temperature of the Earth through the greenhouse effect, predicting that the burning of fossil fuels was large enough to cause significant global warming. Most remarkably he did all of this before 1900, which is pretty depressing because it shows that we already were beginning to understand scientifically some of the deleterious consequences of burning fossil fuels well over a hundred years ago and we did almost nothing about it.

6. AGING AND MORTALITY

I. Night Thoughts During the Hour of the Wolf

According to the ancient Romans, the Hour of the Wolf means the time between night and dawn, just before the light comes, and people believed it to be the time when demons had a heightened power and vitality, the hour when most people died and most children were born, and when nightmares came to one.[12]

Just as growth is an integral part of life, equally so are aging and death. The fact that almost everything dies plays a central role in the evolutionary process because it allows new adaptations, designs, and innovations to emerge and flourish. From this viewpoint, it's not only "good" but also crucial that individuals, whether organisms or companies, die—even if they themselves may not be quite so joyous about it.

This is the curse of consciousness. We all know we are going to die. No other organism is burdened with the enormity of the conscious knowledge that it has a finite lifetime and that its individual existence is eventually and inevitably coming to an end. No creature, whether a bacterium, an ant, a rhododendron, or a salmon, "cares" or even "knows" about dying; they live and they die, participating in the continual struggle for existence by propagating their genes into future generations and playing the endless game of the survival of the fittest. So do we. But over the last few thousand years, we have emerged as the consciousness and conscience of the evolutionary process and begun the extraordinary adventure of contemplating its meaning by bringing ideas of morality, caring, rationality, the soul, the spirit, and the gods to the universe.

I had a minor epiphany when I was sixteen years old. Some school friends persuaded me to join them in going to a small arts cinema in the West End of London to see a film much touted by the intelligentsia at that time. This was Ingmar Bergman's extraordinary film *The Seventh Seal*. It is of Shakespearean grandeur and depth. It tells the story of a medieval knight, Antonius Block, who on his return journey home to Sweden from fighting in the Crusades encounters the personification of Death, who has come to take his life. In his attempt to avoid, or at least delay, the inevitable, Block proposes that they play a game of chess; should he win, his life will be spared. He, of course, eventually loses but only because he is inadvertently tricked into baring his soul to Death, who masquerades as a confessional priest. This allegorical setting provides the stage for delving into the eternal questions concerning the meaning, or pointlessness, of life and its relationship to death. Questions at the very heart of philosophical and religious discourses with which men and women have struggled throughout the centuries are brilliantly depicted by Bergman's genius. Who can forget the final haunting scene in which the black-robed Death leads Antonius and his entourage on an iconic *danse macabre* silhouetted across a distant hillside to meet their inevitable fate?

What an impression this made on an innocent, unconscious, adolescent sixteen-year-old. I think this was my first truly serious inkling that there was more to life than money, sex, and football and began my long-term interest in questions of metaphysics and philosophical thought. I began to voraciously read all of the usual suspects from Socrates, Aristotle, and Job to Spinoza, Kafka, and Sartre, and from Russell and Whitehead to Wittgenstein, A. J. Ayer, and even Colin Wilson, though barely understanding anything of what any of them were saying (especially Wittgenstein, by the way). What I did learn, however, was that although extraordinary men had struggled with the really big questions for a very long time, there actually were no answers. Just more questions.

It speaks to the profundity of Bergman's masterpiece that almost sixty years later the film still makes the same powerful impression, now possibly

more nuanced and poignant, on a slightly jaded seventy-five-year-old man as he approaches the final years. At a critical stage in the film Death very reasonably asks Antonius: "Do you never stop questioning?" to which he emphatically responds, "No. I never stop." And neither should we. The fascination with death, coupled with the incessant questioning and search for any meaning to life, permeates human culture but has mostly been manifested and formalized in the multiplicity of religious institutions and experiences that humans have invented. Science has generally placed itself outside of such philosophical meanderings. However, many scientists have seen the quest for understanding and unraveling "the laws of nature," the passion for wanting to know how things work and what they are made of, as an alternative journey in coming to terms with these big questions even if they themselves are neither "religious" nor particularly "philosophical." Somewhere along the line, I realized that I was one of them, finding in science, or at least in physics and mathematics, some version of the spiritual sustenance that seems to be a universal need. I eventually came to recognize that science was one of the few, if not the only, frameworks that could possibly provide credible answers to some of the big questions.

Once upon a time, science was referred to as *natural philosophy,* implying a somewhat broader connotation than the way we think of it today with a greater connection to philosophical and religious thought. It is no accident that the full title of Newton's famous book the *Principia,* which introduced his universal laws of nature that revolutionized science, is (in English) *The Mathematical Principles of Natural Philosophy.* Although Newton held heretical views such as rejecting the classical doctrines of an immortal soul, the existence of devils and demons, and the worship of Christ as God, which he viewed as idolatrous, he saw his work as the revelation of God as a prime mover. Commenting on the *Principia,* he stated: "When I wrote my treatise about our Systeme I had an eye upon such Principles as might work with considering men for the beliefe of a Deity and nothing can rejoyce me more than to find it useful for that purpose."

The modern scientific method as an outgrowth of natural philosophy rarely invokes such reflections, yet it has proven to be extraordinarily powerful in providing profound and consistent answers to many of the most vexing fundamental questions about the "universe" that have puzzled human beings from time immemorial. How did the universe evolve, what are stars made of, where did all the different animals and plants come from, why is the sky blue, when will the next eclipse occur, and so on and so forth. We understand an enormous amount about the physical universe around us and in many cases with exquisite detail, and we have done it without having to invoke ad hoc or arbitrary arguments that are often the hallmark of religious explanations. Left unanswered, however, are many of the deep questions concerning the very nature of who and what we are as human beings endowed with consciousness and the ability to reflect and reason. We continue to grapple with the nature of mind and consciousness, with psyche and self, with love and hate, and with meaning and purpose. Perhaps all will eventually be understood from the firing of neurons and the complex network dynamics of our brains, but as D'Arcy Thompson proclaimed a hundred years ago, I suspect not. There will always be questions—that is the essence of the human condition—and like Antonius Block, we will never stop asking them even if it is to the great frustration and annoyance of Death. And somehow intertwined with all of this lies the challenge and paradox of understanding aging and mortality, and coming to terms with our collective and individual uneasiness with the finiteness of our own existence.

II. Dawn and a Return to the Sunlight

Having said all of this, I now want to return to the science itself. My intention is by no means to give a comprehensive overview of this somewhat morbid topic, whether from a metaphysical or scientific perspective, but rather to relate it to the scaling and network framework that has been developed in previous chapters. As with growth, I want to show how it provides another important example of how this perspective can be used to give new insights into a fundamental problem in biology and biomedicine by developing a quantitative big-picture theoretical framework for understanding many of the general characteristics of aging and mortality. Furthermore, it is also based on the belief that only by understanding more broadly the mechanistic origins of death and its intimate relationship to life, and their interconnectivity with how other major phenomena in our universe work, can we begin to come to terms with the troubling metaphysical questions that continue to haunt us.

Unlike the predominantly positive images of many life-history events such as birth, growth, and maturity, most of us don't want to confront aging and death. As Woody Allen pithily put it: "I'm not afraid of death, I just don't want to be there when it happens." Much easier would be to be unconscious like an animal or plant and not "be there when it happens." We spend enormous amounts of money trying to prolong our lives and postpone death, even long after we have become debilitatingly frail, and sometimes long after we have become unconscious and are no longer ourselves. In the United States alone we spend more than $50 billion a year on various antiaging products, regimens, and drugs ranging from vitamins, herbs, and supplements to foods, hormones, creams, and exercise aids. The vast majority of medical experts, including the American Medical Association, agree that few of these, if any, are of much proven worth in delaying or reversing the aging process. I should hasten to add that I myself am hardly immune from succumbing to such practices and dutifully take my vitamins, supplements, and other occasional concoctions, though I definitely refrain from doing too much exercise.

We have become obsessed with extending our life spans regardless of cost, whereas it may make greater sense to put the emphasis on maintaining and extending *health span*—that is, living a fuller life in an appropriately healthy body with a reasonably healthy mind and dying when these systems are clearly no longer fully functional. How we conduct ourselves in these matters and how we approach death are intensely personal decisions for which there are no easy answers, and I would not presume to make judgments on an individual's choice. But collectively they raise serious issues for society that need to be addressed, and having a deeper understanding of the aging and dying processes and their relationships to a healthy life should play a role in how we deal with them.

Closely related to this is the ongoing search for a mythical elixir of life, the *elixir vitae*, usually conceived of as a magic potion that bestows immortality on the person who drinks it. It has arisen in many ancient cultures and is often associated with medieval alchemists. It turns up in many myths, with its latest incarnation being in the guise of the Philosopher's Stone ("Sorcerer's Stone" in the United States) in the popular Harry Potter books.

Its modern incarnation has now encroached into the scientific community via several very well-funded programs devoted to the extension of life span. Some of these are pretty hokey, but in recent years several serious established scientists have become engaged in the search for a modern holy grail of life. It is perhaps telling that these projects are overwhelmingly funded from private sources rather than from traditional federal funding agencies such as the National Science Foundation or the National Institute on Aging (which is part of the National Institutes of Health). It is also not surprising that some of the most prominent of these are funded by Silicon Valley tycoons. After all, they have revolutionized society and not unreasonably want both themselves and their hugely successful companies to go on living forever and are willing to spend their money in trying to do so. Among the more prominent ones are Larry Ellison, the founder of Oracle, whose foundation has spent hundreds of millions of dollars on aging research; Peter Thiel, a cofounder of PayPal, who has invested millions in biotech companies oriented toward solving the problem of aging; and Larry Page, a cofounder of Google, who started Calico (the California Life Company), whose focus is on aging research and life extension. And then there's the health care mogul Joon Yun, who, though he didn't make his fortune in classic high-tech, is based in Silicon Valley and is the sponsor of the $1 million Longevity Prize "dedicated to ending aging" through his foundation, the Palo Alto Institute.

Although I remain quite skeptical that any of these will achieve any significant success, they are worthy efforts and a great example of American philanthropy at work, regardless of the motivation. And some of them are undoubtedly first-rate research programs that will deliver some very good and important science even if they don't achieve their stated goals of discovering the elixir of life or even significantly extending life span. In any case, I hope that I'm wrong and that one of these efforts brilliantly succeeds and that life span can indeed be significantly extended *without* compromising health span.

One of the great ironies of this continuing war against death is that we have actually made spectacular progress in extending our life spans over the past 150 years *without* any explicit dedicated program to do so. Prior to the Industrial Revolution and up until the middle of the nineteenth century, life spans across the globe remained fairly constant. The life expectancy at birth prior to 1870 is estimated to have been a mere thirty years when averaged across the entire globe, rising to thirty-four by 1913, but then more than doubling to over seventy years by 2011. Even though there are large variations across different nations with quite disparate standards of living and health care availability, the same dramatic story is repeated everywhere. For instance, in England, where some of the best mortality statistics have been kept since the sixteenth century, the average life span remained at roughly thirty-five years from about 1540 to about 1840, when it began to rise, reaching fifty-two in 1914, the year my father was born, then to about sixty-three in 1940, the year I was born, and has now climbed to more than eighty-one years. Even in some of the very poorest nations this remarkable phenomenon has been repeated: the average life span in Bangladesh was about twenty-five years in 1870 but is now about seventy. A powerful way of expressing this remarkable phenomenon is to note that every country in the world now has a higher life expectancy than the *highest* life expectancy in any country in 1800. This is truly fantastic. What is so fascinating about this achievement is that it was accomplished without any dedicated global, national, or philanthropic private program to extend life. It just happened all on its own without anyone discovering a magic pill, an elixir of life, or fiddling with anyone's genes. What has been going on?

Well, you very likely know or can easily guess the answer. First, a major contributor was the astonishing decrease in infant and child mortality. In the developed world, we tend to forget how extraordinarily prominent child mortality was until relatively recently. Until the mid-nineteenth century, anywhere from one quarter to one half of the children born in European countries didn't make it to their fifth birthday. For instance, of Charles Darwin's ten children, one died after just a few weeks, one after a year and a half, and another, his first daughter, Anne, didn't survive beyond ten years. And Darwin lived an entitled upper-class life with access to all possible amenities and aids including the best possible health care. Can you imagine what it was like for the vast underprivileged majority who were members of the laboring class? By the way, Darwin was particularly attached to Anne, and her tragic death triggered his break with Christianity and led to his coming to terms with the terrible personal realization that death was an integral part of the eternal evolutionary dynamic. Seventy-five years later two of my own grandparents' eight children died in a pattern not so dissimilar from the Darwins': one after only a few weeks, the other, coincidentally also called Anne, died at age ten when she succumbed to St. Vitus' dance, a childhood disease that was not uncommon a hundred years ago. Nowadays it is referred to by the less picturesque name of Sydenham's chorea and affects only about 0.0005 percent of children in the United States.

This is typical and a good example of the huge change that has made childhood mortality a relative rarity in developed and developing countries and significantly reduced it in underdeveloped ones. As was discussed earlier, the coming of the Enlightenment and the Industrial Revolution heralded rapid advances in medicine and huge improvements in health care, both of which were major contributors to an exponentially increasing urban population and to increasing standards of living. Improved housing, public health programs, immunization, antiseptics, and, most important, the development of sanitation, sewage systems, and access to clean running water played an enormous role in overcoming and containing childhood diseases and infections.

All of these accomplishments are the result of the fascinating dynamic that was set in motion by the increasing migration of the populace to urban environments and the development of greater social responsibility, with the city as the provider of basic rights and services. Despite the Dickensian image of destitution and widespread poverty, which was certainly highly prevalent, the increasing access to these basic services led to the decrease in infant and childhood mortality and to a rapid rise in life span, and consequently to a rapid increase in population. Fewer were dying young and more were living longer, a dynamic that has continued unabated to this day. *The city as the engine for social change and increasing well-being is one of the truly great triumphs of our amazing ability to form social groups and collectively take advantage of economies of scale.*

The decrease in infant and child mortality has played a huge role in increasing average life span. For instance, in 1845 average life expectancy at birth in England was only about forty years, but if you made it to five years old, you could expect to live for another fifty years and die at age fifty-five. So if we remove child mortality from the statistics, then the expected life span in 1845 increases by more than ten years. It is interesting to compare this with today's situation. Life expectancy at birth in England is now about eighty-one years. At five years old it increases only marginally to eighty-two, a gain of only one year, reflecting the extremely low rate of infant and child mortality.

Even after factoring out the huge decrease in infant and child mortality it is clear that there has still been a huge increase in the average life span over the past 150 years. Furthermore, we learn that care has to be taken in interpreting the statistics when thinking about combating aging and extending longevity. Obviously all of those infants and children who over the centuries tragically died before even reaching puberty did not die from some quirk of the aging process. Their fate was primarily determined by the inadequacy of the environment they lived in rather than their basic biology. What we learn is that if a child survives to some reasonable age, then he or she has a very good chance of living for considerably longer than the overall average. So, for example, if you made it to twenty-five in 1845, then your life expectancy jumps from forty to a respectable sixty-two. On the other hand, if you made it to eighty, then you were likely to make it only to eighty-five before dying. But this is not so different from today's situation: if you are eighty today, you will probably "only" live to eighty-nine. And perhaps even more surprising is that this is not so different from the experience of our hunter-gatherer ancestors many thousands of years ago. They, too, were dominated by infant mortality, but once that is factored out, they could expect to live to sixty or seventy years old.

On a personal side note, I was gratified to learn from these averages that having reached seventy-five years old, I can expect to live almost twelve years longer and die at the astonishing age of almost eighty-seven, which is far longer than I had ever thought. If true and I am able to stay healthy, it gives me time to finish this book, to see how my children blossom as they approach midlife, possibly even see grandchildren growing up, see the Santa Fe Institute continue to flourish and receive a $100 million endowment, and, the most unlikely of these, see Tottenham Hotspur win the Premier League and even more unlikely the Champions League. My wonderful wife of more than fifty years, Jacqueline, who is now seventy-one years old, should, according to these averages, expect to live to almost eighty-eight, so she'll have more than four years without me driving her nuts in my dotage.

This, of course, is just the stuff of fantasy, as we are using highly coarse-grained averages to say something about individuals, which has all of the pitfalls of extrapolating from the average to the particular. On the other hand, it does give you some idea of the general trends and where you stand with respect to them and provides a very approximate baseline from which to fantasize. Actually, these statistics also play an important role in your life because they are routinely used by insurance companies and mortgage lenders to determine whether you are a safe bet and how much they are going to charge you.

Let's return to the discussion on the statistics of old age and take it one step further: Suppose you had reached one hundred years old in 1845; then it would come as no surprise to learn that statistically you could only expect to live for less than another two years more—to be precise, a year and ten months. Not terribly long. Similarly, if you were one hundred years old today, then you would likewise expect to live for a little more than two more years, actually two years and three months. This represents a gain of just five months in your longevity over your predecessor of 150 years ago, despite the extraordinary advances in health care, medicine, and standards of living that occurred in the meantime.

This exemplifies what all the fuss and bother is about regarding attempts to inhibit aging and mortality. As you continue to increase in age, the time remaining before you die becomes increasingly shorter, eventually becoming vanishingly small. This leads to the notion of a maximum possible age that a human being could conceivably live, which turns out to be less than around 125 years. Very few people have even approached this age. The oldest verified person ever was the Frenchwoman Jeanne Calment, who died in 1997 at the remarkable age of 122 years and 164 days. Just to get a sense of how exceptional this is, the next oldest verified person was the American Sarah Knauss, who lived more than three years fewer than Jeanne, dying at the age of 119 years and 97 days. The next super-champs of long life lived almost two years fewer than Sarah, while the oldest person still alive today is the Italian Emma Murano, who is "only" in her 118th year.

The search for life extension can therefore be boiled down to two major categories: (1) The conservative challenge: how can the rest of us continue the upward march toward a longer life and approach the extraordinary achievements of Jeanne Calment and Sarah Knauss? (2) The radical challenge: is it possible to extend life span beyond the apparent maximum limit of approximately 125 years and live, for instance, to 225 years? In a very real sense we're already achieving the first, whereas it is the second that raises serious scientific questions.

A great deal of research has now been done on centenarians and supercentenarians (those who live beyond 110) to try to discover the origins of their extreme longevity. They represent the extreme tail of age distribution—it is estimated that there are at most only a few hundred of such people alive today. They are curious outliers, and their existence and life history are the source of endless fascination as we seek hints as to how we should conduct our lives or what genes we should have been born with if we wish to live a very long life. Many books and articles are written about them, but it has been hard to distill their collective life histories and genomic composition into a precise formula for increasing longevity.[13] Lots of fairly obvious platitudes are mentioned that are not so different from what your mother advised you when you were growing up—and possibly continues to do even now—like eating green vegetables, not indulging in too many sweets, relaxing and minimizing stress, keeping moderately fit, maintaining a positive attitude and living in a supportive community, and so on. In this regard it's amusing to take a brief look at the life of the super-champion of longevity, Jeanne Calment.

She was born and lived her entire life in Arles in the south of France. Her only child, a daughter, died at age 36 of pneumonia and her only grandchild also died at the age of 36, but of a car accident. Consequently she had no direct heirs. She smoked cigarettes from the age of 21 to 117 and lived on her own until her 110th birthday, walking unaided until she was 114. She was not particularly athletic or much concerned about her health. When asked, she ascribed her longevity to a diet rich in olive oil, which she also rubbed into her skin, port wine, and the nearly one kilogram (2 lbs.) of chocolate she ate every week. Make of this what you will. By the way, you probably recognize Arles as the place where Vincent van Gogh went in order to develop his distinctive style of painting, living for a time with Paul Gauguin. Calment remembers meeting him when she was 13 on his visits to her uncle's shop to buy canvas for his paintings. She vividly remembered him as "dirty, badly dressed, and disagreeable."

The concept of a maximum life span is an extremely important one, as it implies that without some major "unnatural" intervention (which is what the elixir believers are seeking), natural processes inextricably limit human life span to about 125 years. Below, I will address what these limiting processes are and present a theoretical framework based on the network theory for determining this number. Before doing so, however, I want to show how classic *survivorship curves* provide powerful evidence in support of the concept of a maximum human life span.

A survivorship curve simply represents the probability that an individual will live to a given age and is determined by plotting the percentage of survivors in a given population as a function of their age. Its converse is called a *mortality curve* and is the percentage of people who have died at a given age, representing the probability that an individual will die at that age. Biologists, actuaries, and gerontologists have coined the term *mortality* or *death rate* to denote the number of deaths in a population that occur in some given period of time (a month, say) relative to the number that are still alive.

The general structure of survivorship and mortality curves is pretty obvious: most individuals survive the earliest years, but gradually a larger and larger percentage die until a point is reached where the probability of surviving eventually vanishes while the probability of dying reaches 100 percent. A great deal of statistical analysis has been done on such curves across different societies, cultures, environments, and species. One of the surprising results that emerges is that the mortality rate for most organisms remains approximately unchanged with age. In other words, the relative number of individuals that die in *any* time period is the same at any age. So, for example, if 5 percent of the surviving population dies between ages five and six, then 5 percent of the surviving population will also die between ages forty-five and forty-six and between ninety-five and ninety-six. This sounds nonintuitive, but if we put it a different way it will make more sense. A constant mortality rate means that the number of individuals that die in some time period is directly proportional to how many have survived up until that time. If you go back to the earlier discussion on exponential behavior in chapter 3, you will discover that this is precisely the mathematical definition of the *exponential function*, which I will discuss in much greater detail in the following chapter. Here it says that survivorship follows a simple exponential curve, meaning that it becomes exponentially less likely that an individual from the original population will survive the older it gets, or equivalently, that it becomes exponentially more likely that an individual will die the older it gets.

This is precisely the rule that many decay processes in the physical world follow. Physicists use the term *decay rate*, rather than mortality rate, to quantify the decay of radioactive material in which "individual" atoms change their state by emitting particles (alpha, beta, or gamma rays) and "die." Decay rates are typically constant so that the amount of radioactive material decreases exponentially with time, just as the number of individuals does in many biological populations. Physicists also use the term *half-life* to characterize decay rates: this is the time it takes for half of the original radioactive atoms to have decayed. Half-life is a very useful metric for thinking about decay processes in general and has migrated into many fields, including medicine, where it is used to quantify the time efficacy of drugs, isotopes, and other substances that the body processes.

In chapter 9 I will use this language to discuss the mortality of companies and show the surprising result that they, too, follow the same exponential decay law, so that their mortality does not change with age. Indeed, the data show that the half-life of publicly traded companies in the United States is only about ten years. So in just fifty years (five half-lives) only $(\frac{1}{2})^5 = \frac{1}{32}$ or about 3 percent are still posting sales. This begs the fascinating question as to whether the same general dynamics underlies the surprising commonality in the mortality of organisms, isotopes, and companies. We will return to speculate about this later.

But first, back to humans. Until the middle of the nineteenth century, the survivability curves for human beings remained pretty much unchanged, approximating those of other mammals in following an exponential curve. We lived and died according to a constant mortality rate, so the chances of living a very long life were exponentially small. Nevertheless, there was still a finite but exceedingly small probability of becoming a supercentenarian, with the occasional person living to one hundred and beyond. With the huge changes brought on by urbanization and the Industrial Revolution we began to live longer and free ourselves from the shackles of a constant mortality rate. In Figure 25 you can readily see that our survivability curve has progressively shifted from an exponential decay toward developing a flatter and flatter shelf with a shoulder that monotonically shifts toward longer life spans, reflecting the increase in survivability of people at all ages. Also readily apparent is the sharp decrease in infant and child mortality and the continuous shift of the average toward longer and longer life spans.

However, also notice that even as the shoulder shifted toward increasing life spans and people began to live longer, the curve nevertheless always eventually dropped off and approached approximately the *same* value. So despite these enormous gains and the continuous evolution toward ever-increasing average longevity, the end point of the curves, where the probability of surviving vanishes and the probability of dying is 100 percent, always remained the

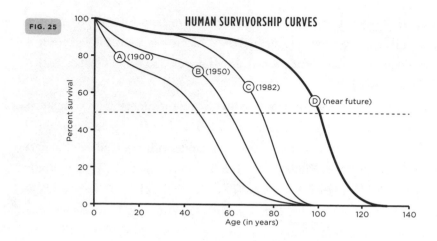

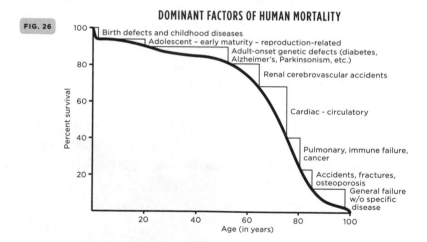

(25) Human survivorship curves showing the rapid shift from classic exponential decay (constant mortality rate) prior to the early nineteenth century toward an increasingly more rectangular shape as the average life span progressively increased due to major changes indicated on the graph. Regardless of this progress, maximum life span has remained at about 125 years. (26) Major causes of mortality at various ages.

same: they all converge on about 125 years. This dramatically and convincingly illustrates the existence of a maximum biological life span.

Figure 26 is an attempt to break down the progressive increase in longevity to its various causes. The biggest contributor has been improvement of housing, sanitation, and public health programs, again illustrating the central role that cities and urbanization have played. In parallel with this, it's equally interesting to deconstruct mortality into its leading medical causes. In order, these are (1) cardiovascular and heart disease, (2) cancer (malignant neoplasms), (3) respiratory disease, and (4) stroke (cerebrovascular diseases). The pattern is quite similar across the globe. An intriguing way of quantifying these is to ask what the gain in life expectancy would be if each of these specific causes was eliminated. A sampling of these is shown in Table 4 taken from data analyzed by the U.S. Centers for Disease Control (CDC) and the World Health Organization (WHO). You can see, for instance, that if all heart and cardiovascular disease were cured, life expectancy at birth would increase by only about six years. Perhaps more surprising is to learn that if all cancer were cured, life expectancy at birth would increase by only about three years—and, at age sixty-five, only by a little less than two years.

There are two important points I want to emphasize arising from these statistics: (1) The leading causes of death are overwhelmingly associated with damage, whether in organs and tissue (as in heart attacks or stroke) or in molecules (as in cancer)—infectious diseases play a relatively minor role. (2) Even

TABLE 4   Estimated Gain in Life Expectancy
if a Given Disease Was Cured

CATEGORY OF DEATH	POTENTIAL GAIN IN LIFE EXPECTANCY (YEARS) IF ELIMINATED
Cardiovascular: all cardiovascular diseases	6.73
Cancer: malignant neoplasms, including neoplasms of lymphatic and hematopoietic tissues, AIDS, etc.	3.36
Diseases of the respiratory system	0.97
Accidents and "adverse effects" (health-care-induced deaths)	0.92
Diseases of the digestive system	0.46
Infectious and parasitic diseases	0.45
Firearm deaths	0.4

if every cause of death were eliminated, all human beings are destined to die before they reach 125 years old, and the vast majority of us will do so well before we reach that ripe old age.

III. Daylight

A great deal has been written about the biology and physiology of aging but rarely from the more quantitative, mechanistic viewpoint that I am trying to emphasize here.[14] So in this spirit, I want to review some of the salient features of aging that need to be quantitatively explained by any theory of longevity and show how these might provide hints as to what the generic underlying mechanisms might be.

Most of the discussion thus far has been about human beings, but I now want to extend it to other animals in order to connect it to scaling laws and the theoretical framework introduced earlier. The discussion will be in the spirit of a coarse-grained description, so there are undoubtedly outliers and even exceptions to some of these statements. This is especially true for the case of aging and mortality because, unlike most other traits, these are not directly selected for in the evolutionary process. Natural selection only needs to ensure that the majority of individuals in a species survive sufficiently long to produce enough offspring to maximize their evolutionary fitness. Once this has happened and they have performed their evolutionary "duty," how much longer they live is of much less importance, so large variations in individual and species life spans can be expected. Thus human beings have evolved to live for at least forty years so that they can produce ten or so children, at least half of whom will survive to maturity and beyond. It's perhaps no accident, then, that this is the age of a woman's menopause. However, to ensure that enough of us reach this age and reproduce accordingly, we have evolved to be sufficiently "overengineered" that statistically many of us are able to live much longer.

Automobiles provide an interesting comparison. For various socioeconomic and technological reasons cars have "evolved" to last for at least 100,000 miles provided they are reasonably well maintained. Depending on fluctuations in the manufacturing process and the degree of maintenance and repair, some cars may live for much longer. Indeed, with enough obsessive mainte-nance, repair, and replacement of parts, cars can be kept alive for a very long time. Human beings have achieved something analogous to this by eating and living well, having regular annual medical tune-ups, keeping hygienic, and occasionally having various body parts replaced. However, it's unlikely that we can do for ourselves what we can potentially do for an automobile and preserve an individual human being indefinitely because, unlike automobiles, which are simple, we are highly complex adaptive systems—and, in particular, are not just the linear sum of replaceable parts.

Here's a summary of some of the significant properties of aging and mor-tality that need to be explained by any theory:

1. Aging and death are "universal": all organisms eventually die. A corol-lary to this is that there is a maximum life span and a corresponding vanishing survival rate.

2. Semiautonomous subsystems of organisms, such as our various organs, age approximately uniformly.

3. Aging progresses approximately linearly with age. For instance, Figure 27 shows how organ functionality degrades with age.[15] Plotted is the percentage of maximum capacity for various vital functions indicating a linear decline with age beginning almost immediately after maturity at about age twenty. It's slightly depressing to see that on average we are physically optimal (100 percent) for only a very few years and that be-ginning around age twenty it's literally downhill all the way. Notice also that we build up to our maximum capacity relatively quickly during our growth period. Later, I will suggest that this aging process is under way even during our earliest years prior to maturity but is hidden by the overwhelming predominance of growth. The aging process effectively begins as soon as you are conceived. Bob Dylan got it right when he sang that "he not busy being born is busy dying."

4. Life spans scale with body mass as a power law whose exponent is ap-proximately ¼. As anticipated, there is a large variance in the data, partly because there are no controlled life-history experiments on lon-gevity for mammals, including us. Some of the data are garnered from

reports on wild animals, some from zoos, some from domesticated animals, some from research laboratories, each with very different environmental and lifestyle conditions. In addition, some are reports of just one or two animals of a species, and some from large cohorts. Although this lack of control is problematic, there are clear trends and consistencies in the data that statistically point to approximate ¼ power scaling.

5. The number of heartbeats in a lifetime is approximately the same for all mammals, as shown in Figure 2 in the opening chapter.[16] Thus, shrews have heart rates of roughly 1,500 beats a minute and live for about two years, whereas heart rates of elephants are only about 30 beats a minute but they live for about seventy-five years. Despite their vast difference in size, both of their hearts beat approximately one and a half billion times during an average lifetime. This invariance is approximately true for all mammals, even though there are large fluctuations for the reasons I outlined above. The greatest outlier from this intriguing invariance is us: for modern human beings, on average our hearts beat approximately two and a half billion times, which is about twice the number for a typical mammal. However, as I have already emphasized, it is only in the past one hundred years that we have been living this long. Over the entire history of humankind up until relatively recently we lived for approximately half as long as we do now and, like the vast majority of mammals, followed the approximately invariant one and a half billion heartbeats "law."

6. Related to this is another invariant quantity: the total amount of energy used in a lifetime to support a gram of tissue is approximately the same for all mammals and, more generally, for all animals within a specific taxonomic group.[17] For mammals it's about 300 food calories per gram per lifetime. A more fundamental way of expressing this is to note that the number of turnovers during a lifetime of the respiratory machinery responsible for the production of energy in cells is approximately the same for all animals within a specific taxonomic group. For mammals this is about ten thousand trillion times (10^{16}) and translates into the invariance of the number of ATP molecules (our fundamental currency of energy) produced in a lifetime to support one gram of tissue.

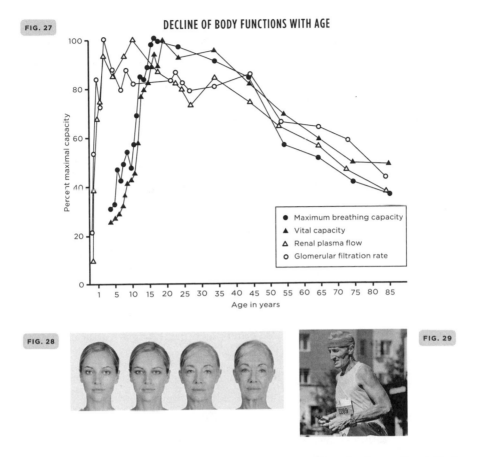

FIG. 27

DECLINE OF BODY FUNCTIONS WITH AGE

- ● Maximum breathing capacity
- ▲ Vital capacity
- △ Renal plasma flow
- ○ Glomerular filtration rate

Percent maximal capacity

Age in years

FIG. 28

FIG. 29

Change in various organ functions with age: percentage of maximal capacity plotted against age. Notice the rapid rise during growth, reaching the maximum around age twenty, after which there is a steady linear decline. Despite this steady decline, a healthy, active life is still possible until late into old age.

Quantities that do not change when other parameters of the system change play a special role in science because they point the way to generic underlying principles that transcend the detailed dynamics and structure of a system. The conservation of energy and the conservation of electric charge are two famous examples of this in physics: no matter how complicated and convoluted the evolution of a system might be as it transforms and interchanges energy and electric charge, the total amount of energy and the total electric charge remain the same. Thus if you add up all of the energy and all of the electric charge in a system at some initial time, then these maintain the same value at any later time no matter what has happened—provided, of course, that you have not added energy or charge from the external environment. To take an extreme example: the total mass-energy in the universe today is exactly the same as it was at the Big Bang more than 13 billion years ago when it was just a compact minuscule point, despite the subsequent evolution of all the galaxies, stars, planets, and life-forms.

The existence of approximate invariant quantities as well as of scaling laws in the complex process of aging and death provides important hints that these processes are not arbitrary and suggests that there might be coarse-grained laws and principles at play. Even more tantalizing is that the scaling laws of longevity have the same quarter-power structure as all other physiological and life-history events.

Before exploring this further, it is instructive to compare some of this with the longevity of automobiles. Unfortunately, there have been surprisingly few scaling analyses of automobiles and other machines, especially regarding their longevity. However, Thomas McMahon, an engineer at Harvard, analyzed data on internal combustion engines, ranging from those used in lawn mowers and automobiles to airplanes, and showed that they follow simple isometric Galilean cubic power scaling laws, discussed in chapter 2. For instance, the horsepower rating of these engines (the analog to their metabolic rate) scales *linearly* with their weight, so to double their power output, you have to double their weight. Thus, unlike organisms, engines do not exhibit an economy of scale as their size increases. McMahon also found that their RPMs (their heart rates) scale with an inverse *cubic* power of their weight.[18]

These are in distinct contrast to the ¼ power scaling laws obeyed by organisms, which result from their optimized fractal-like network structures: their metabolic rates (their horsepower) scale with an exponent of ¾ and their heart rates (their RPMs) with an exponent of -¼. The fact that internal combustion engines have no complex network structures and do *not* follow ¼ power scaling is supporting evidence for the underlying network theory for the origin of ¼ power scaling in biology. Because manufactured engines satisfy classic cubic scaling, one might speculate that their life span increases with the cube root of their weight rather than with a quarter power. Unfortunately, there isn't sufficient data available to test this. However, qualitatively it does predict that bigger automobiles should last longer. In fact, *all* of the top ten longest-lasting vehicles are either large trucks or SUVs, while only three regular-size sedans make it into the top twenty. If you're simply looking for longevity, buy big: the Ford F-250 is tops, with the Chevrolet Silverado second and the Suburban third.

Cars are now typically expected to last for about 150,000 miles. In fact, like the humans who made them, their longevity has increased dramatically over a relatively short period of time, having almost doubled over the past fifty years. Just to get an idea of what this implies, suppose that when averaged over its lifetime, a typical car moves at 30 miles per hour and that its "heart rate" is 2,500 RPMs, then the total number of "engine beats" during a 150,000-mile lifetime is approximately a billion. Amusingly, this is not so different from the number of times a mammalian heart beats in its lifetime. Is this just a coincidence, or is it telling us something about the commonality of mechanisms responsible for aging?

IV. Toward a Quantitative Theory of Aging and Death

All the evidence points to the origins of aging and mortality as being the result of the "wear and tear" processes that unavoidably arise simply from being alive. Like all organisms, we metabolize energy and resources in a highly efficient way in order to combat the continuous fight against the inevitable production of entropy in the form of waste products and dissipative forces that cause physical damage. As we begin to lose the multiple localized battles against entropy we age, ultimately losing the war and succumbing to death. Entropy kills. Or as the great Russian playwright Anton Chekhov poignantly remarked, "Only entropy comes easy."

A central feature of how life is sustained is the transportation of metabolic energy through space-filling networks across all scales to service and feed cells, mitochondria, respiratory complexes, genomes, and other functional intracellular units, as symbolized on page 104. However, these very systems that sustain us are continually causing damage and degrading our bodies. Just as the flows of cars and trucks on highways, or the flow of water through pipes, lead to continuous wear and tear resulting in damage and decay, so it is with the flows in our networks. However, there is a crucial difference: in organisms the damage with the most serious consequences occurs at the cellular and intracellular levels, which are terminal units of these networks where energy and resources are exchanged as, for instance, between capillaries and cells.

Damage occurs across multiple scales through many different mechanisms associated with physical or chemical transport phenomena, but loosely speaking it can be separated into two categories: (1) Classic physical wear and tear due to the viscous drag in the flow, analogous to the wear and tear resulting from ordinary friction when two physical objects move over each other like wearing out your shoes or the tires of your car. (2) Chemical damage from free radicals, which are by-products of the production of ATP in respiratory metabolism. A free radical is any atom or molecule that has lost an electron and consequently has a positive electric charge, making it highly volatile. Most of this kind of damage is caused by oxygen radicals that react with vital cellular components. Oxidative damage to DNA may be particularly deleterious, because in nonreplicating cells such as in the brain and musculature, it causes permanent damage to transcriptional, and perhaps most important, regulatory regions of the genome. Although the detailed role and extent of oxidative damage in aging remains unclear, it has stimulated a mini-industry of antioxidant supplements such as vitamin E, fish oil, and red wine as some of those elixirs of life to combat aging.

The great power and virtue of having a general quantitative theory for understanding the structure and dynamics of these networks and, in particular, for the energetic flows through them, is that it provides an analytic framework for calculating many other subsidiary quantities, such as the growth curves in the previous section and damage rates relevant to aging and mortality that I am going to discuss here. This coarse-grained framework is very general and can incorporate any model of aging based on generalizations of "damage" mechanisms associated with generalized physical or chemical transport phenomena discussed above. The details of the damage mechanism are not important for understanding many of the general features of aging and mortality because the most relevant damage occurs in invariant terminal units of networks (capillaries and mitochondria, for example) whose properties do not appreciably change with the size of the organism. Consequently, the damage per capillary or mitochondrion is approximately the same regardless of the animal.

Because these networks are space filling, meaning that they service all cells and mitochondria throughout the body of the organism, damage occurs approximately uniformly and relentlessly throughout the organism, explaining why aging is approximately spatially uniform and progresses approximately linearly with age. This is why at age seventy-five every part of your body has deteriorated to pretty much the same degree, as indicated in Figure 27. At a more detailed level, this implies that aging within each organ is approximately uniform even though different organs may age at moderately different rates because they have slightly different network characteristics, especially in regard to their potential for repair.

Because larger animals metabolize at higher rates following the ¾ power scaling law, they suffer greater production of entropy and therefore greater overall damage, so you might have thought that this would imply that larger animals would have shorter life spans in obvious contradiction to observation. However, we saw in chapter 3 that on a cellular or per unit mass of tissue basis metabolic rate and therefore the rate at which damage is occurring at the cellular and intracellular levels *decreases* systematically with increasing size of the animal—another expression of economy of scale. Furthermore, as already emphasized, the most significant damage occurs at the terminal units of networks in capillaries, mitochondria, and cells, and their metabolic rates *decrease* with the size of the organism following power law scaling with an exponent of ¼. Cells in larger animals are systematically processing energy at a slower rate than cells in smaller ones. *So at the critical cellular level cells suffer systematically less damage at a slower rate the larger the animal, and this results in a correspondingly longer life span.*

Recall that this down-regulation of terminal units is a result of the hegemony of the network and is the origin of the general economy of scale with increasing size. This is also reflected in the number of terminal units increasing only as the ¾ power of mass, rather than linearly; this played a crucial role in deriving growth curves and for explaining why growth eventually ceases. Because terminal units are invariant, the total damage rate is just proportional to their total number and so also increases only as the ¾ power of mass and is therefore directly proportional to metabolic rate.

Driven by metabolism, the accumulation of damage relentlessly degrades the entire organism. To combat this continual damage our bodies have potent repair mechanisms, which are also powered by cellular metabolism and therefore constrained by the same networks and scaling laws. Consequently their inclusion in the calculation of the total amount of irreversible damage does not change the mathematical structure of the equation but does, of course, affect its overall magnitude. Repair is expensive, and faithfully repairing every single damage event is prohibitively expensive, if not impossible given the enormous number of damages that are continually occurring. The overall scale of repair is primarily determined by the evolutionary requirement that organisms live long enough to produce sufficient offspring to compete in the gene pool.

Aging therefore progresses approximately uniformly, gradually leading to the onset of mortality. This is the result of cumulative unrepaired damage having become sufficiently macroscopic that the functionality of the organism is increasingly compromised, as demonstrated in the graph of Figure 27. Ultimately the organism is no longer able to function, resulting in death from "old age." A small fluctuation or perturbation such as a mild heart flutter is sufficient to end viability. In most cases, however, death occurs earlier than this from multiple causes related to the weakening of specific organs and/or the immune and cardiovascular systems, for example, because of the accumulation of damage in them. All of the dominant causes of mortality listed in Table 4 fall under this umbrella, except of course death arising from externalities and deleterious environmental conditions such as accidents, firearms, and pollutants. Consequently, a calculation of life span based on the origins of death that I've outlined actually sets a limit on *maximum* possible life span.

This can be estimated assuming that the ultimate threshold for death is reached when the fraction of damaged cells (or molecules such as DNA) relative to the total number in the organ or body reaches a critical value, which is approximately the same for all organisms of the same taxonomic group (all mammals, for example). In other words, the total number of damages is proportional to the total number of cells and therefore to the body mass. We simply ask how long it takes for this number of damages to have happened, knowing from the metabolic rate the rate at which damage is occurring and that, on average, each cellular damage event is caused by approximately the same invariant amount of energy. The total number of damages incurred in a lifetime is just the damage *rate* (that is, the number of damage events per unit time which is proportional to the number of terminal units) multiplied by the life span, and this has to be proportional to the total number of cells, and therefore to body mass. Consequently, life span is proportional to the total number of cells divided by the number of terminal units. But the number of terminal units scales with mass with a ¾ power exponent, while the number of cells scales linearly, resulting in life span scaling as the ¼ power of mass, consistent with data.

Notice that just as we saw when discussing growth, the mismatch between the scaling of the sources of energy and therefore the sources of damage (the terminal units) and the scaling of the sinks of energy (cells that need to be sustained) has enormous consequences. In the one case, it ensures that we cease growing, and in the other it ensures that larger animals have expanded lifetimes. And all of this follows from the constraint of the networks.

V. Tests, Predictions, and Consequences: Extending Life Span

A. TEMPERATURE AND EXTENDING LIFE SPAN: Because metabolic rate is proportional to the number of terminal units and these are where most damage occurs, we can relate life span directly to metabolic rate. This results in an alternative expression for life span as the ratio of the body mass to its metabolic rate. In other words, life span is *inversely* proportional to the metabolic rate per unit mass of the organism and therefore inversely proportional to the average metabolic rate of its cells. We saw just above when discussing the metabolic theory of ecology that this systematically scales with body mass following quarter-power scaling and as an exponential with temperature.

This provides an interesting test of the theory by explaining why life spans systematically and predictably increase exponentially as temperature decreases, as shown in Figure 24. This implies that life span can in principle be extended by lowering the body temperature because this lowers cellular metabolic rate and therefore the rate at which damage is incurred. This is a very large effect: I remind you that a modest 2°C decrease in body temperature can result in a 20 percent to 30 percent increase in life span.[19] So if you were able to artificially lower your body temperature by just 1°C (that's about 1.8°F) you could enhance your life span by about 10 percent to 15 percent. The hitch is that you would have to do this for your entire life in order to reap the "benefits." But more saliently, significantly lowering body temperature may well have many other deleterious, potentially life-threatening outcomes. As I have stressed earlier, changing just one component of a complex adaptive system without fully understanding its multilevel spatiotemporal dynamics usually leads to unintended consequences.

B. HEARTBEATS AND THE PACE OF LIFE: These data also confirm the approximate ¼ power scaling of life spans with mass. Because the theory for the cardiovascular system predicts that heart rates *decrease* with a ¼ power of mass, the dependence on mass cancels out when we multiply heart rates by life spans: the *decrease* in one exactly compensates for the *increase* in the other, resulting in an invariant, a quantity that's the same for all mammals. But multiplying heart rates by life spans just gives the total number of heartbeats in a lifetime, so the theory predicts that this should be the same for all mammals, consistent with the data shown in Figure 2 in chapter 1. This argument can be extended to the fundamental level of the respiratory complexes, the basic units inside mitochondria where ATP is manufactured, to show that the number of times the reaction takes place producing ATP is the same for all mammals.

As I have already remarked, large animals live long lives slowly, whereas small ones live short lives fast, but in such a way that their biomarkers such as the total number of times their hearts beat remain approximately the same. When rescaled according to ¼ power scaling the life-history events of all mammals collapse to the same trajectory, an example of which is the universal growth curve shown in Figure 19. Maybe all mammals experience the sequence, pace, and longevity of life as being pretty much the same? A lovely thought.

Once upon a time, when we were "just" another mammal, we did, too. With the coming of social communities and urbanization we evolved into something different, substantially deviating from the constraints that kept us in harmony with nature. Our effective metabolic rate has increased a hundredfold; we've doubled our life span and diminished our fecundity. I'll return to discuss these extraordinary changes in later chapters using the same conceptual framework to understand how this happened.

C. CALORIC RESTRICTION AND EXTENDING LIFE SPAN: We just saw that life span decreases inversely with cellular metabolic rate. Because this systematically decreases with increasing mass across animals, less damage is incurred per cell, resulting in larger animals living longer. However, within species an individual like each of us can decrease its cellular metabolic rate simply by eating less, resulting in less metabolic damage per cell and potentially in an increase in its life span. This strategy is called *caloric restriction*. It has a long, somewhat controversial, history and has been the focus of many studies across a range of animals. Many of these have shown significant benefits, but others have found little effect and the situation remains a bit murky. Almost all of the investigations show some signs of decreased aging whether or not life span is enhanced. It's difficult to carry out long-term controlled experiments and impossible to execute them on human beings, and many of these studies suffer from poor design. I am somewhat biased toward them because I believe in the theory and the concept that lowering metabolism decreases damage, slows the aging process, and increases maximum life span.

Roughly speaking, the theory predicts that maximum life span, and by extrapolation average life span, increases inversely with caloric intake. Taken at face value, the theory predicts that if you consistently decrease your food intake by 10 percent (a couple of hundred calories a day) you could live for up to 10 percent longer (up to ten years more). Figure 30 shows data on caloric restriction from experiments on mice performed in the 1980s by Roy Walford, who was a pathologist at the UCLA School of Medicine and the leading advocate of caloric restriction for extending life span.[20] The data are presented in the form of survival curves for cohorts of mice under different levels of food intake. The effects are indeed dramatic and are consistent with the prediction of a 10 percent increase in life span resulting from a 10 percent caloric restriction but actually show a smaller effect than the predicted doubling of life span resulting from a massive halving of the caloric intake—life span increased by about 75 percent rather than the predicted 100 percent. However, the trends and general shape of the relationship between life span and caloric intake are consistent with the theory.

Given the relative simplicity of the theory, the agreement is surprisingly good. Coupled with the success of its other predictions (including the rate of aging, the allometric scaling of life span, and its temperature dependence) the theory provides a credible coarse-grained baseline for developing a more detailed quantitative theory for understanding aging and mortality. It gives formulae for the rate of aging and maximum life span in terms of generic "universal" biological parameters that show, for instance, how the scale of one hundred years arises from microscopic molecular scales and why mice live for only a very few years. This provides the scientific basis for asking questions about what parameters can be manipulated to extend life and arrest aging, if that is the goal. For example, combining the scaling laws with Figures 23–30 gives quantitative estimates of how long life can be extended by changing body temperature or eating less.

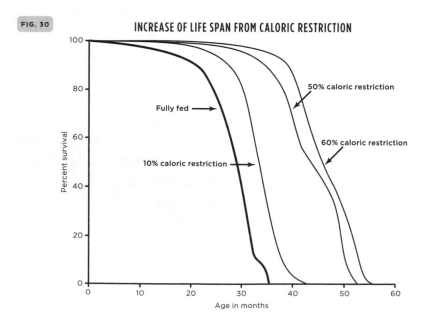

Survival curves for mice under various degrees of caloric restriction showing a concomitant increase in life span.

Furthermore, because this theory is just one piece of a much larger unified framework integrating much of life history, it can also help address the crucial question of what some of the unintended consequences of manipulating life span might be. Naively trying to intervene in the "natural" processes of aging and mortality, whether genetically, physically, or by magic potions, could, and in fact does, have potential deleterious consequences for health and lifestyle. Without a quantitative theoretical framework, such manipulations are potentially dangerous and irresponsible.

Before completing this section, I would be remiss not to mention that Roy Walford, who played a seminal role in aging research, was a man of many talents. Among his many exploits, he gained a certain early notoriety when he and a fellow graduate student used statistical analysis to determine which roulette wheels at a casino in Reno, Nevada, were accidentally biased. By betting heavily on the ones that were most unbalanced, they cleaned up. The casinos eventually realized what was happening and banned them from betting. From his winnings Walford funded his medical school education as well as a year of sailing a yacht in the Caribbean.

5

FROM THE ANTHROPOCENE
TO THE URBANOCENE

A Planet Dominated by Cities

1. LIVING IN EXPONENTIALLY EXPANDING UNIVERSES

One of the most startling and profound discoveries of the twentieth century was to learn that on the cosmic scale we are living in an exponentially expanding universe. An equally profound, but much less heralded, discovery was the realization that on the earthly scale we are also living in an exponentially expanding universe, this one socioeconomic. Although it hardly grabs the same kind of attention, this accelerating socioeconomic expansion has had and will continue to have a significantly more profound effect on your life, your children's lives, and their children's lives than all of the wonders and paradoxes of the exponentially expanding cosmic universe and its archetypal mythologies of dark matter, dark energy, and the Big Bang.

The most obvious manifestation of the exponential rate at which our social and economic lives have been expanding is provided by the huge population explosion that has occurred over the past two hundred years or so. After two million years of slow, steady growth the number of human beings living on the planet is estimated to have eventually reached the billion mark around 1805. But following the Industrial Revolution, the world's population exploded. The

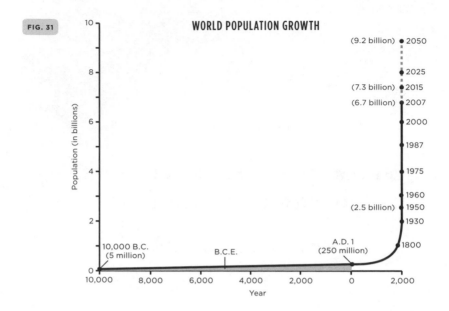

FIG. 31

WORLD POPULATION GROWTH

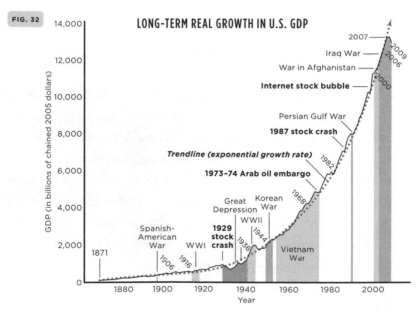

FIG. 32

LONG-TERM REAL GROWTH IN U.S. GDP

(31) The extraordinary faster-than-exponential growth of the world's population since the beginning of the *Anthropocene* ten thousand years ago. The huge rise beginning around 1800 signals the beginning of the Industrial Revolution and the *Urbanocene*. (32) Concomitant with the Urbanocene is the rapid expansion of the economy as represented by the rise in the U.S. GDP since 1800. Despite the many crises, booms, and busts, it is described extremely well by a pure exponential, the dotted line.

transition can be characterized as a shift from traditional hand production to massive industrial machinery and factory production. To a large extent it was stimulated by the invention of large-scale manufacturing processes fueled by the discovery of new methodologies for exploiting the energy stored in huge iron and coal deposits. The subsequent access to apparently unlimited energy and human resources, engendered by the rise of capitalism and individual and corporate entrepreneurship and innovation, marked a major transition in human affairs. The Industrial Revolution was the socioeconomic equivalent of the Big Bang. Having taken two million years to reach a population of a billion, the next billion took only another 120 years, and the one after that took less than 35. Doubling that took just 25 years, reaching 4 billion in 1974, and at present, only 42 years later, we have almost doubled that again with a global population now in excess of 7.3 billion people. Thus, until quite recently, doubling time has been systematically decreasing, reflecting a *faster*-than-exponential growth. Just this year alone we are adding another 80 million, the equivalent of an entire Germany or a Turkey, and are on track to reach a possible 12 billion at the beginning of the next century.

The first pictures of the whole Earth taken from space were inspirational in giving us a completely new psychological perspective on who we are, where we came from, and what sustains us. It was a startling revelation to see a photograph of one's mother for the first time, the one that has given birth to all 7.3 billion of us, bathed in the glorious light of our grandparent, the sun (below left). At the time perhaps no one appreciated this more than the writer and futurist thinker Stewart Brand, who passionately felt that the image of the

whole Earth from space would be a powerful symbol, evoking a sense of shared destiny among all people living on the planet. He relentlessly lobbied NASA to release the first images in 1967, which he then used on the cover of his highly influential *Whole Earth Catalog*, one of the great icons of the 1960s and '70s.

Equally revelatory are the more recent pictures of Mother Earth taken at night when she is not bathed in sunlight (page 211, bottom right). Had it been technically possible to take such a photograph a couple of hundred years ago, it would have appeared black and revealed nothing. Even fifty years ago it would have looked relatively dim. Not so today. Now we have spectacular pictures taken by NASA satellites showing the Earth covered in what looks like a magnificent filigree web of glistening Christmas lights. The brilliance of these "night-lights" is, of course, the explicit result of the exponential population explosion and the extraordinary technical and economic achievements that have accompanied it. And, overwhelmingly, these lights are generated by cities, reflecting the bewildering rate at which we have been urbanizing. As a symbol of twenty-first-century *Homo sapiens* it encapsulates the very essence of the concept of the *Urbanocene* and of *scale,* making it a fitting jacket for this book.

Our recent global population explosion is truly an astonishing achievement, especially when it is realized that, despite huge pockets of poverty, the overall quality of life as measured by health, longevity, and income has by and large increased in concert with this growth when averaged across the entire globe. Traditionally, population growth correlates with increasing socioeconomic and financial indices, so much so that we not only take exponential expansion for granted but have effectively raised it to an axiom. Our entire social and economic paradigm is geared to the continued drive to sustain open-ended exponential growth.

Furthermore, the 7.3 billion of us already on board and the several billion that will soon be joining us over the coming decades need to be fed, clothed, educated, and cared for. Almost all of us want houses, cars, and smart phones, we want to be entertained with televisions, videos, and films in comfortable surroundings, and many of us want to travel and to have access to education and the Internet. Regardless of the diversity of our actions, material desires, and well-being, we all want to have meaningful and fulfilling lives. Together we make up the fantastic tapestry of life, all contributing, participating, benefiting, and suffering, to varying degrees, from the variety of social and economic processes that human beings have invented. But none of this has happened nor can it be sustained without a continuous supply of energy and resources. As presently constituted, our continued success requires the supply of coal, gas, oil, fresh water, iron, copper, molybdenum, titanium, ruthenium, platinum, phosphorus, nitrogen, and much, much more to be available at an exponentially increasing rate.

2. CITIES, URBANIZATION, AND GLOBAL SUSTAINABILITY

Perhaps our greatest invention has been the very stage upon which this panoply of socioeconomic interactions, mechanisms, and processes that drive exponential expansion has been played out, namely the city. This is the view expressed by the urban economist Edward Glaeser in his book *The Triumph of the City*.[1] For concomitant with the population explosion over the past two hundred years has been the exponential urbanization of the planet. The city is the ingenious mechanism we have evolved for facilitating and enhancing social interaction and collaboration, two necessary components of successful innovation and wealth creation. Population and urban growth are, of course, very closely interrelated, each feeding on the other, resulting in our extraordinary dominance of the planet.

The term *Anthropocene* has been suggested as the name for this most recent epoch in the history of our planet during which human activities have significantly affected the Earth's ecosystems. This process began more than ten thousand years ago with the discovery of agriculture and the subsequent transition from mobile hunter-gatherers to sedentary communities and eventually to the emergence of the first cities. Up until that time we were still predominantly "biological," meaning that we were an integrated component of the Earth's multifaceted ecology—just another mammal, so to speak—in meta-equilibrium with all of the other creatures and organisms that comprise the seemingly unbounded diversity of nature. Accordingly, our global population was then just a few million people, reflecting our dynamic interrelationship with the "natural" environment, and the Earth was still fundamentally "pristine."

But eventually along came the Industrial Revolution. Although much of the Earth's landscape had already been significantly modified by human activities prior to its coming, this dramatic series of unprecedented events heralded the beginning of a profound transition to a state of explosive, faster-than-exponential expansion that has led to unforeseen changes in the planet's ecology, environment, and climate over an astonishingly short period. Consequently, some have proposed that we should date the beginnings of the Anthropocene from the Industrial Revolution, while others suggest it should be as recent as the middle of the twentieth century. Others have even suggested that it began more than ten thousand years ago coinciding with the beginning of the Holocene, the geological epoch that began when the Earth warmed up, leading to the development of agriculture and modern man.

I am very enthusiastic about explicitly acknowledging our profound impact on the planet by naming a new epoch but would prefer that we reserve Anthropocene for the entire period going back several thousand years to when we first began to significantly diverge from being predominantly biological to becoming predominantly social by increasing our effective metabolic rate. In that spirit we should also acknowledge that we have already made a sharp transition out of the purely Anthropocene to what could be considered yet another epoch characterized by the exponential rise of cities which now dominate the planet. To designate this much shorter and intense period that began with the Industrial Revolution I would like to introduce a new term and suggest the name *Urbanocene*. Given how profound this change has been and that its future dynamic will determine whether this amazing socioeconomic enterprise will continue to prosper, or whether it is destined to decay and die, I want to set the scene by repeating some of what I wrote in the opening chapter.

As we move into the twenty-first century, cities and global urbanization have emerged as the source of the greatest challenges the planet has faced since humans became social. The future of humanity and the long-term sustainability of the planet are inextricably linked to the fate of our cities. Cities are the crucible of civilization, the hubs of innovation, the engines of wealth creation and centers of power, the magnets that attract creative individuals, and the stimulant for ideas, growth, and innovation. But they also have a dark side:

they are the prime loci of crime, pollution, poverty, disease, and the consumption of energy and resources. Rapid urbanization and accelerating socioeconomic development have generated multiple global challenges ranging from climate change and its environmental impacts to incipient crises in food, energy, and water availability, public health, financial markets, and the global economy.

Given this dual nature of cities as, on the one hand, the origin of our major challenges and, on the other, the generator of creativity and ideas and therefore the source of their solutions, it becomes a matter of some urgency to ask whether there can be a "Science of Cities," by which I mean a conceptual framework for understanding their dynamics, growth, and evolution in a quantitatively predictable framework. This is crucial for devising a serious strategy for achieving long-term sustainability, especially given that the overwhelming majority of human beings will be urban dwellers by the second half of this century, many in megacities of unprecedented size.[2]

None of the problems, challenges, and threats we are facing is new. All of them have been with us since at least the beginnings of the Industrial Revolution. Urbanization is a relatively new global phenomenon, which was not perceived as serious until very recently because cities were very much subdominant relative to the total population. It is only because of the exponential rate of urbanization that the problems they create have now begun to feel like an impending tsunami with the potential to overwhelm us. Fifty years ago, even fifteen years ago, most of us weren't conscious of global warming, long-term environmental changes, limitations on energy, water, and other resources, health and pollution issues, and the stability of financial markets. And if we were, we presumed they were temporary aberrations that would eventually disappear. This is certainly open to debate, with most politicians, economists, and policy makers taking a fairly optimistic view that our ingenuity will triumph. It is in the very nature of an exponential that the future becomes the present at an increasingly more rapid pace, so much so that by the time a problem has arisen it's often too late to address it successfully. Given the general attitude toward this silent threat of exponential expansion, I want to digress and explain its implications, because few in power and policy seem to appreciate it.

3. DIGRESSION: WHAT EXACTLY IS AN EXPONENTIAL ANYWAY? SOME CAUTIONARY FABLES

In discussing the expansion of the universe since the Big Bang or the huge socioeconomic changes that have taken place on the planet since the beginning of the Industrial Revolution, I have been using the phrases "exponential growth" and "exponential expansion" assuming they are well-understood terms. Indeed, I have invoked "exponential" somewhat cavalierly without carefully explaining what it means and what it implies. I may well be underestimating the knowledge and understanding of the general public, but it is not unusual to hear the word "exponential" being used by well-educated journalists, media gurus, politicians, and corporate leaders in ways that indicate that they do not fully understand its meaning or appreciate its formidable implications. In fact, I often feel that if they did we would have a much easier time convincing them of the urgent need to think carefully and strategically about the challenges of long-term sustainability. So at the risk of being pedantic, and because the concept is so important and plays such a critical role in the book, I want to make a short detour to elaborate on its meaning and implications.

"Exponential" is another one of those technical words like "momentum" or "quantum" that originated in a scientific context with a precisely defined meaning but have entered the common jargon because they connote a useful concept that is not already adequately communicated in our everyday language. Colloquially, the phrase "growing exponentially" is generally understood to convey the idea of very fast growth. For example, in my dictionary the first meaning ascribed to "exponential" is "rapid growth." In actuality, exponential growth starts out rather slowly, even innocuously, before smoothly transitioning to what might be termed rapid growth. But it's much more than that.

A population that is growing exponentially is defined mathematically as one in which the rate of increase in its size (per minute, per day, or per year, for instance) is directly proportional to the size of the population that's already there. Thus the growth rate itself grows even faster the bigger the population. So, for example, when the size of a population undergoing exponential growth doubles, the rate at which it is increasing also doubles, which means that it grows faster and faster the bigger it gets, effectively feeding back and running away with itself. Left unchecked, both the population *and* its growth rate would eventually become infinitely large.

You're quite familiar with this kind of growth in your daily life even though it's not usually referred to as exponential growth, for an equivalent way of saying that the rate of increase per unit time is directly proportional to what's currently there is to say that the relative or *percentage growth rate* is constant, which sounds quite innocuous. This is none other than classic compound interest used by banks for calculating the rate of return on your investments. So when presidents, finance ministers, prime ministers, and CEOs announce that their countries or organizations are growing at a rate of 5 percent this year, or when your bank tells you that the rate of return on your savings is 5 percent, implicit in what they are saying is that these are growing exponentially and that the *absolute* growth rate next year is going to be 5 percent larger than it was this year. So if nothing changes, everybody gets progressively richer and more prosperous. Even when the president grimly announces that the economy only grew at 1.5 percent this quarter and gets a lot of negative feedback that the economy is "sluggish," he is still implying that the economy grew exponentially and that we are still on track to grow faster and faster the bigger we get, just at a slower rate. Under a constant percentage growth rate everybody is still getting richer and more prosperous, so no wonder we are hooked on the open-ended steroidlike exponential growth drug. It is truly a real high and is an explicit manifestation of the huge success of our economic dynamic.

The growth of a system, whether an economy or a population, is often expressed in terms of a quantity called the *doubling time,* which is simply the time it takes for the size of the system to double. Exponential growth is characterized by having a *constant* doubling time, which also sounds fairly harmless until one realizes that it implies, for example, that it would take the same time for a population to double from ten thousand to twenty thousand, thereby adding just ten thousand people, as it would for it to double from 20 million to 40 million, thereby adding a humongous 20 million people. Amazingly, the doubling time of the global population has actually been getting systematically shorter and shorter as was indicated above: it took 300 years from 1500 to 1800 for the population to double from 500 million to a billion, but only 120 years to double to 2 billion, and only another 45 to double again to 4 billion. This is shown in Figure 31. So until relatively recently we have been increasing at an *accelerating* pace that is actually *faster* than a pure exponential! Although this acceleration has begun to slow down over the past 50 years, we are still effectively growing at an exponential rate.

Rather than presenting more definitions and dry statistics, I want to relate a couple of amusing stories that illustrate these ideas much more vividly. The surprising attractions and pitfalls of exponential growth have been well known for a long time, especially in the East, where compound interest has been understood and used since ancient times. This is illustrated in one of the great epic poems in world literature, the *Shahnameh,* written about one thousand years ago by the revered Persian poet Ferdowsi. It is the world's longest epic poem and took thirty years to write. Around the time it was being written chess was introduced to Persia from India, where it had been invented. With its rise to popularity, Ferdowsi memorialized it using the chessboard as a means of illustrating the implications of exponential growth. Here's a version of the story:

When the inventor of chess showed the game to the king, the ruler was so taken by it that he asked the inventor to name his reward for creating such a marvelous and challenging game. The man, who was mathematically inclined, asked the king for what seemed to be an extremely modest reward in the form of grains of rice. However, these were to be apportioned in the following manner: he would receive 1 grain of rice on the first square of the chessboard, 2 grains on the second, 4 on the third, 8 on the fourth, 16 on the fifth, and so on, doubling the amount for each progressive square. The king, though somewhat offended by such an apparently measly response to his very generous offer, reluctantly accepted the inventor's request and ordered the treasurer to count out the grains of rice as prescribed by the inventor. However, when the treasurer had not completed the assignment by the end of the week, the king called him to task and asked him the reason for his extreme tardiness. The treasurer responded by telling the king that it would take more than the entire sum of all the assets in his kingdom to give the inventor his reward.

Let's examine why the treasurer's response is not only correct but is in fact a gross *underestimate* of the size of the reward. It's actually a pretty simple argument. Recall that a chessboard is made up of 64 squares (8 × 8). The instructions for determining the reward are that there is to be just 1 grain of rice on the first square, 2 on the second, 4 on the third, and so on. So on the eighth square, for example (the one in the top right-hand corner of the board), there are $2 \times 2 \times 2 \times 2 \times 2 \times 2 \times 2 = 128$ grains. However, by the time we get to the last square, that is the 64th, which is in the bottom right-hand corner of the board, the number of grains of rice is determined by multiplying 63 2s together (that is, $2 \times 2 \times 2 \times 2 \times 2 \times 2 \ldots 63$ times). This is a truly astronomical number: if you carry this out on the calculator on your laptop or smart phone you will quickly confirm that it's 9,223,372,036,854,775,808, which is just a little under 10 million trillion grains of rice! This amount of rice would form a pile larger than Mount Everest.

This illustrates the fantastic power and ultimate absurdity of unchecked exponential growth. It also illustrates some of its unsuspecting characteristics: it starts out surprisingly slowly but once unleashed gets completely out of control, gobbling up everything before it. Furthermore, at any one time the size of an exponentially growing population is larger than the sum of all the individuals that existed before it. For example, the number of grains on any single square is always larger than the sum of all of the grains on all of the other previous squares added together. Thus there are more people alive on the planet today than all of the people that have lived from the beginning of the exponential explosion until now. So reaching a potentially unsustainable population or a seemingly "infinite" one comes upon the system quite unexpectedly as our next cautionary tale aptly demonstrates. As I will discuss later, in naturally occurring communities that go through a stage of exponential expansion, such as forests and bacterial colonies, there are typically natural feedback mechanisms that lead to ecological limits to growth often related to competitive forces and environmental resource constraints.

Which brings me to my second cautionary tale, which contains a Talmudic sort of question. The story is in the form of a fictional thought experiment inspired by the real-life process of growing bacterial colonies. Suppose we want to prepare a sample of an antibiotic, such as penicillin, and start the process with a single bacterium which, for the sake of argument, we know splits into two identical daughter bacteria every minute. So after one minute we have 2 bacteria, each of which splits into 2 new bacteria after another minute, giving us 4. Another minute later we have 8, then 16, and so on, doubling the number with each successive minute. The analogy to the exponential growth of rice grains on the chessboard is obvious. Suppose we start the growth process at eight o'clock in the morning and have carefully calculated that there are just sufficient nutrients to ensure that the container is completely filled with bacteria by exactly twelve noon. Here's the question: at what time between eight a.m. and noon is the container half full?

Those who get the wrong answer typically suggest a time over halfway between 8:00 and 12:00, such as 10:30 or 11:15. The correct answer, which comes as a surprise to some, is 11:59, just one minute before noon. I'm sure you've got it: since the population doubles every minute, it must have been half of its final size just one minute before the process ends at noon, which is 11:59 a.m.

I'd like to take this little thought experiment one step further by paradoxically working backward: 1 minute before noon the container is half full, 2 minutes before noon it's only ¼ full ($\frac{1}{2} \times \frac{1}{2}$), 3 minutes before noon it's ⅛ full ($\frac{1}{2} \times \frac{1}{2} \times \frac{1}{2}$), and so on. At 11:55 a.m., just 5 minutes before noon, the container is only ¹⁄₃₂ full ($\frac{1}{2} \times \frac{1}{2} \times \frac{1}{2} \times \frac{1}{2} \times \frac{1}{2}$), that is, it is only about 3 percent full with the bacteria barely being visible. Proceeding in this way, a similar calculation shows that at 11:50 a.m., with only 10 minutes remaining, the container is only 0.1 percent full and therefore looks empty. So for almost the entire lifetime of this little universe the container appears empty with almost nothing happening in it, even though the colony is continuing to grow exponentially during the entire time. Only in the last few minutes, corresponding to a tiny fraction of its entire existence and just before this bacterial universe comes to an oblivious end, is there any visible action in the container.

Now let's look at this from the point of view of the bacteria living in the colony. Even after 100 generations, which is equivalent to 100 minutes in "real" time, corresponding to roughly 2,000 years in "human time" (assuming 20 years for each human generation), life is wonderful, food is abundant, and the community is continuing to expand and colonize their little universe. Even after 200 generations, all seems to be going famously well; and even after 235 generations it still looks pretty good, though a few of the bacteria may already have become conscious of the "boundaries" of their universe and perceived for the first time that food is beginning to get a little scarce. Very shortly thereafter at 239 doublings, when the population has reached the absurdly huge number of 10^{71} (that is, a trillion multiplied by itself a million times), are things beginning to look very bad for everyone, and indeed, just one generation later, it's all over!

Although the details of this little fable are not quite right—bacterial doubling times are typically more like thirty minutes rather than one minute, and more important, effects such as the production of toxic wastes and consequent cellular mortality have been ignored—the basic message and implications of unrestrained exponential growth are for real. Page 222 shows an illustration of the growth trajectory and life cycle of real bacterial colonies that you would find in any elementary ecology textbook. As you can see, it traces out the story that I just told: rapid growth, followed by stagnation and collapse. Crucial in this is that the system is closed, meaning that the resources available to the colony have been kept finite as in the test tube story I just related above. Pointedly, this is analogous to the closed situation we have engineered for ourselves on Earth, with our almost total reliance on fossil fuels rather than remaining open and being powered externally by the sun. While exponential growth is a

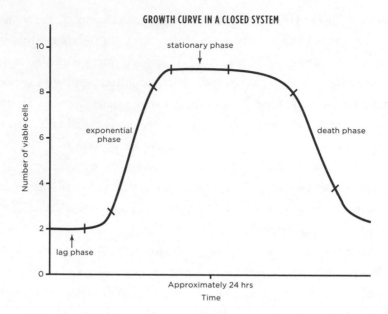

remarkable manifestation of our extraordinary accomplishments as a species, built into it are the potential seeds of our demise and the portent of big troubles just around the next corner.

4. THE RISE OF THE INDUSTRIAL CITY AND ITS DISCONTENTS

I have presented these somewhat provocative cautionary tales both to illustrate the meaning and implications of open-ended exponential growth and to set the stage for a later discussion on global sustainability. It's hard not to view them anthropocentrically as the epic saga of modern man that is destined to end tragically and relatively precipitously in much the same unsuspecting way that it did for the bacteria. Are these stories realistic metaphors for what we've been doing over the last two hundred years? Should we be preparing for the worst, or at the very least changing our profligate ways? Or are these just mythical fables that are misleading in their simplicity and will humans continue on their journey toward an even more glorious future of health, wealth, and prosperity?

Questions such as these have fueled a highly spirited ongoing debate that began very soon after the Industrial Revolution set the wheels of exponential growth in motion and has continued right up to the present time. The transition from working the land and artisanal hand production to automated machines and the creation of factories for the mass production of goods, the technological innovation and increased productivity in agriculture, the introduction of new chemical manufacturing and iron production processes, the improved efficiency of water power, and the increasing use of steam fueled by the change from renewable wood energy to fossilized coal energy, all contributed to an inevitable migration of more and more people away from a traditional rural existence to the rapidly expanding urban centers that were perceived as providing greater opportunities for employment. This process continues unabated across the globe to this day.[3]

The huge changes wreaked by the Industrial Revolution created many wealthy entrepreneurial manufacturers and factory owners and saw the rise of a sizable and increasingly influential middle class, but the fate of the newly urbanized working class, whether in factories or mines, was pretty abysmal. Think of the Dickensian image of Oliver Twist's London: a city pervaded by crime, pollution, disease, and destitution with a large working-class population living in misery. Slums, which arose as a response to rapid population growth and industrialization, were notoriously overcrowded with unsanitary and squalid living conditions.

In many ways it was Manchester, the hub of the booming textile industry and consequently a major driver of British ambitions to "rule the waves" in order to ensure supplies of raw materials such as cotton, that was the true symbol of the Industrial Revolution. It was the world's first industrial city and grew by a factor of six from a population of little more than 20,000 in 1771 to 120,000 in 1831, and to more than two million by the end of the nineteenth century nearly seventy years later. The evolution of Manchester provided the template that has been repeated innumerable times across the globe right up to the present day, from Düsseldorf and Pittsburgh to Shenzhen and São Paulo.

Looking back on megacities of the past such as London or New York, we recognize that they suffered from much the same negative image often associated with megacities of today, such as Mexico City, Nairobi, or Calcutta. This is how Manchester's textile workers were described 150 years ago: "the notorious fact is that well constitutioned men are rendered old and past labour at forty years of age, and that children are rendered decrepit and deformed, and thousands upon thousands of them slaughtered by consumptions [tuberculosis], before they arrive at the age of sixteen." Nevertheless, despite the shameless exploitation and terrible dehumanizing living and working conditions, these cities became highly mobile, rapidly evolving diverse societies, offering huge opportunities that ultimately resulted in many of them becoming drivers of the world's economy. Much the same could be speculated about megacities emerging today in Africa, Asia, and other parts of the world. To quote the American architect and urban planner Andrés Duany: "In 1860, the capital city of Washington, with a population of 60,000, had unlighted streets, open sewers, and pigs roaming about its principal avenues. This condition was worse than the worst of our current cities. There is hope."

I cannot resist inserting a small personal note when writing about the rise of Victorian megacities and the plight of the "working poor." Although I was born in a rural area of England, the county of Somerset, I have family roots in the East End of London and by a strange twist of fate ended up attending the last several years of my high school there. The East End is a product of the rapid expansion of London during the nineteenth century and became one of the poorest, most overcrowded areas of the city and, consequently, a breeding ground for disease and crime. The infamous Jack the Ripper was perhaps the most notorious East End criminal of all time. In keeping with this dubious tradition my desk mate who sat next to me in class for my first two years at that high school eventually became the most wanted criminal in Britain. At that time much of the East End was still referred to as a slum and still oozed a Dickensian ambience, especially in the winter months when the days were short, the skies dark gray, and classic swirling pea-soup fogs descended on the city— the perfect setting for a Sherlock Holmes mystery.

During the summers up through college I worked as a laborer in local breweries. My first experience was in the summer of 1956 when I was fifteen years old. It was at the old Taylor Walker's brewery in Limehouse, at that time a particularly disreputable part of the East End adjacent to the docks on the north bank of the Thames. Limehouse has featured in many books and films and had remained relatively unchanged from its Victorian past. In 1956 it still had a reputation for crime, though the opium dens of Dickens's *The Mystery of Edwin Drood* were long gone.

The brewery, which dated from 1730, was built in 1827 and partially revamped in 1889. It was a typical Victorian redbrick factory building with poor lighting and ventilation and pretty awful working conditions that had remained much the same for more than a hundred years. My job consisted of continuously and mindlessly loading crates of beer bottles onto a vertical conveyor belt that took the used bottles to be cleaned and refilled with beer. Every five seconds or so another heavy crate had to be fed to this ancient iron machine, relentlessly for nine and a half hours a day, five and a half days a week (this included an hour's overtime each day, as well for the half day on Saturday), with an hour's break for lunch and a fifteen-minute break in the morning and afternoon. This was the hardest work I have ever done in my life (except possibly for string theory and helping to guide the Santa Fe Institute through the market meltdown of 2008). I would return home (almost an hour away) exhausted, eat a huge meal, and be asleep by 8:30 p.m. and up at 6:30 the next morning.

During the breaks, a man with a grubby old body-length leather apron emerged from the pages of Dickens carrying a large dirty iron bucket with an old beaten-up pewter mug attached to it by an iron chain. The bucket contained Taylor Walker's cheapest brown ale, and we were privileged to have as much of it to drink as we wished using the pewter mug. Needless to say, the mug was neither cleaned nor rinsed between uses. God knows what diseases I caught, and I certainly didn't mention any of this to my mum! For all of this I got paid 1s 11¾d an hour, which is almost a tenth of a pound (10p in today's coinage), or roughly US15¢. This is not quite as bad as it sounds; inflated to today's value it corresponds to approximately £2.18 (or roughly $3) an hour.

This was actually pretty good for a fifteen-year-old, and I earned enough during those two or three months of the summer to fund a modest hitchhiking holiday and to enjoy what London could offer an adolescent boy for the rest of the year. But if I had been a thirty-year-old man with a wife and three children, goodness knows how I would have made ends meet, even if I were being paid twice as much. Conditions and prospects were extremely poor, though no doubt they were an improvement from fifty or one hundred years earlier when the workday was twelve hours long, six days a week, and children were routinely used as labor in mines and manufacturing. I was politically conservative but like many before me was strongly affected by what I witnessed in the big city at both ends of the economic spectrum. From Marx and Engels to George Bernard Shaw and the Bloomsbury set, and to Clement Attlee and his postwar colleagues of the British Labour Party, many thinkers from comfortable upper-middle-class backgrounds were shocked by the poverty and deprivation they saw in the East End of London, the mills of Lancashire, and the collieries of South Wales.

It's easy to forget that harsh, unsanitary working conditions were the norm for many of the laboring class long before the Industrial Revolution. All of the sins we associate with the Industrial Revolution and urbanization were just as prevalent in preindustrial society, whether it was child labor, dirty living conditions, or long working hours. Indeed, it was the improvements rendered by science and the Enlightenment that eventually contributed to the large decrease in infant and child mortality and therefore to the rapid increase in the population growth rate. Life may appear worse for an urban industrial worker relative to his agricultural counterpart partially because of the perceived dehumanizing harshness of factories and mines relative to working on the land but also because the scale and extent of the problem was so much larger as a result of exponential expansion. Similar arguments persist today, when many believe that life was so much better when we all supposedly lived in small villages and towns where there was a sense of community and relatedness that today seems absent in the hubbub of the modern city. I will return to this later when I discuss the dynamics of cities and show how an accelerating pace of life is integral to the idea of an open-ended growing economy that all of us have come to depend upon, whether we live in fast-track cities or sleepy rural villages.

5. MALTHUS, NEO-MALTHUSIANS, AND
THE GREAT INNOVATION OPTIMISTS

Thomas Robert Malthus is usually credited with being the first person to recognize the potential threat posed by open-ended exponential growth and connect it to the challenge of resource limitation and availability. Malthus was an English cleric and scholar and an early contributor to the newly emerging fields of economics and demography and their implications for long-term political strategy. He published a highly influential essay in 1798 titled *An Essay on the Principle of Population* in which he declared that "the power of population is indefinitely greater than the power in the earth to produce subsistence for man." His argument was that the population "multiplies geometrically," meaning that it increases at an exponential rate, whereas the ability to grow and supply food increases only "arithmetically," meaning that it increases at a much slower linear rate, so the size of the population will eventually outstrip the food supply, leading to catastrophic collapse.

Malthus concluded that in order to avoid such a catastrophe and ensure a sustainable population, some form of population control was needed. This would either arise through "natural" causes such as an increase in disease, starvation, and war or, more preferably, by changing social behavior, particularly that of the working poor, whose reproductive rates he perceived as the apparent cause of the problem. Being a devout Christian he didn't much like the idea of contraception and preferred the concept of moral restraint such as abstinence, delayed marriages, and restricting marriage for those in dire poverty or with health or mental defects. Sound familiar? Given his deeply felt religious and moral beliefs, it's highly debatable whether Malthus would have been an enthusiastic supporter of mass sterilization or the freedom to obtain abortions had they been available at that time. However, he would surely have been a huge supporter of the one-child policy adopted by China. Needless to say, modern neo-Malthusians have no such religious scruples concerning artificial birth control, abortion, or possibly even sterilization programs if they are on a voluntary basis.

An unfortunate consequence of Malthus's analyses was that they were interpreted as effectively blaming the poor for their own predicament because they persisted in reproducing too rapidly. It was relatively easy therefore to conclude that it was this rather than exploitation by capitalists that was the origin of their poverty and their generally abysmal condition. As a further consequence of this thinking, traditional Malthusians believed that paternalistic charity for the poor, whether governmental or philanthropic, was counterproductive in that it resulted only in increasing their numbers, therefore contributing further to the exponential growth in the number of dependent poor, the end result being that this would eventually bankrupt the country. In its various modern incarnations this argument also sounds quite familiar. These ideas inevitably set off a great debate that resonated throughout the next two centuries and in a broader context has continued unabated up to the present time.

In some ways it is quite surprising that the debate has not been put to rest, because Malthus's ideas were heavily criticized and summarily dismissed almost from the moment he proposed them by many of the most influential social and economic thinkers across the entire political spectrum . . . and for good reasons, as we shall see. Over the last two hundred years widespread criticism has come, and continues to come, from diverse quarters ranging from Marxists and socialists to libertarian free market enthusiasts, and from social conservatives and feminists to human rights advocates. A classic criticism that I find particularly amusing was that of Marx and Engels, who dismissed Malthus as a "lackey of the bourgeoisie," sounding like a parody of themselves from a Monty Python skit.

On the other hand, Malthus's ideas have influenced many important thinkers, even if some didn't entirely agree with everything he said. These have included the great economist John Maynard Keynes as well as Alfred Russel Wallace and Charles Darwin, the originators of the theory of natural selection. In more recent years with the burgeoning concerns over global sustainability, Malthus's ideas have been broadened to include questions of resource limitation in general (not just food), with less emphasis on the poor and even on population growth, but rather on general questions of the environment, climate change, and the recognition that these issues transcend geography and economic class.

However, among mainstream economists and social thinkers, Malthusian theory with its implications of imminent catastrophe has become a dirty word. Most believe that the basic assumptions of the theory are fundamentally wrong, and there is plenty of evidence to back them up. Perhaps of greatest importance is that in complete contradiction to Malthus's expectations, agricultural productivity did not increase linearly with time but instead tracked population growth and increased exponentially. Furthermore, there has been a steady decrease in human fecundity accompanied by a steady increase in the average standard of living. As average wages increase and access to contraception becomes more accessible, workers insist on reproducing less rather than more.

I have met scant few economists who do not automatically dismiss traditional Malthusian-like ideas of eventual or imminent collapse as naive, simplistic, or just plain wrong. On the other hand, I have met scant few physicists or ecologists who think it's nuts to believe otherwise. The late maverick economist Kenneth Boulding perhaps best summed it up when testifying before the U.S. Congress, declaring that "anyone who believes exponential growth can go on forever in a finite world is either a madman or an economist."

Most economists, social scientists, politicians, and business leaders typically justify their optimistic views by appealing to the canonical mantra of "innovation" as the magic wand that will keep us exponentially afloat. They rightly point out that it has been our extraordinary inventiveness and openness to change and innovation, driven in large part by a free market economy that has continued to fuel exponential growth and increase standards of living. The original argument of Malthus was wrong because of the unforeseen technological advances in agriculture stimulated by the spirit and discoveries of the Enlightenment and the Industrial Revolution. These led to inventions such as the thresher, the binder, the cotton gin, the steam tractor, and the wrought iron plow with steel cutting edges, as well as advances in crop rotation and the increasing use of commercially produced fertilizer. These contributed enormously to the greater efficiency of production by increasing yields and mechanizing processes that for the previous ten thousand years had primarily been done by hand. In 1830 it took almost three hundred hours of human labor to grow one hundred bushels of wheat; by 1890 this was reduced to less than fifty hours. Today it takes less than a few hours.

Our own time has seen an extraordinary continuation of this amazing revolution in food productivity as agriculture has become increasingly industrialized. Food production in the developed world is dominated by huge agribusiness consortia who use science and technology to maximize yields and optimize distribution. This mechanization of food production in which meat, fish, and vegetables are effectively manufactured in production lines in huge factories as if they were cars or television sets has been rapidly spreading across the globe, providing sustenance for billions of people at affordable prices.

Just to get a sense of the scale of the change, consider, for example, that in 1967 there were roughly a million pig farms in the United States, whereas today there are only about 100,000, and over 80 percent of the pigs presently being produced are from these dedicated factory farms. Four companies alone now produce 81 percent of the cattle, 73 percent of the sheep, 57 percent of the pigs, and 50 percent of the chickens consumed in the United States. Globally, 74 percent of the world's poultry, 43 percent of its beef, and 68 percent of its eggs are produced this way. Consequently, not much more than 1 percent of the U.S. population now works in agriculture compared with about a quarter of the population in the 1930s, when on average each farm worker supplied food to about eleven consumers. Today that number is closer to one hundred. This tremendous gain in efficiency and the huge decrease in the need for agricultural labor has been a major driver behind the exponential growth in the population of cities.

It's hard not to be convinced by the "innovation" argument when thinking about long-term sustainability. Just think of the amazing array of new gadgets, machines, artifacts, processes, and ideas that have been generated over the past two hundred years, let alone the last twenty. From airplanes, automobiles, computers, and the Internet to the theory of relativity, quantum mechanics, and natural selection, it has been an unimaginable exponential joyride—a planet on steroids with an endless supply of more wonders than Ali Baba or Horatio could ever have dreamed of.

According to the World Bank, one of the major Millennium Development

Goals set by the United Nations in 2000, namely, to cut the 1990 poverty rate in half by 2015, was attained five years ahead of schedule in 2010. Furthermore, on average people today are living longer and at a higher standard of living than ever before. That's one side of the coin. The other is that half of the world's population is still living on less than $2.50 a day, with up to a billion people lacking adequate access to clean drinking water or enough food to eat. It seems that despite all of our marvelous advances, the Malthusian threat is still lurking in the background.

This was forcefully argued almost fifty years ago in the best-selling popular book *The Population Bomb*, published in 1968 by the ecologist Paul Ehrlich.[4] It began by throwing down the gauntlet and provocatively proclaiming that:

The battle to feed all of humanity is over. In the 1970s hundreds of millions of people will starve to death in spite of any crash programs embarked upon now. At this late date nothing can prevent a substantial increase in the world death rate. . . .

Similar dire predictions, such as "I don't see how India could possibly feed two hundred million more people by 1980," were made and a series of draconian suggestions, including forced sterilization, were proposed for mitigating the impending catastrophe.

Shortly thereafter in 1972 a study carried out by Dennis Meadows and Jay Forrester at MIT titled *The Limits to Growth* was published.[5] It focused on exploring how finite resources will affect a continuation of exponential growth and what the potential consequences of "business as usual" might be. The research was sponsored by an organization called the Club of Rome, a confederation of distinguished "world citizens, sharing a common concern for the future of humanity" and including former heads of state, diplomats, scientists, economists, and business leaders from around the globe. The study was the first serious attempt to model possible scenarios for the sustainability of the planet using available data combined with computer simulations on food production, population growth, industrialization, nonrenewable resources, and pollution. As such, it is the forerunner of subsequent attempts to seriously model the planet's future, including the recent modeling of climate change.

Like Malthus's essay and Ehrlich's book, *The Limits to Growth* received a lot of attention in the popular media and was equally provocative in stimulating debates about the future of the planet. And like those, it received considerable criticism, especially from economists for its failure to include the dynamics of innovation.

A leading critic was the well-known economist Julian Simon, who expressed a fairly extreme version of the view held by many economists that the spectacular growth we have witnessed over the past two hundred years will be sustained "forever" by human ingenuity and our continued ability to innovate. In fact, Simon argued in his 1981 book *The Ultimate Resource* that a larger population is actually better because it stimulates even more technological innovation, inventiveness, and ingenuity, thereby leading to new ways of exploiting resources and increasing standards of living.[6]

As we move into the twenty-first century this vision of a cornucopian "horn of plenty," with its image of limitless barrels of fish being continuously replenished by the magic, not of divine intervention, but of the free expression of human ingenuity and the boundless possibilities of a free market economy, has reemerged as a significant component of corporate and political conceptual thinking. Indeed, Simon's views have effectively been embraced by many in the academic, business, and political communities. A succinct summary of this view was articulately expressed by the economist Paul Romer, one of the founders of *endogenous growth theory*, which holds that economic growth is driven primarily by investment in human capital, innovation, and knowledge creation.[7] Romer declares that "every generation has perceived the limits to growth that finite resources and undesirable side effects would pose if no new recipes or ideas were discovered. And every generation has underestimated the potential for finding new recipes and ideas. We consistently fail to grasp how many ideas remain to be discovered. Possibilities do not add up. They multiply." Put slightly differently, this proclaims that ideas and innovation increase multiplicatively (that is, exponentially) and not arithmetically (that is, linearly) in tandem with the exponential growth in population and that the process is open-ended and effectively limitless.

On the other hand, the last few decades have also seen a resurfacing of the spiritual successors of *The Population Bomb* and *The Limits to Growth* with the rise of the environmental movement and the development of a serious concern for the future of the planet. Closely related to this is a deep concern for the impact of unregulated corporate and political ambition, which has stimulated a perceived need for "corporate social responsibility." Bridging the divide and reducing the continuous tension between rampant capitalism, packaged as the engine of innovation and ingenuity that fuels growth and prosperity for all, versus the doom-and-gloom concerns of environmentalists and those who heed the warning signals of climate change and potential economic collapse, has emerged as one of the major political challenges of the twenty-first century.

While it may not be unreasonable to hold the view that collective human ingenuity and innovation facilitated by a free market economy is the secret for maintaining long-term open-ended growth in defiance of potential collapse, I find it somewhat baffling that this is often coupled with a denial or, at the very least, a deep skepticism concerning some of its inevitable consequences. Like many who advocate "innovation" as the panacea for meeting future global socioeconomic challenges, Simon was a vocal skeptic when it came to believing that human activity caused global environmental damage or was the origin of serious health concerns, whether from climate change, pollution, or chemical contamination. The spirit and substance of the Second Law of Thermodynamics and its manifestation in terms of entropy production represent the dark side of open-ended exponential growth. Independent of how superbly innovative we are, ultimately everything is driven and processed by the use of energy, and the processing of energy has inevitable deleterious consequences.

6. IT'S ALL ENERGY, STUPID

Globally we use an enormous amount of energy, somewhere in the range of about 150 trillion kilowatt-hours a year. This is one of those astronomical numbers like the annual budget of the United States, whose magnitude and meaning are extremely difficult for most of us mere mortals to comprehend, making our eyes glaze over when we hear it. Commenting on the U.S. budget, the leader of Senate Republicans from 1959 to 1969, Senator Everett Dirksen, was reputed to have said, "A billion here, a billion there, and pretty soon you're talking about real money." And this at a time when the budget was thirty times smaller than its present size of around $3.5 trillion. This represents about $10,000 for every man, woman, and child in the United States, which gives us a much better sense of its size.

To get a similar sense of what the scale of the global energy consumption means and to put its enormous magnitude into some sort of approachable perspective, here are a couple of comparisons that might help. I pointed out in the opening chapter that the 2,000 food calories a day you need to stay alive is equivalent to almost 100 watts, the power of a lightbulb. You are extraordinarily efficient in your use of energy relative to anything that is man-made. Your dishwasher, for example, requires more than ten times more energy per second than you do just to wash dishes, while your car uses it at a rate in excess of a thousand times more just to move you around. When you add up all the energy an average human being on the planet uses to fuel all of the machinery, artifacts, and infrastructure integral to modern life it comes to about thirty times the rate of our natural energy requirements.

To put it slightly differently, the rate at which we need to process energy to sustain our standard of living remained at just a few hundred watts for hundreds of thousands of years, until about ten thousand years ago when we began to form collective urban communities. This marked the beginning of the Anthropocene, in which our effective metabolic rate began its steady rise to its present level of more than 3,000 watts today. But this is just its average value taken across the entire planet. The rate at which energy is used in developed countries is far higher. In the United States it is almost a factor of four larger, at a whopping 11,000 watts, which is more than one hundred times larger than its "natural" biological value. This amount of power is not a lot smaller than the metabolic rate of a blue whale, which is more than one thousand times larger in mass than we are. Thinking of us as an animal using thirty times more energy than we "should" given our physical size, the effective human population of the planet accordingly operates as if it were much larger than the 7.3 billion people who actually inhabit it. In a very real sense, we are operating as if our population were at least thirty times larger, equivalent to a global population in excess of 200 billion people. If the most optimistic of cornucopian thinkers are correct and the world's population reaches 10 billion by the end of the century, all living at a standard comparable to that of the United States, the subsequent effective population would then exceed a trillion people.

This exercise not only gives a sense of scale for how much energy we use but also illustrates just how far out of ecological equilibrium we have come relative to the rest of the "natural world." Of equal significance is that this huge increase of energy consumption has taken place over an infinitesimally short period of time when measured by evolutionary standards, so any systemic adjustment or adaptation to its impact has been almost impossible to accommodate. It has been estimated, for instance, that when we were still an integral component of the natural world operating at an energy level of a few hundred watts and had not yet discovered agriculture, our global population was only around 10 million people. It has now grown to be effectively twenty thousand times larger over such a short period of time that it has led to a major disruption of the dynamical evolutionary balance of nature and therefore to potentially catastrophic ecological and environmental consequences.

This is all pretty sobering, but it's even more so when one is reminded of the inevitable inefficiencies in our use of energy and the subsequent production of entropy resulting in pollution, low-grade heat, and environmental damage and destruction. Of the annual world energy consumption, which is up by a factor of almost two from its value in 1980, roughly one third goes to waste. For example, only about 20 percent of the energy in gasoline is actually used to keep a car moving. A major role of innovation is to decrease such inefficiencies by refining extant technologies, inventing new ones, or developing new ways of organizing their uses. We are witnessing increasing public and corporate consciousness of the challenges of energy consumption, waste, and inefficiency as government programs and taxation policies encourage new ways of thinking and addressing these issues. There is no doubt that significant progress has been and will continue to be made, but can it be enough? It is a matter of faith that the free market system geared to open-ended growth, even when tempered by governmental intervention, stimulation, and regulation, can find a meta-stable balance between making significant profits and solving the problem of sustainability. The primary function of business, after all, is not to increase efficiencies, but to make profits.

Life on our planet evolved and has been sustained by transforming energy directly from the sun into biological metabolic energy that fuels organisms. This amazing process has continued successfully for more than two billion years and, as such, can confidently be referred to as being "sustainable" despite continuous changes in the dramatis personae representing new innovative life-forms resulting from natural selection. A crucial element in how life has been sustained is that the energy source, namely the sun, was external, reliable, and relatively constant. It shone every day and any variation in its output occurred over long enough periods of time for adaptations to accommodate to the change.

This ongoing, ever-evolving, quasi-steady state very slowly began to change with our discovery of fire, which is the chemical process that releases the sun's energy stored in dead wood. When coupled with the invention of agriculture, this began the transition to the Anthropocene as we emerged from a purely biological organism to our present state as an urbanized socioeconomic creature no longer in meta-equilibrium with the "natural" world. The truly dramatic and revolutionary departure from almost three billion years of sustainable business as usual came about in just the last two hundred years when our discovery and exploitation of the sun's energy stored underground in coal and oil heralded the beginning of the Urbanocene. Fossil fuels were, and still are, perceived like the sun itself as an almost limitless source of energy whose subsequent release sparked the Industrial Revolution.

From a scientific perspective the truly revolutionary character of the Industrial Revolution was the dramatic change from an *open* system where energy is supplied externally by the sun to a *closed* system where energy is supplied internally by fossil fuel. This is a fundamental systemic change with huge thermodynamic consequences, because in a closed system the Second Law of Thermodynamics and its requirement that entropy always increases strictly applies. We "progressed" from an external, reliable, and constant source of energy to one that is internal, unreliable, and variable. Furthermore, because our dominant source of energy is now an integral component of the very system it is supporting, its supply is hostage to continually changing internal market forces.

Powered by fossil fuels, our socioeconomic accomplishments in just two hundred years surpass anything that natural selection, powered directly by the sun, has ever done biologically over such a short period of time. But there is a potentially heavy price to pay for letting the fossil energy genie out of the bottle and we have to either learn to live with it, if possible, or put it back in the bottle whence it came.

An example of the consequences of the Second Law is the warming of the atmosphere due to the release of energy stored underground in fossil fuels onto the surface of the planet. This is strongly enhanced by the production of gases such as carbon dioxide and methane as entropic by-products from burning these fuels, leading to the well-known greenhouse effect in which heat gets trapped in the atmosphere. I will not be discussing in any detail how the rates of physical and chemical processes depend on temperature, other than to reiterate that they scale *exponentially* rather than as a power law. Consequently, the processes that govern the weather and the life history of plants and animals are *exponentially sensitive to small changes in the temperature* at which they operate. I remind you that a 2°C rise in average temperature leads to a whopping 20 percent increase in these rates. So small changes in the ambient temperature over relatively short periods of time that are not sufficiently long for adaptive processes to develop can lead to huge ecological and climatological effects. Some of these may be positive, but many will be catastrophic. Regardless, however, of the sign of the effect, significant changes are upon us, and we desperately need to understand their origins and consequences and forge strategies for adaptation and mitigation.

The crucial question is not whether these effects are anthropogenic in origin because they almost certainly are, but rather to what extent they can be minimized without leading to rapid discontinuous changes in our physical and economic environment and ultimately to the potential collapse of the global socioeconomic fabric. Hence my bewilderment at those in the general public including political and corporate leaders who reject the cautionary exhortations of scientists, environmentalists, and others, and why I am continually baffled by their lack of action. Yes, we should all delight in and promote the huge successes and fruits of the free market system and of the role of human ingenuity and innovation, but we should also recognize the critical roles of energy and entropy and together act strategically to find global solutions to their deleterious consequences.

Despite the obviously central role that energy has played in bringing us to this point in the history of the planet and, in particular, in the socioeconomic development of modern human societies, you will be hard-pressed to find even a sentence or two about it in any classic textbook on economics. Remarkably, concepts like energy and entropy, metabolism, and carrying capacity have not found their way into mainstream economics. The continued growth of economies, markets, and populations over the past two hundred years coupled with a parallel increase in standards of living are, not surprisingly, seen as testament to the success of classic economic thinking and as a rejection of neo-Malthusian ideas. There has been no need to think seriously in terms of energy as an underlying driver of economic success or of population growth, let alone to consider entropy as its inevitable consequence. Nor has there been the need to consider the possibility that resources may actually be limited, nor that there might be underlying physical constraints that would question open-ended growth. Until now.

These issues have been conceptually circumvented by invoking the almost magical role that innovation and human ingenuity have played, and presumably will continue to play, in keeping the entire enterprise viable, especially if stimulated by a relatively free market economy. Just as the concept of a mysterious, almost limitless, dark energy is invoked to "explain" why the physical universe continues to expand exponentially, so the specter of an almost limitless supply of innovative ideas is invoked to explain why the socioeconomic universe will continue to expand, overcoming all obstacles in its course.

In addition, there also seems to be an unexpressed presumption that ideas, the seeds of innovation, cost nothing—after all, they are "just" neural processes in the brains of human beings, and collectively we can produce an almost infinite number of them in our heads. But like everything else, ideas and the innovations they inspire require energy, and lots of it, to support the smart individuals who are thinking and to provide the appropriately stimulating environments and collective experiences we institutionalize in places such as universities, laboratories, parliaments, cafés, concert halls, and conference venues.

The very essence of this is embodied in the concept of the city and of urban life. This was aptly expressed by the famous anthropologist Margaret Mead, who observed: "The city as a center where, any day in any year, there may be a fresh encounter with a new talent, a keen mind or a gifted specialist—this is essential to the life of a country." Indeed, cities have evolved as the engine we invented for enhancing and facilitating social interaction, thereby stimulating idea creation and innovation. Smart and ambitious people are drawn to cities, and this is where our new ideas are incubated, where entrepreneurship flourishes, and where wealth is created. Supporting all of this is extremely expensive, so it is naive to dissociate ideas from energy—one cannot flourish without the other. Of the trillions of thoughts, ideas, speculations, and proposals for new machines, new products, and new theories, only an infinitesimal minority ever lead to any significance. Almost all fall by the wayside, even though as a totality they all contribute a necessary background noise and weltanschauung for new and innovative phenomena to arise and blossom. All of this requires huge amounts of energy: *ex nihilo nihil fit*—nothing comes from nothing.

A possible science of sustainability requires understanding global dynamics as a complex evolving adaptive system composed of many interlocking and interacting subsystems that are themselves complex adaptive systems, all evolving together potentially under energy, resource, and informational constraints. We need to understand how the dynamics of innovation, technological advances, urbanization, financial markets, social networks, and population dynamics are interconnected and how their evolving interrelationships fuel growth and societal change—and, as manifestations of human endeavors, how they are all integrated into a holistic interacting systemic framework . . . and whether such a dynamically evolving system is ultimately sustainable.

The arguments of those like Malthus, Paul Ehrlich, and the Club of Rome may be flawed, but their conclusions and implications may well have validity. In any case, they did us a great service by raising some of the most important existential issues humankind has to face as we move almost blindly into the twenty-first century. Although the population bomb has been swept under the rug, questions of sustainable energy supply and its potential deleterious consequences have been brought to consciousness and are now being seriously debated.

Viewed from the standpoint of the enormous flux of energy that is reliably delivered on time every day by the sun to the Earth, there is no energy problem. To get a sense of the relative scales involved, consider the following: the total amount of energy delivered by the sun to the Earth is approximately a million trillion (10^{18}) kilowatt-hours a year, compared with our "measly" needs (on this scale) of 150 trillion (1.5×10^{14}) kilowatt-hours we collectively use each year. So on the scale of what the Earth receives from the sun, our energy use represents only about 0.015 percent of what is in principle actually available to us. To put it another way: more energy is delivered by the sun in just one hour than is used by the entire world in a single year. Indeed, the scale of solar energy is so vast that in one year it is about twice as much as will *ever* be obtained from all of the Earth's nonrenewable resources of coal, oil, natural gas, and uranium combined. So from this point of view, there is no energy problem—at least in principle.

Consequently, the long-term strategy for sustained global energy availability is clear: we need to return to the biological paradigm where most of our energy needs are supplied directly from the sun but in such a way as to maintain and expand what we have so far accomplished. We urgently need to develop the technology that will enable us to harness affordable abundant energy from the sun, primarily from direct radiation but also indirectly from wind, tidal forces, and wave motion. What a marvelous challenge to our professed ingenuity and powers of innovation. Here's the opportunity for dynamic and charismatic political and corporate leadership to forge the way to a sustainable global energy future based on the dynamics of entrepreneurship, the free market system, and government stimulation. Surely, with a track record of remarkable inventions that include the steam turbine, the telephone, the laptop computer, the Internet, quantum mechanics, and the theory of relativity, this should be a piece of cake. However, one of the more mysterious aspects of the twenty-first century is that those who seem most vocal in promoting and celebrating innovation and the free market economy as the engine of sustainability seem so reluctant to recognize the urgency of the challenge and to champion research and development in exploiting the almost infinite power of solar energy.

The lack of progress until relatively recently is quite astonishing given that the basic technology for developing solar energy has been known for more than a hundred years. In 1897, an American engineer, Frank Shuman, built a proof of principle device for utilizing energy from the sun by showing that it could power a small steam engine. His system was eventually patented in 1912, and in 1913 he constructed the world's first solar thermal energy power plant, which was built in Egypt. It generated only about 50 kilowatts (about 65 horsepower) but was able to pump more than 5,000 gallons of water a minute (about 22,000 liters a minute) from the Nile onto adjacent cotton fields. Shuman was an enthusiast and advocate for solar energy and in 1916 was quoted in *The New York Times* as saying:

> *We have proved the commercial profit of sun power . . . and have more particularly proved that after our stores of oil and coal are exhausted the human race can receive unlimited power from the rays of the sun.*

Given how long ago this statement was made, Shuman's observation was remarkably prescient, even if it is yet to be realized. The discovery and development of cheap oil in the 1930s discouraged the advancement of solar energy, and Shuman's vision and basic design were effectively forgotten until the first energy crises of the 1970s. It is, however, encouraging that technologies such as photovoltaic cells have now been developed that make Shuman's dream a possible reality as the price of renewables begins to become competitive with traditional fossil fuel energy generation.

Another fundamental difference between fossil fuel and solar energy lies in the underlying physical mechanism for how their energy is generated. The process of burning fossil fuels releases energy stored in chemical bonds that hold the atoms and molecules of coal, oil, or gas together. All molecules, whether they are the building blocks of your body, your brain, your house, or your computer, are held together by the forces of electromagnetism and, as such, are characterized by energies whose magnitude is in the range of electron-volts (eV), which is the unit conventionally used to measure them. An electron-volt is infinitesimal on the scale of the energies we have been considering: 1eV is equivalent to only about a three hundred trillion trillionth of a kilowatt-hour (1eV = 3×10^{-26} kWh), so in terms of these atomic units, the amount of energy we consume each year is about 5×10^{39}eV. You can loosely think of this as saying that each year we break down this many molecules to supply us with our energy needs.

On the other hand, the sun, which consists primarily of hydrogen and helium, is fueled by nuclear energy stored in the bonds that hold nuclei together. It is released as radiation when hydrogen nuclei fuse together to form helium nuclei. This is called nuclear fusion and is the fundamental physical mechanism for how the sun shines, providing us energy in the form of the light and heat that gave rise to all life on the planet. It remains the sole source of energy for all life on Earth, except for us over the past few thousand years since we discovered its power stored in fossil fuels.

The scale of nuclear energy is about a million times larger than that of chemical electromagnetic energy that is released in the burning of fossil fuels: nuclear processes involve energies in the range of millions of electron-volts (MeV) rather than electron-volts characteristic of molecular chemical reactions. It is this huge enhancement factor that makes the idea of exploiting nuclear energy so attractive: the same amount of matter can produce approximately a million times more energy from the nuclei of its atoms than from its molecules. Thus instead of having to use 500 gallons (about 2,000 liters) of gasoline a year to run your car, you would need only the equivalent of a few grams of nuclear material, about the size of a small pill.

The promise of "unlimited" energy from nuclear power stations producing energy using the same physics that powers the sun is a fantastic idea. When it was first conceived right after the Second World War in the heady days following the development of the atomic bomb, there was tremendous optimism that nuclear energy would soon displace fossil fuels as our primary source of energy. I certainly recall as a young teenager in the 1950s reading articles in newspapers proclaiming that by the time I was grown up and a family man, electricity would be so cheap there would be no need to meter it. Typical of this euphoria were proclamations such as that by the Nobel Prize–winning nuclear chemist Glenn Seaborg, who was chair of the U.S. Atomic Energy Commission, that "there will be nuclear powered earth-to-moon shuttles, nuclear powered artificial hearts, plutonium heated swimming pools for SCUBA divers, and much more."

Unfortunately, using nuclear fusion to generate economically competitive energy has proven to be quite elusive and technologically extremely challenging despite vigorous international efforts to make it viable. Instead, nuclear power has been successfully developed using nuclear fission in which energy is released when heavy nuclei (of uranium) dissociate into lighter products, a process analogous to conventional chemical production of energy from fossil fuels. At present about 10 percent of the world's electricity is generated using fission, with France leading the way with over 80 percent of its electricity coming from nuclear reactors.

Like conventional fossil fuel power plants, energy produced by nuclear reactors is internal to the entire global system and consequently suffers from similar issues regarding the production of entropy and deleterious by-products. Although nuclear power, like solar, is not a significant source of greenhouse gases and therefore not a driver of possible climate change, its by-products can be extremely deleterious because their energy scale is so much higher (by a factor of a million). Consequently, radiation resulting from nuclear processes can be extremely damaging to molecules and therefore to organic tissue, leading to serious health issues of which cancer is the best known. To a large extent, our atmosphere protects us from similar radiation from the sun, but in reactors on Earth this is a major challenge. In addition, there is the issue of safe and reliable storage and disposal of the waste products from nuclear reactions, which remain radioactive for thousands of years.

Despite huge efforts to ensure the safety of nuclear reactors, there have now been enough accidents to have dampened the enthusiasm for their utilization as an alternative energy source to fossil fuels, even though the number of direct fatalities has been extremely small. The reverberations from the disaster at the Fukushima nuclear power plant in Japan in 2011 have led to a dramatic decrease in the present and projected future reliance on nuclear energy world-wide. Although fossil fuels have caused hundreds of thousands, if not millions, of deaths and an enormous number of health problems, they are still perceived by many as being more desirable than the potential risks coming from nuclear reactors. Questions of long-term safety and quantitative assessments of the entropic consequences of energy production and utilization are a highly complex and controversial social, political, psychological, and scientific issue. How many deaths are caused directly by energy production, how many indirectly, what health issues are considered dangerous, and what are their long-term consequences? How do we compare different technologies? What metrics should we be using?

Just to give a sense of the sorts of comparisons we have to make, consider the following. We are surprisingly tolerant of death and destruction arising from "unnatural, man-made" causes when they occur on a continual and regular basis, but are extremely intolerant when they occur suddenly as discrete events even though the numbers involved are much smaller. For instance, each year more than a million and a quarter people die from car accidents world-wide, which is comparable to the number who die from lung cancer, the most common cause of cancer death. Nevertheless, the fear and anxiety about dying from cancer seems to be far greater than the concern about being killed in an automobile accident, and this is reflected in the large discrepancy in the resources we devote to addressing each of these problems. It is interesting to compare both of these to the number of people who have died directly from nuclear accidents. Even when integrated over the entire lifetime of all nuclear power plants this is less than one hundred people, and most of them died in the Chernobyl disaster in the USSR in 1986, while none died at Fukushima. On the other hand, many thousands may have contracted cancer and died, or will die prematurely, because of radiation exposure from such accidents, particularly from Chernobyl. But this should be "balanced" by the estimated 50 million people who are injured, maimed, or disabled each year by car accidents.

And so the arguments go back and forth as we try to compare apples and oranges, grappling for the appropriate metrics to help us make these difficult decisions and comparisons as we try to prioritize the global energy portfolio that will play a central role in determining how society evolves in the coming decades. Adding to our difficulties are the psychosocial imponderables such as the almost universal love affair with the automobile and the almost universal fear of a major nuclear power disaster, which is difficult to disentangle from the universal fear of nuclear bombs. My intention here is not to give a complete review of the pros and cons of energy options, but rather to present a few simple examples of the sorts of quantitative statistics that we need to ponder in debating these issues. We need to think quantitatively and to develop the underlying science that is needed for addressing these challenges so as to inform rational political decisions.

Regardless of whether one believes in the innovative capacity of human beings to solve the problems of nuclear energy, whether fusion or fission, or the challenge of affordable and reliable solar technology sufficient to support the energy needs of 10 billion people, or to reverse the amount of carbon we are pumping into our atmosphere, we are still left with the long-term problem of entropy production. Apart from its many other issues, the nuclear option, like that of traditional fossil fuels, keeps us trapped in the paradigm of a closed system, whereas the solar option has the critical capacity for potentially returning us to a truly sustainable paradigm of an open system.

6

PRELUDE TO
A SCIENCE OF CITIES

1. ARE CITIES AND COMPANIES JUST VERY LARGE ORGANISMS?

The success of the network-based theory for understanding scaling laws and providing a big-picture conceptual framework for quantitatively addressing diverse questions across a broad spectrum of biology naturally leads to the question of whether this framework could be extended to understand other networked systems such as cities and companies. Superficially these have much in common with organisms and ecosystems. After all, they metabolize energy and resources, produce waste, process information, grow, adapt and evolve, contract disease, and even develop what could be characterized as tumors and growths. In addition, they age, and in the case of companies, almost all of them eventually die, whereas for cities only extremely few ever do, an enigma we shall come to consider later.

Many of us blithely use phrases like the "metabolism of a city," the "ecology of the marketplace," the "DNA of a company," and so on, as if cities and companies were biological. Even as far back as Aristotle we find him continually referring to the city (the *polis*) as a "natural" *organic* autonomous entity. In more recent times an influential movement in architecture has arisen called *Metabolism,* which was explicitly inspired by analogy with the idea of biological regeneration driven by metabolic processes. This views architecture as an integral component of urban planning and development and as a continually evolving process, implying that buildings should be designed ab initio with change in mind. One of its original proponents was the well-known Japanese architect Kenzo Tange, the 1987 winner of the Pritzker Prize, considered to be the Nobel Prize of architecture. I find his designs, however, to be surprisingly *in*organic, dominated by right angles and concrete and somewhat soulless, rather than having the curvaceous, softer qualities of an organism.

Writers, too, have often expressed organic visions of cities. An extreme example is Jack Kerouac, one of the charismatic founders of Beat poetry and literature in the 1950s, who whimsically wrote, "Paris is a woman but London is an independent man puffing his pipe in a pub." However, it is in business that the concepts and language if not the actual science of ecology and evolutionary biology have seized the imagination, especially in Silicon Valley. The concept

Clockwise from top, images of cities: Steel and concrete skyscrapers in São Paulo, Brazil; the "organic" city of Sana'a, Yemen; the integration of town and country, Melbourne, Australia; the profligate use of energy in Seattle.

of a *business ecosystem* has become a standard buzzword to connote a sort of Darwinian survival of the fittest in the marketplace. It was introduced in 1993 by James Moore, then at Harvard Law School, in an article he wrote titled "Predators and Prey: A New Ecology of Competition," which won the McKinsey award for article of the year.[1] It's a fairly standard ecological narrative, with individual businesses replacing animals in the evolutionary dynamics of natural selection. In keeping with much of the traditional literature on understanding companies, it's entirely qualitative with no quantitative predictive power. Its great virtue is that it emphasizes the role of community structure, the importance of systemic thinking, and the inevitable processes of innovation, adaptation, and evolution.

So are all of these references to biological concepts and processes just qualitative metaphors in the same way we loosely use scientific jargon like "quantum leap" or "momentum" to describe phenomena that are difficult to capture in conventional language, or do they express something deeper and more substantive, connoting that cities and companies are indeed just very large organisms following the rules of biology and natural selection?

These were the kinds of general ruminations I was contemplating when I began informal discussions in 2001–2 with colleagues at the Santa Fe Institute whose backgrounds were in the socioeconomic sciences. Fortuitously, Sander van der Leeuw, a well-known anthropologist then at the University of Paris who later moved to run the School of Sustainability at Arizona State University, was on sabbatical leave at SFI, and David Lane, who had previously run the SFI economics program, was often on-site. David was a well-known statistician who, inspired by SFI, had switched to economics. He had been chair of statistics at the University of Minnesota but had moved to the University of Modena in Italy, where he began a program oriented toward understanding innovation, especially in the manufacturing sector that had been the lifeblood of northern Italy. (You probably recognize Modena for its marvelous balsamic vinegar, not to mention as the home of Ferrari, Lamborghini, and Maserati. On my first visit there, David introduced me to their traditional balsamic vinegar, a remarkable elixir to be distinguished from the tamer stuff many of us nowadays use on our salads, but which costs a good deal more than some of the most expensive bottles of wine I've ever bought.)

Despite my skepticism, David and Sander convinced me that extending the network-based scaling theory from biology to social organizations was indeed a worthwhile project. They became the prime movers in putting together a broad program that covered our joint interests ranging from innovation and information transfer in both ancient and modern societies to understanding the structure and dynamics of cities and companies, all from a complexity perspective. The program was called the Information Society as a Complex System (ISCOM) and was generously funded by the European Union. Shortly thereafter Denise Pumain, a well-known urban geographer at the Sorbonne in Paris, joined our collaboration and the four of us each ran one component of the project. I assembled a new multidisciplinary collaboration centered on SFI whose first goal was to ask whether cities and companies manifest scaling and, if so, to develop a quantitative principled theory for understanding their structure and dynamics.

As in much of life, it is often instructive to take a retrospective look at what one had proposed to do long after it's over. For instance, if I look back at the list of attendees at one of our early workshops, very few of them eventually became ongoing members of the collaboration. This is not unusual at the beginning of a program such as this that proposes to broach new questions that transcend disciplinary boundaries. At the outset all kinds of people with diverse backgrounds who are well versed in expertise that might be pertinent to the program are invited to participate in the hope that synergies will happen, sparks will fly, and a real sense of purpose and excitement about prospects for something new will be generated. Many find, however, that even if they are fascinated by the intellectual challenges and potential outcomes of the proposed project, it simply isn't compelling enough to sacrifice the time to get fully involved and reset the priorities of their own research agendas. Others discover that they really aren't that interested after all, or that it's unlikely that anything of substance will come out of the effort. Eventually, however, by word of mouth, by serendipitous connections and informal discussions, and by osmosis and diffusion, an evolving group of researchers emerges whose members are to varying degrees willing to commit to a longer-term involvement with the challenge and who will actually do the substantive work over the ensuing years. Such was the process that led to the scaling and social organization component of ISCOM.[2]

Although its scope and emphasis broadened as progress was made, the vision of the proposal remained pretty much intact over the years. Its motivation was originally expressed as follows: "Because of the obvious analogy with social network systems, such as corporate and urban structures, it is both natural and compelling to investigate the possibility of extending the same sort of analyses used for understanding biological network systems to social organizations," with an added emphasis that "the flow of information in social organizations is as significant as the flow of matter, energy and resources." Many questions were asked, including "What is a social organization? What are the appropriate scaling laws? What constraints must be satisfied by the architecture of the structures that channel social flows of information, matter and energy? In particular, are the relevant constraints all physical, or might there also be social and cognitive constraints that must be taken into account?"

New York, Los Angeles, and Dallas superficially look and feel quite different from one another, as do Tokyo, Osaka, and Kyoto, or Paris, Lyon, and Marseille, yet their differences are relatively small compared with the differences we perceive between whales, horses, and monkey, which, as was shown earlier, are actually scaled versions of one another following simple power law scaling relationships. These hidden regularities are manifestations of the physics and mathematics of the underlying networks that transport energy and resources in their bodies. Cities are sustained by similar network systems such as roads, railways, and electrical lines that transport people, energy, and resources and whose flow is therefore a manifestation of the metabolism of the city. These flows are the physical lifeblood of all cities and, as with organisms, their structure and dynamics have tended to evolve by the continuous feedback mechanisms inherent in a selective process toward an approximate optimization by minimizing costs and time: regardless of the city, most people on average want to get from A to B in the shortest possible time at the cheapest cost, and most businesses want to do likewise with their supply and delivery systems. This suggests that despite appearances, cities might also be approximate scaled versions of one another in much the same way that mammals are.

Cities, however, are much more than just the physicality of their buildings and structures connected and serviced by various transport systems. Although we tend to think of cities in physical terms—the beautiful boulevards of Paris,

Road networks in Los Angeles and the New York City subway network; hidden are the other infrastructural networks, such as water, gas, and electricity.

the London Underground, the skyscrapers of New York, the temples of Kyoto, and so on—cities are much more than their physical infrastructure. In fact, the real essence of a city is its people—they provide its buzz, its soul, and its spirit, those indefinable characteristics we viscerally feel when we are participating in the life of a successful city. This may seem obvious, but the emphasis of those who think about cities, such as planners, architects, economists, politicians, and policy makers, is primarily focused on their physicality rather than on the people who inhabit them and how they interact with one another. It is all too often forgotten that the whole point of a city is to bring people together, to facilitate interaction, and thereby to create ideas and wealth, to enhance innovative thinking and encourage entrepreneurship and cultural activity by taking advantage of the extraordinary opportunities that the diversity of a great city offers. This is the magic formula that we discovered ten thousand years ago when we inadvertently began the process of urbanization. Its unintended consequences have resulted in an exponentially increasing population whose quality of life and standard of living have on the average also been increasing.

Like almost everything else involving the psychosocial world of human beings, William Shakespeare understood our fundamental symbiotic relationship with cities. In his rather gruesome political drama *Coriolanus*, a Roman tribune named Sicinius rhetorically remarks, "What is the city but the people?" to which the citizens (the plebs) emphatically respond, "True, the people are the city." Which I translate to mean: cities are emergent complex adaptive social network systems resulting from the continuous interactions among their inhabitants, enhanced and facilitated by the feedback mechanisms provided by urban life.

2. ST. JANE AND THE DRAGONS

No one is more identified with viewing cities through the collective lives of their citizens than the famous urban writer-theorist Jane Jacobs. Her defining book, *The Death and Life of Great American Cities,* had an enormous influence across the globe on how we think about cities and how we approach "urban planning."[3] It's required reading for anyone interested in cities whether a student, a professional, or just an intellectually curious citizen. I suspect that every mayor of every major city in the world has a copy of Jane's book sitting somewhere on his or her bookshelf and has read at least parts of it. It's a wonderful book, extremely provocative and insightful, highly polemical and personal, very entertaining and well written. Although published in 1961 and explicitly focused on major American cities of that period, its message is very much broader. In some ways it is perhaps more widely relevant now than it was then, especially outside of the United States, as many cities have followed variations of the classic American urban trajectory with the dominance and challenge of the automobile, the mall, the growth of suburbia, and the consequent loss of community.

Ironically, Jane had no fancy academic credentials, not even an undergraduate degree, nor did she engage in traditional research activities. Her writings are more like journalistic narratives, based primarily on anecdotes, personal experience, and a deep intuitive understanding of what cities are, how they work, and how they "should" work. Despite the explicit focus on "great American cities" in her book, one gets the impression that most of her analyses and commentaries are based on her personal experiences of New York City. She was extraordinarily intolerant of urban planners and politicians, and vicious in her attacks on traditional urban planning, especially in regard to its apparent lack of recognition that people, not buildings and highways, were primary. These classic quotes from her writing are typical of her critical attitude:

> *The pseudoscience of planning seems almost neurotic in its determination to imitate empiric failure and ignore empiric success.*

> *In this dependence on maps as some sort of higher reality, project planners and urban designers assume they can create a promenade simply by mapping one in where they want it, then having it built. But a promenade needs promenaders.*

> *There is no logic that can be superimposed on the city; people make it, and it is to them, not buildings, that we must fit our plans. . . . We can see what people like.*

> *His aim was the creation of self-sufficient small towns, really very nice towns if you were docile and had no plans of your own and did not mind spending your life with others with no plans of their own. As in all Utopias, the right to have plans of any significance belonged only to the planner in charge.*

The "his" in this last quote is a reference to Sir Ebenezer Howard, the inventor of the concept of the "garden city." This has had a powerful influence on town planning throughout the twentieth century, providing the idealized model of suburbs around the world. Howard was a visionary utopian thinker who was strongly affected by the plight and exploitation of the working class in nineteenth-century Britain. Howard's vision of the garden city was a planned community with distinct areas of residences (housing), factories (industry), and nature (agriculture) in prescribed proportions that he deemed ideal for providing the best of urban and country living. No slums, no pollution, plenty of fresh air with room to breathe and to live the good life. The integration of town and country was seen as a step toward a new civilized society, a curious marriage between libertarianism and socialism. His garden cities would be largely independent but managed cooperatively by the citizens who had an economic interest in them, though they would not own the land.

Unlike most utopian dreams, Howard's vision struck a resonant chord among both liberal thinkers and hard-core investors. He was able to form a company that raised sufficient private investment to build two garden cities from scratch north of London: Letchworth Garden City (1899), whose present population is 33,000, and Welwyn Garden City (1919), now with 43,000 inhabitants. However, in order to accomplish this dream in the real world, many of his ideals had to be sacrificed or seriously compromised, including his rigid top down design plans that Jane Jacobs railed against. Nevertheless, his basic philosophy of a planned "town and country" community has persisted to this day and has left its mark not only on the many variants of garden cities that have since sprung up around the world but also in the design concept of almost every suburban development of every city. An interesting special case of this on a large scale is Singapore. Even as the city has grown to become a major global financial center with more than five million inhabitants and has continued to build the usual ostentatious steel and glass skyscrapers, its saving grace is that it has maintained the dream of being a garden city on a grand scale. This is primarily due to its visionary top-down leader the late Lee Kuan Yew, who had the foresight in 1967 to require that Singapore develop as a "city in a garden" with abundant vegetation, open green spaces, and a sense of tropical lushness despite its chronic shortage of land. Singapore may not be the world's most exciting city, but its ambience of greenness is palpable.

Ironically, Howard's actual design of these garden cities was not at all organic. Their layout and organization are the very essence of simple Euclidean geometry, with the only curves in sight being perfect circles connected by perfect straight lines—the very opposite of the apparently jumbled mess of organically evolved cities, towns, and villages. No fractal-like boundaries, surfaces, or networks à la Mandelbrot appear in Ebenezer Howard's vision of a garden city—an example of which is shown on page 256. This move away from organic geometry became the signature of modernist movements in both architecture and urban planning during the twentieth century. It is perhaps no better exemplified than by the hugely influential Swiss-French architect and urban theorist Charles-Édouard Jeanneret-Gris, universally known as Le Corbusier, and the philosophy often referred to as "form following function." Part of his motivation in taking this pseudonym, which was derived from his mother's family name, was to demonstrate that anyone could reinvent themselves.

Like Ebenezer Howard, Le Corbusier was greatly influenced by the squalid living conditions in city slums and sought efficient ways to ameliorate the plight of the urban poor. Out of this concern emerged his bold proposal to

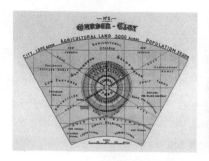

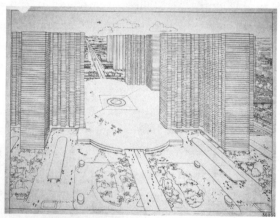

Top left: An example of Ebenezer Howard's plans for a garden city; top right: the new city of Masdar in Abu Dhabi; middle and bottom left: examples of Le Corbusier's design for a new city.

erase much of central Paris (and for that matter, Stockholm) and replace it with multiple high-density concrete, glass, and steel high-rises crisscrossed with railway lines, highways, and even airports. It was all pretty stark and spartan and even a bit sinister, mimicking a turn to the right in his political thinking during the turbulent years of the 1930s. This is reflected in phrases he used such as "cleaning and purging" the city, or developing "a calm and powerful architecture," and in his insistence that buildings be designed without ornamentation. Thank goodness that his grandiose plans were not acted upon so that we can still enjoy some of the more decadent urban embellishments of central Paris and Stockholm.

Le Corbusier had an enormous influence on architects and urbanists across the globe, as evidenced by the dominance of rigid steel and concrete structures that adorn the central districts of all of our major cities. Just as Howard's urban design philosophy has left its indelible mark on suburban city life, so has Le Corbusier's left its indelible mark on our downtown cityscape. This is especially evident in the design of new capital cities such as Canberra, Chandigarh, and Brasilia. A particularly interesting case is that of Brasilia, whose civic buildings were designed by the architect Oscar Niemeyer, who was much influenced by Le Corbusier although he qualified his admiration by declaring:

I am not attracted to straight angles or to the straight line, hard and inflexible, created by man. I am attracted to free-flowing, sensual curves. The curves that I find in the mountains of my country, in the sinuousness of its rivers, in the waves of the ocean, and on the body of the beloved woman. Curves make up the entire Universe, the curved Universe of Einstein.

To which, by the way, he might have added Mandelbrot and fractals. It is ironic that despite this admirable declaration, Brasilia became emblematic of what a city should not be. It has often been characterized as a "concrete jungle"—stark and soulless even though, harking back to the influence of Ebenezer Howard, it does have many open green spaces and parks. After visiting Brasília shortly after it was inaugurated in 1960, the avant-garde French writer-philosopher Simone de Beauvoir echoed Jane Jacobs by rhetorically asking:

What possible interest could there be in wandering about? . . . The street,
that meeting ground of . . . passers-by, of stores and houses, of vehicles
and pedestrians . . . does not exist in Brasília and never will.

Fifty years later, emerging from the shackles of its original plan, the city, now boasting more than two and a half million inhabitants, has slowly begun to evolve organically and develop a "meeting ground" integrated with a more humanistic livable environment. Meanwhile in 1989, just two years after Kenzo Tange received the Pritzker Prize, it was awarded to Oscar Niemeyer. Another more recent Pritzker Prize winner, Norman Foster, has also tried his hand at designing a city ex nihilo, in this case in the harsh desert environment of the Gulf States. This is the much-publicized city of Masdar in Abu Dhabi, which is envisioned to be a showcase for a sustainable, energy-efficient, user-friendly high-tech community by taking advantage of abundant solar energy enabled by sexy advances in IT. It's a bold and exciting plan, even if it is rather a strange beast. It is planned to have about fifty thousand inhabitants by around 2025 at a cost of about $20 billion. Its main business is expected to be the high-tech research and manufacturing of environmentally friendly products to be supported by an influx of an additional sixty thousand commuters from Abu Dhabi itself. Perhaps the most bizarre aspect of Masdar is that its boundaries have been designed to be about as inorganic and unimaginative as possible—they form an exact square. Yes, a square city.

It is hard not to perceive Masdar as effectively a large private suburban residential industrial park rather than a vibrant diverse autonomous city. In many ways its philosophy is derivative of Ebenezer Howard's garden city concept brought into the high-tech culture of the twenty-first century, except that it appears to be designed for the privileged rather than for the working poor. Nicolai Ouroussoff, who was the architecture critic for the *New York Times* from 2004 to 2011, suggested that Masdar is the epitome of a gated community: "the crystallization of another global phenomenon: the growing division of the world into refined, high-end enclaves and vast formless ghettos where issues like sustainability have little immediate relevance." It's too early to tell whether Masdar will become a real city or remain just a grandiose upscale "gated community" stuck out in the Arabian desert.

The tension between form and function, between town and country, between organic evolutionary development and the parsimony of unadorned steel and concrete, and between the complexity of fractal-like curves and surfaces and the simplicity of Euclidean geometry, remains an ongoing debate with no simple resolution or any easy answer. Indeed, many modern architects have explored, struggled, and experimented with many sides of these ongoing tensions, as for example Niemeyer's testament rejecting the "hard and inflexible" and embracing "free-flowing, sensual curves" versus the actuality of some of his soulless concrete buildings. Think of the organic grace of Eero Saarinen's TWA terminal building at Kennedy airport in New York City, or Frank Gehry's whimsical concert hall in Los Angeles and his magical museum in Bilbao, Spain, or Jørn Utzon's wonderful Sydney Opera House, and even that weird phallic symbol in London, dubbed "the gherkin," built by the very same Foster who is building a square city in the desert. At the extreme end of this spectrum and in marked opposition to the admonitions of Le Corbusier and his disciples has been a smattering of remarkable architects like Antoni Gaudí in Spain or Bruce Goff in the United States. Both seemed unbounded in their imaginative visions and were willing to embrace the fantasy of organic structures as witnessed by Gaudí's magnum opus, the extraordinary Sagrada Familia cathedral in Barcelona, or Goff's Bavinger House in Norman, Oklahoma, inspired by the famous Fibonacci sequence of numbers manifested in nautilus shells, sunflowers, and spiral galaxies.

All of these innovative examples are of individual structures, but there is no real equivalent in the design of entire cities, nor in urban development beyond variations on the garden city theme. However, in the 1980s a movement called the New Urbanism arose that was an attempt to combat some of the issues inherent in an automobile and steel and concrete–dominated society where people become alienated from one another and where commuting long distances to work becomes the norm. The movement advocated a return to diverse, mixed-use neighborhoods architecturally as well as socially and commercially, with an emphasis on community structure through designs that enhanced pedestrian use and public transportation. Much of the thinking was inspired by the critical writings of the great urbanists Lewis Mumford and Jane Jacobs, who had reminded us that cities are people and not just infrastructure in the service of the automobile and corporate concrete-and-steel high-rises.

Jane Jacobs gained her fame and notoriety during the 1950s and '60s battling plans to run a four-lane limited-access highway through Greenwich Village in New York City, where she then lived. This was the height of the period of "urban renewal" and "slum clearance" in which massive, unattractive high-rise public housing projects were erected along with major four-lane highways running through downtown city areas with little regard for the urban fabric or the human scale. The man behind all of this in New York City was Robert Moses, the powerful mastermind who reshaped and rejuvenated the infrastructure of the city over a period of almost forty years. Although he accomplished many important things for New York, including the construction of bridges and expressways connecting Manhattan to the other boroughs, he did it at the expense of destroying many traditional neighborhoods.

A major piece of Moses's vision was to construct the Lower Manhattan Expressway designed to go directly through Greenwich Village, Washington Square, and SoHo. Jane Jacobs led the fight to stop this extraordinary encroachment, claiming that it would destroy an essential feature of the city. After a long and bitter struggle she eventually won the battle. She was much vilified during the process, and not just by politicians and developers, but by many urban planners and practitioners, including Lewis Mumford, who saw her as a flaky sentimental reactionary blocking progress and the future commercial success of New York City. In keeping with the spirit of Le Corbusier, Moses's plans also called for multiple city blocks to be razed and replaced with upscale high-rises. Although this was carried out in many areas of the city, it went by the wayside in Greenwich Village though it did result in the development of Washington Square Village, a project owned by New York University, which was ultimately used for faculty housing. I have actually had the pleasure of staying there for short periods during extended visits to NYU. I thoroughly enjoyed it—not because I particularly enjoyed living in a typical modern high-rise apartment complex but because it gave me immediate access to the wonderfully exciting life of Greenwich Village, SoHo, and Little Italy. These are populated with all those crazy people who contribute to the urban buzz and proliferation of galleries, restaurants, and diverse cultural activities that help make New York such a great city—all of which the prophet Moses might well have unwittingly destroyed had it not been for Saint Jane the savior. New York and the rest of us should be eternally grateful to her.

Many cities across the globe have suffered from this vision of urban renewal and slum clearance, all carried out with the very best of intentions and often for good reason. However, all too often the sense of community is neglected, to say nothing of the plight of those being displaced, leading to untold unintended consequences. In too many cases seemingly exitless highways have cut through traditional neighborhoods, leading to isolated islands literally cut off from the major arteries of the city. Together with the construction of bland high-rise apartment complexes, these islands have often bred alienation and crime. In the United States it is a testament to the rallying call of Jane Jacobs that the massive highways built fifty years ago that ran through the downtown areas of major cities such as Boston, Seattle, and San Francisco have now been torn down. It is not so easy, however, to resurrect old neighborhoods and community structures that had evolved over many decades, but cities are very resilient and adaptive and will no doubt evolve something new and unexpected.

As a footnote to this piece of urban history, it is ironic that NYU's long-term strategic plans include a proposal to redevelop the Washington Square Village complex by demolishing those very same high-rise apartment buildings and restore the area to its original structure—*plus ça change, plus c'est la même chose.*

In an interview[4] in 2001 Jane Jacobs was asked:

What do you think you'll be remembered for most? You were the one who stood up to the federal bulldozers and the urban renewal people and said they were destroying the lifeblood of these cities. Is that what it will be?

To which she replied:

No. If I were to be remembered as a really important thinker of the century, the most important thing I've contributed is my discussion of what

*makes economic expansion happen. This is something that has puzzled
people always. I think I've figured out what it is.*

Alas, she was wrong. She is, in fact, primarily remembered for her fight to
preserve the integrity of lower Manhattan and for her insights into the nature
of cities and how they function, including recognizing the critical roles of diversity and community in creating a vibrant urban socioeconomic ecology. In
more recent years she has been beatified "as a really important thinker of the
century" by much of the urban planning community but also by the broader
intelligentsia and cognoscenti. Unfortunately, her contributions to economics
per se for which she wanted to be remembered have not fared nearly so well
and have barely been acknowledged. She wrote several books on urban economics and on economics itself, focusing primarily on questions of growth
and the origins of technological innovation.

A major point throughout her writing is that macroeconomically, cities are
the prime drivers of economic development, not the nation-state as is typically
presumed by most classical economists. This was a radical idea at the time
and almost entirely ignored by economists, especially as Jane was not a card-
carrying member of the clan. Obviously the economy of a country is strongly
interrelated with the economic activity of its cities, but, like any complex adaptive system, the whole is greater than the sum of its parts.

Almost fifty years after Jane's hypotheses about the primacy of cities in
national economies were articulated, many of us who have come to study cities
from a variety of perspectives have arrived at some version of her conclusions.
We live in the age of the Urbanocene, and globally the fate of the cities is the
fate of the planet. Jane understood this truth more than fifty years ago, and
only now are some of the experts beginning to recognize her extraordinary
foresight. Many writers have picked up this theme, including the urban economists Edward Glaeser and Richard Florida, but none has been as forthright
and bold as Benjamin Barber in his book with the provocative title *If Mayors
Ruled the World: Dysfunctional Nations, Rising Cities.*[5] These are indicative of
a rising consciousness that cities are where the action is—where challenges
have to be addressed in real time and where governance seems to work, at least
relative to the increasing dysfunctionality of the nation-state.

3. AN ASIDE: A PERSONAL EXPERIENCE
OF GARDEN CITIES AND NEW TOWN

Following the destruction wrought by the Second World War in which millions of houses were destroyed, the socialist government in Britain faced a mammoth housing crisis. The majority of these damaged houses were in working-class areas, so this greatly accelerated the ongoing processes of "urban development" and "slum clearance" that had already been on the agenda before the war, and in which Ebenezer Howard's idea of garden cities was a classic visionary example. By the 1950s and '60s the preferred model for new housing had evolved from the traditional British desire for a single-family house to the more efficient building of high-rise apartment complexes. These were of mixed success and raised many of the issues we've already discussed. Will Hutton, a political economist at Oxford and formerly editor in chief of *The Observer* newspaper, commented as recently as 2007:

> *The truth is that council housing is a living tomb. You dare not give the house up because you might never get another, but staying is to be trapped in a ghetto of both place and mind. . . . [C]ouncil estates need to be less cut off from the rest of the economy and society.*

As part of this new postwar housing program the British government embarked on creating a series of "New Towns" in order to relocate people from poor or bombed-out urban areas. Their design was inspired by the perceived image of garden cities as the wave of the future for the working class with residences in a countrylike setting and factories located in a separate enclave. The first of these was Stevenage, which was designated a "New Town" in 1946 and where I lived for almost a year in 1957–58. So I actually have some personal experience of what living in a garden city is like.

To my great astonishment I had been offered a place at Cambridge University in Gonville and Caius College to begin in the autumn of 1958 when the new academic year was to commence. So toward the end of 1957 I summarily quit my school in the East End of London and got a temporary job in the research labs of International Computers Limited (ICL), also known as the British Tabulating Machine Company, located in Stevenage.

Like any teenager living away from home for the first time, this was a defining experience and I learned a great deal during that period. From the many new vistas that opened up before me, there are three that are of relevance to this narrative. First, and the most obvious, is that working in an innovative research environment that allows and even encourages freedom of thought and movement beats laboring in the confines of a brewery mindlessly feeding a machine with beer bottles.

The second was that Jane Jacobs, who I suspect had never actually been to a garden city despite her damning comments about them, was right. It was many years before I came to know who Jane Jacobs was, but I quickly came to see that, compared with living in a somewhat run-down Victorian row house in lower-middle-class northeast London, Stevenage was like living in a fancy country resort. And that was its problem. Just as Jacobs sarcastically remarked a few years later: it was a "really very nice town if you were docile and had no plans of your own and did not mind spending your life with others with no plans of their own." Although this is pretty harsh, it does capture that sense of boredom, routine, isolation, and benign "niceness" hiding and suppressing inner passions that later became associated with suburbia. Not that Hackney and the East End of London were paragons of urban bliss; nor, by the way, were Greenwich Village, Little Italy, or the Bronx, despite Jane's protestations. It has become fashionable to nostalgically romanticize working-class London and wax eloquently about community in inner city life, but the truth is that it was dirty, unhealthy, rough, and tough, with its own version of architectural dreariness and the potential for loneliness and alienation. However, these were mightily compensated for by action, diversity, and a pulsating sense of people participating in life with ready access to museums, concerts, theaters, cinemas, sports events, gatherings, protest meetings, and all of the marvelous amenities a traditional city has to offer.

These were the early days of commercial computers, and like IBM in the United States, ICL was developing both old-fashioned vacuum tube and new transistor-based machines programmed by the dreaded Hollerith punched cards. We of a certain age all remember them with a kind of nightmarish nostalgia. As a graduate student at Stanford some years later I developed a passionate hatred of those awful cards and the tedious routine of programming in languages with weird names like Fortran and Balgol. Too bad, actually, because it turned me off to computer development and programming forever so that, even though I was reasonably good at it and was present "at the beginning," both in Stevenage and in what became Silicon Valley, I didn't have the foresight to see that computers would be useful for anything beyond doing complicated calculations and analysis. No doubt that's why I ended up being an academic of modest means rather than a wealthy product of the Stanford entrepreneurial IT machine.

The third vista that opened up was a glimpse into the sophistication and potential power of what electrical circuits could accomplish. From a few very simple modular units (resistors, capacitors, inductors, and transistors) connected by wires in clever and complicated ways following very simple rules, something miraculously powerful and "complex" emerged that could perform extraordinary tasks at lightning speeds—this was the electronic computer. This was my introduction to the primitive concept of networks, emergence, and complexity, though none of those words nor any of that language had yet been articulated. Once I entered the life of a student at Cambridge, all of this was completely forgotten. But some of it must have unsuspectingly remained buried deep in my unconscious waiting to emerge forty years later when I began to speculate that networks form the fundamental scaffolding for understanding how our bodies, our cities, and our companies work.

4. INTERMEDIATE SUMMARY AND CONCLUSION

My aim in this brief and somewhat personal digression is not to give a comprehensive critique or balanced overview of urban planning and design but rather to highlight some of its specific characteristics relevant for setting the scene and providing a natural segue into the possibility of developing a science of cities. I am not an expert, nor do I have credentials in urban planning, design, or architecture, so my observations are necessarily incomplete.

One important insight that resulted from these observations was that most urban development and renewal—and in particular almost all newly created planned cities such as Washington, D.C., Canberra, Brasilia, and Islamabad—has not been very successful. This seems to be the consensus among critics, experts, commentators, and the like. Here's the popular travel writer Bill Bryson offering sarcastic comments about Canberra from his book *Down Under:*

> *Canberra: There's nothing to it!*
> *Canberra: Why wait for death?*
> *Canberra: Gateway to everywhere else![6]*

It is notoriously difficult to make an objective judgment about the success of a city. It is not even clear what characteristics and metrics one should use to determine success or failure. Measurements of psychosocial phenomena such as happiness, fulfillment, and the quality of life do not readily lend themselves to reliable quantification, let alone to being modeled. On the other hand, the more concrete characteristics of life such as income, health, and cultural activities clearly do. Much of what has been written about the success of cities is not much more sophisticated than elaborations on the sort of anecdotes I've already quoted and is, at best, intuitive analysis based on narratives much in the style of Jane Jacobs or Lewis Mumford.[7]

There are many academic sociological studies based on interviews and surveys that try to develop a more objective, "scientific" perspective. Urban sociology as a discipline has a long and illustrious but somewhat controversial and often surprisingly parochial history—it was even used by Robert Moses to justify blasting highways through traditional neighborhoods. However, with all of that in mind, it does seem clear that to varying degrees almost all planned cities end up being soulless and alienating, lacking a buzz of popular and cultural activities and with a general dearth of community spirit. Relative to the promises and hype that have usually accompanied the building of a new city or a major urban redevelopment, it's probably fair to say that almost none of them meet their expectations and many could be characterized as failures.

However, cities are remarkably resilient, and being complex adaptive systems are continually evolving. For instance, for many of us, Washington, D.C., was a noncity that we only visited for historical or patriotic reasons, or because we needed to do business with the government. It was pretty deadly and a bit of a concrete jungle dominated by massive government buildings often projecting that eerie sense of a Kafkaesque bureaucracy oddly reminiscent of the old Soviet style.

Look at Washington now—despite its many problems, it has evolved into a highly diverse and vibrant city that has attracted large numbers of ambitious and creative young people enticed by its sense of action and community. The larger metropolitan area now has an expanded economy that is no longer solely dependent on government jobs. And almost magically, those government edifices don't look quite as menacing any longer, softened by the proliferation of many excellent restaurants and gathering places brimming with young people from all over the world. It's taken a long time for Washington to become a "real" city, a place that even Jane Jacobs might have admired. There is hope.

This brings me to another important point. In the grand scheme of things, it probably hasn't mattered very much that these new inorganic planned cities, such as Washington, Brasilia, or even Stevenage, were "failures" in that they weren't exciting places abounding with opportunities for people to live fulfilling lives, expand their horizons, and feel that they are part of a vibrant creative community. Cities evolve and eventually develop a soul, though it might take a long time. Furthermore, in the not-so-distant past there were proportionately many fewer people living in urban environments and many fewer planned cities. However, because urbanization has been expanding exponentially— recall that when averaged over the next thirty years, we are adding the equivalent of a new city of almost a million and a half people to the planet *every week*—this situation has utterly and completely changed.

Now it really matters. To accommodate the continued exponential increase, new cities and urban developments are being built at a truly astonishing rate. China alone will be constructing two to three hundred new cities over the next twenty years, many with populations of over a million, while megacities that already dominate the developing world continue to expand, many of them spawning slums and informal settlements as more and more people flock to cities.

As I remarked earlier, megacities of the past such as London and New York suffered from much the same negative image associated with megacities of today. Nevertheless, they developed into major economic engines offering huge opportunities and driving the world's economy. Here's the problem: cities do indeed evolve, but they take many decades to change and we simply no longer have the time to wait. It took 150 years for Washington, 100 years for London, and more than 50 years for Brasilia, which is still very much a work in progress. Added to this is the sheer scale of the problem. China has embarked on the daunting task of constructing hundreds of new cities to urbanize 300 million rural residents. Out of expediency, these are being built without any deep understanding of the complexity of cities and their connection to socio-economic success. Indeed, most commentators report that many of these new cities, like classic suburbs, are soulless ghost towns with little sense of community. Cities have an organic quality. They evolve and physically grow out of interactions between people. The great metropolises of the world facilitate human interaction, creating that indefinable buzz and soul that is the wellspring of its innovation and excitement and a major contributor to its resilience and success, economically and socially. It is shortsighted and even courting disaster to ignore this critical dimension of urbanization and concentrate only on buildings and infrastructure.

7

TOWARD A SCIENCE OF CITIES

Almost all theories of the city are largely qualitative, developed primarily from focused studies on specific cities or groups of cities supplemented by narratives, anecdotes, and intuition. They are rarely systemic and typically do not integrate issues of infrastructure with those of socioeconomic dynamics. It may be that the sort of quantitative "physics-inspired" theory of cities that I am advocating is simply not conceivable. Cities and the process of urbanization may be just "too complex" to be subjected to laws and rules that transcend their individuality in a useful way. Science at its best is the search for commonalities, regularities, principles, and universalities that transcend and underlie the structure and behavior of any particular individual constituent, whether it be a quark, a galaxy, an electron, a cell, an airplane, a computer, a person, or a city. And it is at its very best when it can do that in a quantitative, mathematically computational, predictive framework, as is the case for electrons, airplanes, and computers, for instance. However, there are many big challenges such as consciousness, the origins of life, the origins of the universe, and indeed cities themselves that potentially cannot be fully addressed in this way, and we have to recognize and be satisfied with limits to our knowledge and understanding. But it is nevertheless incumbent upon us to push the scientific paradigm as far as possible to determine where the boundaries are and not be deterred by the specter of overwhelming complexity and diversity. Indeed, the very question of the boundaries themselves and the potential limits to knowledge and understanding is itself fundamental and important, both philosophically and practically.

It is in this dual spirit of the urgent need for a theory to help address the practical existential questions of long-term global sustainability and the desire to understand an extraordinary fundamental phenomenon of nature for its own sake that the research program on cities and companies was initiated at the Santa Fe Institute. Its origins and early formation were briefly described at the beginning of the previous chapter. This present chapter is devoted to an overview of some of its more salient accomplishments that have contributed toward formulating a possible *science of cities* and to connect them to work of other researchers who have been pursuing related agendas. I will also try to relate it to more traditional ideas and models that have been proposed for understanding various multifaceted aspects of cities and urbanization.

This is a very old subject going back at least to Aristotle. Consequently, a wide variety of perspectives and frameworks have been developed for trying to understand what cities are, how they arose, how they function, and what their future is. Just within academia itself there is a dizzying array of separate departments, centers, and institutes representing a broad spectrum of alternative ways of perceiving cities: urban geography, urban economics, urban planning, urban studies, urbanomics, architectural studies, and many more, each with its own culture, paradigm, and agenda, though rarely interacting with one another. The situation is rapidly changing as new developments are being initiated, many stimulated by the advent of big data and the vision of smart cities, both somewhat naively touted as panaceas for solving all of our urban problems. But tellingly, there are as yet no explicit departments of "urban science" or "urban physics." These represent a new frontier as the urgency to understand cities from a more scientific perspective emerges. This is the context of what I am presenting here, namely using scale as a powerful tool for opening a window onto the development of a quantitative conceptual integrated systemic framework for understanding cities.

The first step in carrying out such a program was to ask if cities are approximately scaled versions of one another in a similar way that animals are. In terms of their measurable characteristics, are New York, Los Angeles, Chicago, and Santa Fe scaled versions of one another and, if so, is their relative scaling similar to the way in which Tokyo, Osaka, Nagoya, and Kyoto scale, despite their very different appearances and characters? Does their scaling manifest any analog of the universality we saw in biology where whales, elephants, giraffes, human beings, and mice are approximately scaled versions of one another, all neatly and quantitatively expressed by the preponderance of quarter-power scaling laws?

Compared with biology, surprisingly little attention had been paid to such questions regarding cities, urban systems, or companies prior to our work. To some extent this is because urban studies have historically been even less generally quantitative than biology, but also because relatively few computational mechanistic models of cities or companies have been proposed, let alone confronted with data.

1. THE SCALING OF CITIES

An early recruit to our collaboration was Dirk Helbing, who was the director of the Institute for Transport and Economics at Dresden University of Technology in Germany when I first met him. Dirk had been trained in statistical physics and had applied these techniques to understand highway traffic and pedestrian crowds. He's now at the prestigious Swiss Federal Institute of Technology in Zurich, usually referred to as the ETH, where he runs a large project called the Living Earth Simulator. This is designed to model global-scale systems from economies, governments, and cultural trends to epidemics, agriculture, and technological developments using big data sets and fancy algorithms.

In 2004 Dirk recruited one of his students, Christian Kuhnert, to investigate how various characteristics of cities scale with city size across European nations. Some of the results of that early investigation are shown in Figure 33, where you can readily see that the data manifest a surprising simplicity and regularity across cities and countries.[1] Plotted in these graphs is perhaps one of the more mundane characteristics of cities, namely the number of gasoline stations as a function of city size. Their number is shown on the vertical axis while the size of the city as measured by its population is shown on the horizontal one. As was done in earlier graphs illustrating scaling phenomena, the data are plotted *logarithmically*, meaning that the coordinates increase sequentially by factors of ten. You don't have to know any mathematics or remember what a logarithm is, or even know very much about cities, to clearly see that there is an extraordinary regularity in how the number of gas stations varies across different cities. To a good approximation, the data closely follow a simple straight line rather than falling randomly all over the graph, clearly indicating that the variation is not arbitrary but follows a highly constrained systematic behavior. The resulting straight line tells us that the number of gas stations increases with population size following a simple power law, entirely reminiscent of what we saw earlier in how biological and physical quantities scale.

Furthermore, the slope of the straight line, which is the *exponent* of the power law, is about 0.85, a little bit higher than the 0.75 (the famous ¾) we saw for the metabolic rate of organisms (Figure 1). Equally intriguing is that this exponent takes on approximately the same value for how gasoline stations scale across all of the countries shown in the figure. This value of around 0.85 is smaller than 1, so in the language developed earlier, the scaling is *sublinear*, indicating a systematic *economy of scale*, meaning that the bigger the city the fewer the number of gas stations needed on a per capita basis. Thus, on average, each gas station in a larger city serves more people and consequently sells more fuel per month than in a smaller one. To put it slightly differently, with each doubling of population size, a city needs only about 85 percent more gas stations—and not twice as many as might naively be expected—so there is a systematic savings of about 15 percent with each doubling. This becomes a very large effect when comparing, for example, small cities of around fifty thousand people to metropolises one hundred times larger with populations in the range of five million. To service one hundred times as many people requires an increase of only about fifty times as many gas stations, so on a per capita basis, the large city needs only about half as many gas stations as the small one.

It may not come as such a big surprise to learn that larger cities require fewer gas stations per capita than smaller ones, but what is surprising is that this economy of scale is so systematic: it is approximately the same across all of these countries, obeying the same mathematical scaling law with a similar

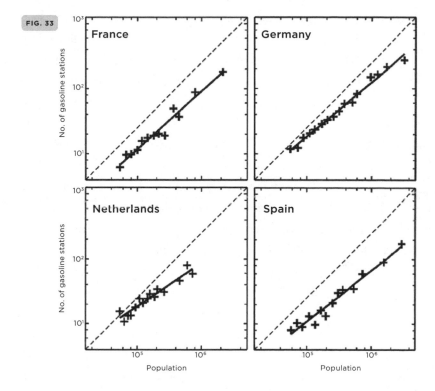

The number of gasoline stations plotted logarithmically against the size of cities in four European nations showing that they all scale sublinearly with a similar exponent. The dotted line has a slope of 1 and indicates linear scaling.

exponent of around 0.85. What is even more surprising is that other infrastructural quantities associated with transport and supply networks, such as the total length of electrical lines, roads, water and gas lines, all scale in much the same way with approximately the same value of the exponent, namely about 0.85. Furthermore, this systematic behavior appears to be the same across the globe wherever data could be obtained. So as far as their overall infrastructure is concerned, cities behave just like organisms—they scale *sublinearly* following simple power-law behavior, thereby exhibiting a systematic economy of scale, albeit to a lesser degree as represented by the different values of their exponents (0.75 for organisms vs. 0.85 for cities).

The extension of this initial probe into whether cities scale to a broader suite of metrics and across a wider range of countries was carried out by a highly talented group of new recruits to the collaboration. These included Luis Bettencourt, whom I first got to know when he was a postdoctoral fellow in astrophysics at Los Alamos working on the evolution of the early universe. He had meanwhile spent a couple of years at MIT before returning to Los Alamos as a staff member in the applied mathematics group. Luis was born, raised, and educated in Portugal, though you'd never know it as he speaks English fluently without any trace of an accent, so much so that when I first met him I thought he was English. He had in fact obtained his doctorate in physics at Imperial College in London, where coincidentally I hold a position in the mathematics department. Luis's fluidity with language is matched by his fluidity with science. He very quickly became engaged with the cities project, gathering and analyzing data from across the globe. He is a passionate adherent of the cause of developing a deep understanding of cities and has now established himself as one of the world's leading experts in this arena.

Luis was joined in this endeavor by another very bright recruit, José Lobo, an urban economist now in the sustainability program at Arizona State University. When we first met, José was a young faculty member in the Department of City and Regional Planning at Cornell University and had been coming to SFI for several years. Like Luis, José brought a real talent for statistics and sophisticated data analysis to our program and, in addition, brought professional expertise in cities and urbanization, a critical component of our collaboration.

Luis and José took the lead in assembling and analyzing extensive data sets covering a broad range of metrics about urban systems across the globe, ranging from Spain and the Netherlands in Europe to Japan and China in Asia and Colombia and Brazil in Latin America. This convincingly verified the earlier analyses showing sublinear scaling of infrastructural metrics and strongly supported the universality of systematic economies of scale in cities. Regardless of the specific urban system, whether Japan, the United States, or Portugal, and regardless of the specific metric whether the number of gas stations, the total length of pipes, roads, or electrical wires, only about 85 percent more material infrastructure is needed with every doubling of city size.[2] Thus a city of 10 million people typically needs 15 percent less of the same infrastructure compared with two cities of 5 million each, leading to significant savings in materials and energy use.[3]

This savings leads to a significant decrease in the production of emissions and pollution. Consequently, the greater efficiency that comes with size has the nonintuitive but very important consequence that on average the bigger the city, the greener it is and the smaller its per capita carbon footprint. In this sense, New York is the greenest city in the United States, whereas Santa Fe, where I live, is one of the more profligate ones. On average, each of us in Santa Fe is putting almost twice as much carbon into the atmosphere as New York. This should not be thought of as somehow reflecting the greater wisdom of New York's planners and politicians, nor as the fault of Santa Fe's leadership, but rather as an almost inevitable by-product of the dynamics underlying economies of scale that transcend the individuality of cities as their size increases. These gains are mostly unplanned, though policy makers in cities can certainly play a powerful role in facilitating and enhancing the hidden "natural" processes that are at work. In fact, this is a large part of what their job is. Some cities are very successful at doing this, while others are much less so. I'll discuss the question of relative performance in the next chapter.

These results are very encouraging and provide powerful evidence in support of the quest for a possible theory of cities. However, of even greater significance was the surprising discovery that the data also reveal that socioeconomic quantities *with no analog in biology* such as average wages, the number of professional people, the number of patents produced, the amount of crime, the number of restaurants, and the gross urban domestic product (GDP) also scale in a surprisingly regular and systematic fashion, as illustrated in Figures 34–38.

Also clearly manifested in these graphs is the equally surprising result that all of the slopes of these various quantities have approximately the same value, clustering around 1.15. Thus these metrics not only scale in an extremely simple fashion following classic power law behavior, but they all do it in approximately the same way with a similar exponent of approximately 1.15 regardless of the urban system. So in marked contrast to infrastructure, which scales *sublinearly* with population size, socioeconomic quantities—the very essence of a city—scale *superlinearly*, thereby manifesting systematic *increasing returns to scale*. The larger the city, the higher the wages, the greater the

FIG. 34

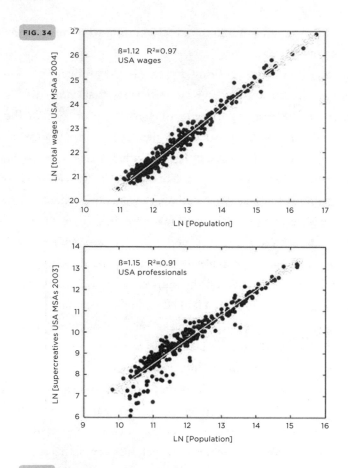

ß=1.12   R²=0.97
USA wages

ß=1.15   R²=0.91
USA professionals

FIG. 35

INNOVATION MEASURED BY PATENTS

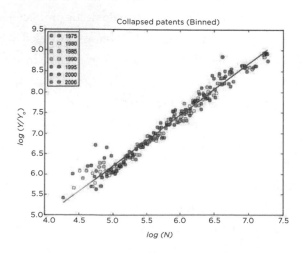

Collapsed patents (Binned)

1975
1980
1985
1990
1995
2000
2006

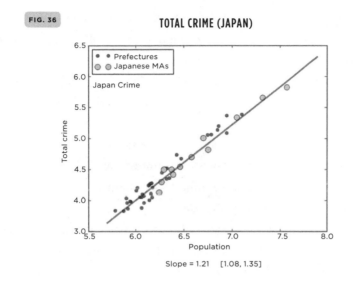

FIG. 36

TOTAL CRIME (JAPAN)

Slope = 1.21   [1.08, 1.35]

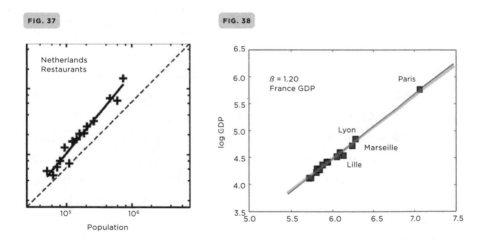

FIG. 37

FIG. 38

Opposite and this page: The scaling of a variety of socioeconomic metrics as a function of population size in a number of different urban systems showing the remarkable similarity of their superlinear exponents (the slopes of all the graphs are close to the same value of 1.15): (34 top) Wages in the United States. (34 bottom) The number of professional people ("supercreatives") in the United States. (35) The number of patents produced in the United States.[4] (36) Crime in Japan. (37) Restaurants in the Netherlands. (38) GDP in France.

GDP, the more crime, the more cases of AIDS and flu, the more restaurants, the more patents produced, and so on, all following the "15 percent rule" on a per capita basis in urban systems across the globe.

Thus the larger the city the more innovative "social capital" is created, and consequently, the more the average citizen owns, produces, and consumes, whether it's goods, resources, or ideas. This is the good news about cities and why they have been so attractive and seductive. On the other hand, cities also have a dark side, so here's the bad news. To approximately the same degree as for the positive indicators, negative indicators of human social behavior also systematically increase with city size: doubling the size of a city not only increases wages, wealth, and innovation by approximately 15 percent per capita but also increases the amount of crime, pollution, and disease to the same degree. Apparently, the good, the bad, and the ugly come together hand in glove as an integrated, almost predictable package. A person may move to a bigger city drawn by more innovation, more opportunity, better wages, and a greater sense of "action," but she can also expect to confront an equivalent increase in garbage, theft, stomach flu, and AIDS.

These results are pretty amazing. We typically think of each city, and especially the one *we* live in, as being unique, with its own history, geography, and culture, having its own special individuality and character that we feel we recognize. Boston not only looks different but also "feels" different from New York, San Francisco, or Cleveland, just as Munich looks and feels different from Berlin, Frankfurt, or Aachen. And they do and are. But who would believe that within their own urban systems they are approximately scaled versions of one another, at least as far as almost anything that you can measure about them is concerned? If you are given the size of a city in the United States, for example, then you can predict with 80 to 90 percent accuracy what the average wage is, how many patents it's produced, how long all of its roads are, how many AIDS cases it's had, how much violent crime was committed, how many restaurants there are, how many lawyers and doctors it has, et cetera. Much about a city is determined simply by its size. There are, of course, outliers and variations around such estimates, and these will be addressed in the following chapter.

Another important point to recognize is that the observed scaling laws are between cities *within the same* national urban system, meaning within the same country. Scaling laws such as those shown in Figures 34–38 don't predict how cities scale *between* different urban systems. The overall scale of the various metrics, such as wages, crime, patent production, and total road lengths, depends on the overall economy, culture, and individuality of each national urban system. For example, the overall scale of crime is much lower in Japan than in the United States, whereas the overall production of patents is higher in the United States. Thus while scaling laws predict how Chicago's metrics scale relative to Los Angeles's and how Kyoto's scale relative to Osaka's, they don't directly predict how Osaka scales relative to Chicago. However, if we know the overall scale of the metrics—which can be deduced from just knowing, for example, how New York's metrics scale relative to Tokyo's, we can then predict how any Japanese city scales relative to any American one.

Usually we tend to view most of these diverse urban metrics and characteristics as being unrelated and independent of one another. For example, we don't think of the number of cases of a particular disease as being connected to the number of patents produced or the number of gasoline stations in a city. Who would believe that wages, patents, crime, and disease all scale with city size in approximately the same "predictable" fashion anywhere in the world? The data convincingly show that despite appearances cities are approximately scaled versions of one another: New York and Tokyo are, to a surprising and predictable degree, nonlinearly scaled-up versions respectively of San Francisco and Nagoya. These extraordinary regularities open a window onto underlying mechanisms, dynamics, and structures common to all cities and strongly suggest that all of these phenomena are in fact highly correlated and interconnected, driven by the same underlying dynamics and constrained by the same set of "universal" principles.

Consequently, each of these urban characteristics, each metric—whether wages, the length of all the roads, the number of AIDS cases, or the amount of crime—is interrelated and interconnected with every other one and together they form an overarching multiscale quintessentially complex adaptive system that is continuously integrating and processing energy, resources, and information. The result is the extraordinary collective phenomenon we call a city, whose origins emerge from the underlying dynamics and organization of how people interact with one another through social networks. To repeat: cities are an emergent self-organizing phenomenon that has resulted from the interaction and communication between human beings exchanging energy, resources, and information. As urban creatures we all participate in the multiple networks of intense human interaction that is manifested in the metropolitan buzz of productivity, speed, and ingenuity, no matter where we live.

It is important to recognize that these patterns of increased productivity and concomitant decreased costs hold true across nations with very different levels of development, technology, and wealth. Although there is much more information on cities in richer parts of the world, extensive data sets are beginning to become more readily available from rapidly developing countries such as Brazil and China. Good data from countries like India and those in Africa remain frustratingly elusive, though this will almost certainly change in the near future. The data thus far analyzed fit the mold, and below I'll refer to some that have already played an important role in helping to establish just how "universal" the nature of systematic scaling is. The GDP for cities in Brazil and China, for instance, closely follow the same superlinear curve that western European and North American cities exhibit, even though they start from a lower baseline. These patterns hold true because the same basic social and economic processes are at work, whether in São Paulo's favelas, under Beijing's smog-filled skies, or along Copenhagen's tidy streets.

Finally, it's worth noting that not all characteristics of cities scale nonlinearly. For example, each person on average has a single home and a single job regardless of city size, so that the number of jobs and number of houses increase linearly with city size. In other words, the exponents of their corresponding scaling curves are very close to 1, as confirmed by data: double the number of people in a city, and you'll find twice as many houses and twice as many jobs. Buried in this are some hidden assumptions and conclusions. Obviously not everyone has a job (especially children and old people), and some people have more than one job; in addition, although almost everyone has a home, not everyone has a house. Nevertheless, most people have a single job, and the average occupancy of houses (the "family," however defined) is approximately the same across cities, so at the coarse-grained level these dominate and a simple linear relationship results.

To summarize: the bigger the city, the greater the social activity, the more opportunities there are, the higher the wages, the more diversity there is, the greater the access to good restaurants, concerts, museums, and educational facilities, and the greater the sense of buzz, excitement, and engagement. These facets of larger cities have proven to be enormously attractive and seductive to people worldwide who at the same time suppress, ignore, or discount the inevitable negative aspects and the dark side of increased crime, pollution, and disease. Human beings are pretty good at "accentuating the positive and eliminating the negative," especially when it comes to money and material well-being. In addition to the perceived individual benefits coming from increased city size, there are huge collective benefits arising from systematic economies of scale. Coupled together, this remarkable combination of increasing benefits to the individual with systematic increasing benefits for the collective as city size increases is the underlying driving force for the continued explosion of urbanization across the planet.

2. CITIES AND SOCIAL NETWORKS

So how is it that urban systems across the globe in countries as diverse as Japan, Chile, the United States, and the Netherlands scale in basically the same way despite having vastly different geographies, histories, and cultures, and despite having evolved independently of one another? It's not as if there was some international agreement among nations that's been operating over the past centuries that required them to build and develop their cities according to these simple scaling laws. It happened without anyone enforcing it, designing it, or policing it. It just happened. So what is the common unifying factor that transcends these differences and underlies this surprising structural and dynamical similarity?

I have already strongly hinted at the answer: the great commonality is the universality of social network structures across the globe. Cities are people, and to a large extent people are pretty much the same all over the world in how they interact with one another and how they cluster to form groups and communities. We may look different, dress differently, speak different languages, and have different belief systems, but to a large extent our biological and social organization and dynamics are remarkably similar. After all, we are all human beings sharing pretty much the same genes and the same generic social history. And no matter where we live on the planet, all of us emerged only relatively recently from being mobile hunter-gatherers to becoming predominantly sedentary communal creatures. The underlying commonality that is being expressed by the surprising universality of urban scaling laws is that the structure and dynamics of human social networks are very much the same everywhere.

With the development of language, human beings acquired the capability of exchanging and communicating new kinds of information on a scale and at a rate that was unprecedented in the entire history of life. A major outcome of this revolution was the discovery of the fruits of economies of scale: by working together, we could build and accomplish more with the same amount of individual effort, or equivalently, we could complete specific tasks faster using less energy per person. Communal activities such as building, hunting, storing, and planning all evolved and benefited from the development of language and the consequential enhanced ability to communicate and think. Furthermore, we developed imagination and brought to consciousness the concept of the *future* and therefore the remarkable ability to plan, to think ahead and construct possible scenarios in anticipation of future challenges and events. This powerful innovation in human cerebral activity was entirely new to the planet and led to extraordinary consequences that have affected not just human beings but almost all of the planet's inhabitants from microscopic bacteria to giant whales and sequoias.

It is true that many other creatures, such as herding animals and especially social insects, also discovered economies of scale, but their accomplishments are relatively primitive and static compared with what humans have achieved. The power of language has allowed us to go well beyond classic economies of scale, such as our cells achieve or our hunter-gatherer predecessors attained to evolve and build on those advantages by adapting to new challenges over periods of time that are vastly shorter than typical evolutionary timescales that had hitherto been required for major innovations to be made. Ants brilliantly self-organized to evolve remarkably robust and hugely successful and sophisticated physical and social structures, but it took them millions of years to do so. Furthermore, they accomplished this more than 50 million years ago and have barely evolved beyond it since. On the other hand, once we had invented verbal language, it took us only tens of thousands of years to evolve from hunting and gathering to becoming sedentary agriculturalists—and even more remarkably, only another ten thousand years to evolve cities, become urbanists, and invent cell phones, airplanes, the Internet, quantum mechanics, and the general theory of relativity.

It is, of course, a matter of judgment as to whether we are better adapted to life than ants are, and a matter of futurology as to whether their cities, economies, quality of life, and social structures ultimately prove more or less sustainable than ours. As things presently stand, my money is definitely on them outlasting us. They are extraordinarily efficient, robust, and stable and they've already been around significantly longer than we have and will very likely be around long after we've gone. Nevertheless, despite all of our faults, and we certainly have plenty of them, my anthropocentric judgment in terms of the quality and meaning of life is unerringly with us.

We have uniquely evolved perhaps the most precious and mysterious property of life, namely consciousness, and with it contemplation and conscience, and these have revealed a glimmer of insight into how we can address some of the most awesome questions facing us. The quintessential anthropocentric process of wonder, thinking, contemplation, reflection, questioning, and philosophizing, of creating and innovating, of searching and exploring, has been enhanced and engendered by our invention of the city as the crucible of civilization, the engine that facilitates creativity and ideas.

If cities are thought of solely in terms of their physicality, as just buildings and roads and the multiple network systems of wires and pipes that supply them with energy and resources, then they are indeed quite analogous to organisms, manifesting similar systematic scaling laws encapsulating economies of scale. However, when humans began forming sizable communities they brought a fundamentally new dynamic to the planet beyond that of biology and the discovery of economies of scale. With the invention of language and the consequent exchange of information between people and groups of people via social networks, we discovered how to innovate and create wealth. Cities are therefore much more than giant organisms or anthills: they rely on long-range, complex exchanges of people, goods, and knowledge. They are invariably magnets for creative and innovative individuals, and stimulants for economic growth, wealth production, and new ideas.

Cities provide a natural mechanism for reaping the benefits of high social connectivity between people conceiving and solving problems in a diversity of ways. The resulting positive feedback loops act as drivers of continuous multiplicative innovation and wealth creation, leading to superlinear scaling and increasing returns to scale. Universal scaling is a manifestation of an essential trait resulting from our evolutionary history as social animals common to all peoples worldwide, transcending geography, history, and culture. It arises from the integration of the structure and dynamics of social networks with the physical infrastructural networks that are the platform upon which the panoply of urban life is played out. Although this is a dynamic beyond biology, it shares a similar conceptual framework and mathematical structure as exemplified by fractal-like network geometries discussed in chapter 3.

3. WHAT ARE THESE NETWORKS?

Recall that the generic geometric and dynamical properties of biological networks that underlie quarter-power allometric scaling are: (1) they are space filling (so every cell of an organism, for instance, must be serviced by the network); (2) the terminal units, such as capillaries or cells, are invariant within a given design (so, for instance, our cells and capillaries are approximately the same as those of mice and whales); and (3) the networks have evolved to be approximately optimal (so, for instance, the energy our hearts have to use to circulate blood and support our cells is minimized in order to maximize the energy available for reproduction and the rearing of offspring).

These properties have direct analogs in the infrastructural networks of cities. For example, our road and transport networks have to be space filling so that every local region of the city is serviced, just as all of the various utility lines have to supply water, gas, and electricity to all of its houses and buildings. It's also natural to extend this concept to social networks: averaged over time, each person interacts with a number of other people as well as with groups of people in the city in such a way that collectively their network of interactions fills the available "socioeconomic space." Indeed, this urban network of socio-economic interactions constitutes the cauldron of social activity and interconnectivity that effectively defines what a city is and what its boundaries are. To be part of a city you have to be an ongoing participant in this network. And, of course, the invariant terminal units of these networks, the analogs of capillaries, cells, leaves, and petioles, are people and their houses.

A challenging and very interesting question is what, if anything, is being optimized in the structure and dynamics of cities. Compared with biological life cities haven't been around very long—only for hundreds of years relative to the millions, if not hundreds of millions, that many organisms have been here. So any drive toward optimality arising from incremental adaptations and feedback mechanisms as cities grow and evolve hasn't had a lot of time to settle down and reach full fruition. This is further complicated by the much more rapid rate at which innovations and changes have taken place in cities relative to typical biological rates of evolution. Nevertheless, market forces and social dynamics are continually at work, so it's not entirely unreasonable to speculate that the evolution of infrastructural networks has moved toward minimizing costs and energy use. For instance, when it comes to transport, most journeys, whether by bus, train, car, horse, or foot, are typically undertaken with the aim of minimizing either travel time or distance, or both. There are undoubtedly huge local inefficiencies in electrical, gas, water, and transport systems, many resulting from historical legacies and economic expediency. Nevertheless, and despite appearances, upgrades, improvements, replacements, and maintenance are continually being undertaken, so when viewed over a long enough period of time there is a clear trend toward approximate optimization of these network systems. The emergence of systematic scaling laws with a common exponent for diverse infrastructural quantities in different urban systems across the globe can be viewed as a consequence of this evolutionary process.

Notice, however, that relative to what we saw in the scaling of most metrics in biology there is a much greater spread of the data around the idealized scaling curves for cities. Compare, for instance, the tightness of the fits to the data for animal metabolic rates, shown in Figure 1, relative to the much greater spread of the data for the average wage in cities shown in Figures 34–38. This greater variance reflects the much shorter time cities have had to organically evolve toward the idealized optimal configuration represented by the scaling curves—the straight lines in the logarithmic plots. The deviations from these lines are a measure of the residual footprint of the unique history, geography, and culture of each individual city and will be discussed in greater detail below. In contrast to the scaling exponent (the 0.85), which is the slope of the straight lines in the logarithmic plots and which is approximately the same for all urban systems, the degree of variation (that is, the spread) of the data around these straight lines (all having similar slopes) is different in different urban systems. This is largely because different nations have devoted different amounts of resources to maintenance, improvements, and innovations of their cities.

In terms of the socioeconomic dynamics of cities we can likewise ask what, if anything, is being optimized in urban *social* networks. This is a tough question to answer definitively, and many scholars have obliquely tried to address it from multiple points of view.[5] If we think of the city as the great facilitator of social interactions or as the great incubator for wealth creation and innovation, it is natural to speculate that its structure and dynamics evolved so as to maximize social capital by optimizing the connectivity between individuals. This suggests that social networks and the entire social fabric of cities and urban systems—that is, who is connected to whom, how much information flows between them, and the nature of their group structure—is ultimately determined by the insatiable drive of individuals, small businesses, and giant companies to always want more. Or, to put it in crass terms, that the socioeconomic machinery that we all participate in is primarily fueled by greed in both its negative and positive connotations as in the sense of the "desire for more." Given the enormous disparities in income distributions that are observed in all cities across the globe, and the apparent drive of most of us to want more despite having plenty, it's not hard to believe that greed in its various forms is an important contributor to the socioeconomic dynamics of cities. To quote Mahatma Gandhi: "The Earth provides enough to satisfy every man's needs, but not every man's greed."

Greed is the pejorative image of this insatiable desire for more, but it also has an extremely important, positive flip side. Metaphorically, it is the social analog of the evolutionary biological drive of animals, including us, to maximize their metabolic power relative to their size. As was discussed in chapter 3, this can be thought of as derivative from the principle of natural selection and underlies the allometric scaling laws that permeate biology. The extension of the concept of the survival of the fittest to the social and political domain has led many thinkers to the controversial concept of Social Darwinism, whose roots go back to Malthus. Regardless of its validity, this idea has been sadly misrepresented, abused, and misused by politicians and social thinkers, sometimes with devastating consequences, to support all sorts of extreme views ranging from eugenics and racism to rampant laissez-faire capitalism.

The desire for more can apply to many things beyond wealth and material assets. It is a hugely powerful force in society that poses enormous moral, spiritual, and psychological challenges at both the individual and collective levels. The desire to succeed, whether in sports, business, or academia—to run the fastest, have the most creative company, or generate the most profound and insightful idea—has been a major underlying societal dynamic that has been instrumental in bringing us the extraordinary standard of living and quality of life many of us are privileged to enjoy. At the same time we have tempered our rampant materialistic greed by evolving altruistic and philanthropic behavior that has been integrated into our sociopolitical structures to protect us from excesses.

With the invention of the city and its powerful combination of economies of scale coupled to innovation and wealth creation came the great divisions of society. Our present social network structures barely existed in their present form until urban communities evolved. Hunter-gatherers were significantly less hierarchical, more egalitarian and community oriented than we are. The struggle and tension between unbridled individual self-enhancement and the care and concern for the less fortunate has been a major thread running throughout human history, especially over the past two hundred years. Nevertheless, it seems that without the motive of self-interest our entrepreneurial free market economy would collapse. The system we have evolved critically relies on people continually wanting new cars and new cell phones, new widgets and gadgets, new clothes and new washing machines, new thrills, new entertainment, and pretty much new everything, even when they already have enough of "everything." It may not be a pretty picture and it doesn't work for everyone, but so far, it's worked remarkably well for most of us, and apparently most of us seem to want it to continue. Whether it can is a topic I'll return to in the last chapter.

Later in this chapter I will elaborate in some detail on the nature of the flows of information, energy, and resources in both social and infrastructural networks and show how they lead to the observed scaling. Much like biological networks, these networks are hierarchical and fractal-like in nature. So, for example, in infrastructural networks the flows in utility lines systematically decrease from central supply units, such as power stations and waterworks, through the pipes and electrical lines of their respective networks to supply individual houses in much the same way that blood flow in the circulatory system decreases in approximately regular geometric proportion from the heart down through the aorta to the capillaries to supply cells. The fractal-like nature of these networks and of the flows through them ensures an efficient distribution of energy and resources and underlies sublinear scaling and economies of scale.

It's actually a little more subtle than this, because cities are not uniform but typically have a number of local hubs of activity that behave semiautonomously, even though they are hierarchically interconnected with one another. These local hubs are often referred to as "central places," following a popular model of urban systems known as *central place theory* that gained great currency with urban planners and geographers after it was introduced in the 1930s by the German geographer Walter Christaller.

4. CITIES: CHRISTALLS OR FRACTALS?

It's a curious theory. Basically it's a static, highly symmetric geometric model for how cities and urban systems are physically configured. It was postulated by Walter Christaller based on his observations of cities in southern Germany, somewhat analogously to the way Jane Jacobs formulated her version of the city from her personal experiences of New York. Little or no attention was paid

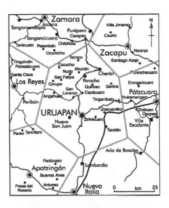

● CITY
○ TOWN
● VILLAGE

Christaller's hexagonal lattice concept of central places together with "real world evidence" from central Mexico supporting the idea.

to quantitative reckoning and testing, systematic analysis and confrontation with data, or mathematical formulation with subsequent prediction; so it's not exactly science, at least the way I have tried to present it here. In spirit, it has much more in common with the rigid inorganic garden city designs of Ebenezer Howard, which were primarily inspired by idealized Euclidean geometric patterns with almost no consideration given to the role of people other than as economic units. It has, nevertheless, many interesting features and has been extremely influential in the design and thinking of cities during the twentieth century.

In a curious play on his name, Christaller posited that urban systems, and by implication individual cities, can be represented as idealized two-dimensional crystalline geometric structures based on a highly symmetric hexagonal lattice pattern that repeats itself at smaller and smaller rescaled granularities, as illustrated above. Hexagons were chosen as the simplest nontrivial shape that could fit together edge to edge so as to completely fill the geographical extent of the city or urban system without gaps in between. These hexagons act as "central places" of commercial activity with smaller hexagonal central places embedded within them. Christaller had been motivated in this design by the observation that towns of a similar size in southern Germany were approximately equidistant from one another (all presumed located at the vertices of a hexagon) and from a larger central city acting as a hub (located at the center of the hexagon). Although this regularity is generally not observed in most urban systems or within cities, and despite its rather contrived and unnatural structure, Christaller's model for the geometry of urban systems incorporates two very important features that it shares in common with organically evolved network structures. It is space filling and self-similar (and therefore hierarchical), even though neither of these terms had yet been invented. His model also incorporated other important general features such as the idea of minimal travel time and distance to obtain services, something I will return to below.

Central place theory remains a major conceptual component in today's urban planning and design, even if its shortcomings are well recognized. During the early 1950s it formed the basis for restructuring the municipal relationships and boundaries of the newly founded Federal Republic of Germany (West Germany), a system that persists today. Somewhat paradoxically, Christaller had joined the Communist Party after the Second World War, having been a member of the Nazi Party during the war when he had worked under the SS. He had devised a grand plan inspired by his theory for reconfiguring the economic geography of the conquered territories of Czechoslovakia and Poland to accommodate German expansion. A further, but tragic, irony in this story is that the German economist August Lösch, regarded as the founder of the field of regional science and who is best known for expanding Christaller's work to make it less static, more mathematical, and more realistic, was an active member of a vocal anti-Nazi Protestant group. He remained in Germany and went into hiding during the war but died from scarlet fever just a few days after it ended. He was just thirty-nine years old.

The actual self-similarity of cities more closely reflects the organically evolved hierarchical network structures of transport and utility systems than the rigid hexagonal crystalline structures of Christaller. The city is not a top-down engineered machine dominated by straight lines and classic Euclidean geometry, but rather is much more akin to an organism with its crinkly lines and fractal-like shapes typical of a complex adaptive system—which it is. This is clear from just a casual look at the growth patterns of a typical city with its ever-expanding filigreed infrastructural network pattern reminiscent of the growth pattern of a bacterial colony, as illustrated on page 291. A careful math-

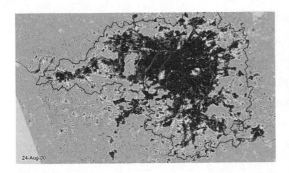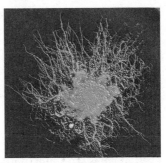

The organic growth of Paris showing the development of a fractal-like geometry (left), and a bacterial colony showing the development of its fractal-like geometry (right).

ematical analysis of such patterns shows that cities are, in fact, approximate self-similar fractals much like biological organisms or geographical coastlines. For example, if the length of the perceived boundary of a city is measured at different resolutions, analogous to what Lewis Fry Richardson did for coastlines, and these are plotted logarithmically, then approximate straight lines result whose slope is the conventional fractal dimension of the city boundary.

As I explained earlier, fractal dimensions are a measure of an object's degree of crinkliness, which some interpret as a measure of its complexity. Stimulated by the explosive interest in fractals and the incipient development of a science of complexity in the 1980s, the distinguished urban geographer Michael Batty carried out extensive statistical analyses on cities to measure their fractal dimensions.[6] Batty and his colleagues as well as others that followed found values in the range of 1.2 but with a large variance up to values close to 1.8. In addition to providing a metric for comparing the complexity of different cities, perhaps one of the more interesting uses of a fractal dimension is as a diagnostic barometer of the health of a city. Typically, the fractal dimension of a healthy robust city steadily increases as it grows and develops, reflecting a greater complexity as more and more infrastructure is built to accommodate an expanding population engaging in more and more diverse and intricate activities. But conversely, its fractal dimension decreases when it goes through difficult economic times or when it temporarily contracts.

These fractal dimensions are a measure of the self-similarity of a city's various infrastructural networks and are derived from analyzing images such as those shown on page 291 at different resolutions. However, a city's fractal nature may not always be so obvious from just looking at its physical manifestation. After all, the street plan of New York City or for that matter almost any American city is typically a regular rectangular grid, which is about as simplistically Euclidean as you can get. This is obviously not as true in Old World cities like London or Rome, whose meandering streets have a more obvious fractal-like organic structure. In either case, hidden beneath the geometry even of cities with rectangular grids lurks a fractality that permeates all cities and is reflected in the universality of the scaling laws.

Let me illustrate this with an example that is for an entire urban system rather than for a specific city, but the point is the same. Page 293 shows a map of the interstate road network system of the United States. Its construction began after the Second World War during the Eisenhower administration, inspired by Hitler's autobahns in prewar Germany and, like the autobahn, strongly motivated by perceived defense needs. Indeed, its official title is the National System of Interstate and Defense Highways. Consequently, the roads were planned to be as straight as possible in order to minimize distance and travel times between major cities, in much the way that the Romans built their roads two thousand years earlier so as to maintain control over their empire. The result, as you can see, is that the interstate system roughly approximates a rectangular grid much like a typical American city, though, of course, geography and local conditions dictate deviations here and there. Overall, however, it's surprisingly regular and does not look much like a classic fractal.

Yet despite appearances, the interstate system is in fact a quintessential fractal when viewed through the lens of the actual traffic flowing on it, rather than when viewed simply as a physical road network. The traffic flow is the very essence of the interstate and is the fundamental reason for its existence. To reveal its fractality, consider for simplicity some port city such as Boston, Long Beach, or Laredo. Trucks leave these ports on a regular basis to deliver goods all over the United States using the interstate road network. The U.S. Department of Transportation keeps careful statistics on such traffic flows so it's easy to add up the total number of trucks traveling in each road section over some specified period of time such as a month. Take Laredo, Texas, as a specific example. Obviously the sections of the interstate directly leaving the city have the most traffic on them, because all of the trucks have to use them to leave. As these trucks travel farther away from the city they fan out across the country by branching off onto other road sections of the interstate and eventually onto roads of the local state system. Consequently, these sections have less and less truck traffic on them the farther removed they are from Laredo delivering goods to more and more distant cities and towns.

This is neatly illustrated in the map of the flow of transport in Texas, where

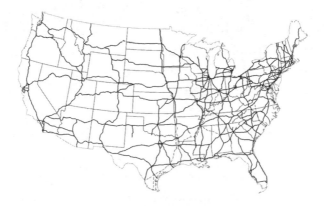

MAJOR FLOWS BY TRUCK TO, FROM, AND WITHIN TEXAS: 2010

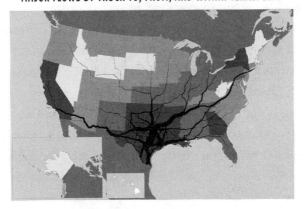

Top: A standard map of the interstate highway system of the United States; bottom: a map of the flow of transport in Texas, revealing the hidden fractal structure concealed in the physical road system. The thickness of the road represents the relative flow of traffic. Many of the thinner segments, the "capillaries," represent noninterstate roads whereas the thicker segments, the main "arteries," are larger roads. Compare this with the cardiovascular blood transport system shown in chapter 3.

the thickness of each road section represents the truck traffic flowing on it that originated in Laredo—in other words, the thicker the road, the greater the number of trucks on it that started in Laredo. As you can readily see, the regular grid of the interstate that you are used to seeing on a map has been transformed into a much more interesting hierarchical fractal-like structure remarkably reminiscent of our circulatory system. Thus in terms of what actually matters, namely the traffic flowing on it, this is really what the road network looks like. The main road leaving Laredo acts like an aorta, the subsequent roads are its arteries, and the terminal roads leading to the various towns and cities where goods are finally delivered are its capillaries. Its heart is the city of Laredo itself, which "pumps" trucks through the interstate circulatory system. This pattern is repeated for each city across the country, so the system is a generalization of our physiological circulatory system in which each city behaves as a pumping heart, or as a "central place" in the jargon of Christaller.

Unfortunately, no one has performed a similar analysis within cities, primarily because we don't have the detailed statistics on traffic flow for every street in the city. The advent of smart cities with the promise of countless detectors on every street corner monitoring traffic will eventually provide sufficient data to carry out similar analyses on all cities to reveal the dynamical structure of their transportation system, much like the map on page 293. This would provide a detailed quantitative valuation of traffic patterns and the attractiveness of specific locations as well as other metrics that are crucial for planning purposes such as in successfully developing new areas of a city or deciding on the placement of new malls or stadiums.

The leading advocate for developing the concept of the fractal city and integrating ideas from complexity theory into traditional urban analysis and planning has been Mike Batty, who runs the Centre for Advanced Spatial Analysis (CASA) at University College London. His work has focused primarily on computer models of the physicality of cities and urban systems. He is enthusiastic about the concept of cities as complex adaptive systems and has consequently become a major proponent of developing a science of cities. His vision is a little different from mine and is summarized in his recent book *The New Science of Cities,* which emphasizes the more phenomenological traditions of the social sciences, geography, and urban planning as against the more analytic, mathematical traditions of physics based on underlying principles that I've been articulating.[7] Ultimately, both approaches are needed if we are to accomplish the huge challenge of understanding cities.

5. CITIES AS THE GREAT SOCIAL INCUBATOR

A city is not just the aggregate sum of its roads, buildings, pipes, and wires that comprise its physical infrastructure, nor is it just the cumulative sum of the lives and interactions of all of its citizens, but rather it's the amalgamation of all of these into a vibrant, multidimensional living entity. A city is an emergent complex adaptive system resulting from the integration of the flows of energy and resources that sustain and grow both its physical infrastructure and its inhabitants with the flows and exchange of information in the social networks that interconnect all of its citizenry. The integration and interplay of these two very different networks magically gives rise to increasing economies of scale in its physical infrastructure and simultaneously to a disproportionate increase in social activity, innovation, and economic output.

The previous section focused on the physicality of cities in highlighting their self-similar fractal nature, manifested in the sublinear scaling of infrastructural metrics such as road lengths and the number of gas stations. These arise in cities by much the same process as they do in organisms in that they are the consequence of the generic properties of optimized, space-filling transport networks that constrain how energy and resources are supplied to all parts of the city. These physical networks are all very familiar to us—roads, buildings, water pipes, electrical lines, automobiles, and gasoline stations are the very visible stuff of urban life—and it's not difficult to imagine how they mimic the physiological networks in our own bodies, such as our cardiovascular system. Not quite so obvious, however, is to visualize the geometry and structure of social networks and the flows of information between the people in them.

The study of social networks is a huge field embracing all of the social sciences, with a long and illustrious history going back to the founding of the discipline of sociology. Although sophisticated mathematical and statistical techniques for analyzing such networks were developed by social scientists out of academic interest as well as for corporate and marketing reasons, the field got a huge boost when physicists and mathematicians began to get interested in complex adaptive systems in the 1990s. This was further enhanced by the emergence of new communication tools ushered in by the revolution in information technology, which led to new kinds of social networks such as Facebook, Twitter, and the like. Together with the advent of smart mobile phones, this has led to an explosion in the amounts and quality of data available for analysis of how people interact.

Over the past twenty years an emerging subfield of *network science* has blossomed in its own right, leading to a deeper understanding of both the phenomenology of networks in general and also the underlying mechanisms and dynamics that generate them.[8] The subject of network science covers an enormous range of topics including classic community organizations, criminal and terrorist networks, networks of innovation, ecological networks and food webs, health care and disease networks, and linguistic and literary networks. Such studies have provided important insights into a broad range of important societal challenges including devising the most effective strategies for attacking pandemics, terrorist organizations, and environmental issues, for enhancing and facilitating innovative processes, and for optimizing social organizations. A lot of this fascinating work has been carried out or stimulated by many of my colleagues associated with the Santa Fe Institute.

It's a Small World: Stanley Milgram and Six Degrees of Separation

You are very likely familiar with the notion of "six degrees of separation." This was articulated by the highly imaginative social psychologist Stanley Milgram in the 1960s and is often referred to as "the small world problem."[9] It arose from answering the fascinating question: how many people on average separate you from any other arbitrary random person in the country? An easy way to conceptualize this is to think of it graphically by first representing each person as a dot on a piece of paper; this is called a *node*. If two such people— each represented by dots—know each other, then a line is drawn between them; this is called a *link*. With this simple prescription you can construct the *social network* of any community, an example of which is given on page 298. So, for example, if we look at the social network of the entire country, it's interesting to ask how many links on average separate two random people. Obviously, for people you know well, which are referred to as your friends but include your family members and coworkers, there is only a single link. For friends of your friends whom you don't know, there are two links separating you. Taking this one step further, there are three links separating you from friends of your friends of your friends whom you don't know ... and so on. You get the idea—this can be straightforwardly extended indefinitely until everyone in the network is accounted for. I live in Santa Fe, New Mexico, and maybe you, the reader, live in Lewiston, Maine, which is more than two thousand miles away. As far as I know, I don't know anyone in Lewiston, and it's very likely that you don't know anyone in Santa Fe. But an interesting question to ask is what is the minimum number of links I have to go through—that is, how many iterations of friends of friends of friends of friends, et cetera—before I eventually connect to you. There are about 350 million people in the United States, so you might think that this number would be something quite large, like fifty, one hundred, or even one thousand. Amazingly, Milgram discovered that, on average, the number of links connecting any two people is approximately just six. Hence the phrase "six degrees of separation"—that's it, we are separated from each other by just six links, so despite appearances we are surprisingly closely connected.

This unexpected result was first analyzed mathematically by the applied mathematician Steven Strogatz and his then student Duncan Watts.[10] They showed that small-world networks typically have an overabundance of *hubs* and a large degree of clustering relative to randomly connected networks. Hubs are simply nodes with an unusually large number of links associated with them. This is familiar from the way airline companies organize their flight schedules using a "hub and spoke" algorithm derived from network theory. For example, Dallas is a major hub for American Airlines, so in order to fly from almost anywhere in the western half of the United States to New York City on American, you have to go through Dallas. This greater degree of clustering because of this hub structure means that small-world networks tend to contain modular subnetworks, called *cliques*, which themselves have high connectivity within them so that almost any two nodes are connected. These sorts of general properties, which are characteristic of social networks, lead to the result that the shortest path between nodes is on average a relatively small number and that this number is essentially independent of the size of the population, so that the six degrees of separation is approximately the same across all communities. Furthermore, it turns out that the modular structure is typically self-similar so that many characteristics of small-world networks satisfy power law scaling.

Steve Strogatz is an eclectic applied mathematician at Cornell University who uses ideas from nonlinear dynamics and complexity theory to analyze and explain a broad range of fascinating problems. For example, he has done some lovely work showing how crickets, cicadas, and fireflies synchronize their behaviors and more recently extended it to show why London's Millennium Bridge was dysfunctional.[11] This latter problem has some interesting lessons for the science of cities, and I want to digress to explain it.

As part of Britain's celebration of the millennium, it was decided to build a new pedestrian bridge across the River Thames connecting landmark sites such as the Tate Modern Gallery and Shakespeare's Globe Theatre on the south bank to St. Paul's Cathedral and the City of London, London's financial

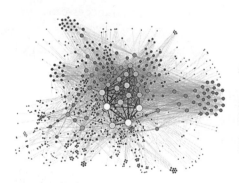

Left: An example of a social network showing *nodes* of individuals connected via single *links*. Note that connections between some individuals require two or more links and that some individuals act as hubs with an overabundance of links. Right: Social networks typically have modular subunits of closely interacting individuals, such as families, groups of very close friends, or workers.

center, on the north bank. The competition for its design was won by some of the usual suspects: the distinguished architect Lord Norman Foster—whom I mentioned earlier as the lead designer of the weird square city of Masdar in the Arabian desert—aided by the well-known sculptor Sir Anthony Caro and the engineering firm of Arup. It's a lovely design and a wonderful addition to London. Surprisingly, it is the only purely pedestrian bridge joining the two halves of the city, and walking across it at any time of day is an exhilarating experience whether you are going to St. Paul's, the Tate, the Globe, or none of the above. Before it opened the designers referred to it as "a pure expression of engineering structure," comparing its design to "a blade of light," with others calling it "an absolute statement of our capabilities at the beginning of the 21st century."

Its opening day on June 10, 2000, was a huge success, with ninety thousand people crossing it, with up to two thousand on it at any one time. Unfortunately, however, the bridge had to be closed two days later and was not reopened for almost another year and a half because of a serious unforeseen design flaw. It turned out that the motion of people walking across the bridge induced it to sway laterally from side to side and this motion was amplified by an unconscious tendency for at least some people to match their footsteps to the sway, thereby exacerbating the oscillations. This not only made it uncomfortable and unnerving but potentially very dangerous.

This is a classic case of a positive feedback mechanism, often manifested as a *resonance*, which has been well known to physicists and engineers for a very long time. We routinely teach it in elementary physics classes and explain its role in how musical instruments and our vocal cords produce sounds, how lasers work, and even how pushing a child in a swing by matching the frequency of pushes with the natural frequency of the swing (its "resonant frequency") makes it go higher and higher. Effectively, this was what the pedestrians were doing when they were walking across the Millennium Bridge. Their natural collective sway was driving the lateral oscillations of the bridge in synchrony with its natural resonant frequency.

The possibility that bridges may be susceptible to potentially threatening resonances hidden in their structure is a well-known phenomenon, so much so that soldiers were traditionally told to break the regularity of their footsteps when marching across bridges. Modern bridges are designed to ensure that this sort of thing doesn't happen. So how did it happen to a sophisticated bridge built at the end of the twentieth century by leading architects, designers, and engineers with all of the requisite knowledge and computing power available at their fingertips?

It seems that when thinking about possible resonances and oscillations of bridges, only *vertical* motions are considered, while the possibility of *lateral horizontal* motion is generally ignored, which I find astonishing. In their defense, the designers of the Millennium Bridge pointed out that this lateral sway is "a phenomenon of which little was previously known in the engineering world." The cost of the bridge was almost $30 million, with an additional $8 million needed to correct the problem. A little bit of science input beforehand—maybe a Steve Strogatz—might have gone a long way to save quite a bit of cash.

The same sentiment applies to the design and development of cities. The Millennium Bridge debacle, like Brunel's *Great Eastern* flop before it, is a relatively "simple" illustration of how supplementing and integrating traditional approaches, no matter how sophisticated they are, with a broad systemic scientific perspective based on underlying principles in an analytic framework might have prevented significant grief and embarrassment and saved quite a bit of money. Developing and building cities is much more challenging and complex than building a bridge or a ship, but the same argument applies. Knowing and being cognizant of the underlying principles and dynamics, seeing the problem in a broad systemic context, thinking quantitatively and analytically, all need to be integrated with the necessarily dominant focus on detail relevant to the specific problem in order to optimize design and minimize unintended consequences.

Steve Strogatz was an external professor at the Santa Fe Institute when he did his work on small-world networks with Duncan Watts. He has written some excellent popular books on mathematics and nonlinear dynamics and has been a science writer for the *New York Times*.[12] Duncan joined SFI as a postdoctoral fellow upon completing his doctorate at Cornell. This coincided with the beginning of my own engagement there and I had the pleasure of sharing an office with him on the days I spent at the institute. Duncan is now an established scientist in his own right and runs a vibrant group at Microsoft devoted to research on online social networks.

One of Watts's projects has been to verify Milgram's six degrees of separation result using huge amounts of data on e-mail messages sent between people so as to determine how many links are needed to connect any arbitrary two of them. This was important because Milgram's work, which was based on traditional letters sent via the regular postal service, had been heavily criticized for its relative sparseness of data and lack of systematic control.

Milgram is equally famous for his extremely provocative and thought-provoking experiments investigating obedience to authority. Having been strongly influenced by the events of the Holocaust and in particular by the trial in 1961 of Adolf Eichmann, one of its main architects, he devised experiments to show how easily any of us can be persuaded by peer and group pressures to perform acts or make statements in violation of our beliefs and conscience. These, too, were heavily criticized not only on scientific and methodological grounds but because of ethical issues surrounding how the participants were perceived to be duped and the emotional stress that this might have caused. Milgram was a young faculty member at Yale at the time, but moved to Harvard shortly thereafter, where he did his work on six degrees of separation. He did not get tenure at Harvard, partially because of the controversy related to the ethical issues of his experiments, and settled permanently back in New York at City University (CUNY).

Milgram grew up in modest circumstances in New York City, the son of immigrant Jewish bakers whom I would have enjoyed meeting given my chronic addiction to good bread. He was a high school friend of another eminent social psychologist, Philip Zimbardo, who became famous for his "prison experiments" at Stanford in the early 1970s. These were inspired by Milgram's obedience to authority research and demonstrated how otherwise normal people (Stanford students in this case) can be induced to perform sadistic acts when playing the role of a prison guard or to exhibit extreme passivity and depression when playing the role of a prisoner. Zimbardo's research came to prominence following the revelations of prisoner abuse by guards at the Abu Ghraib prison during the Iraq War.[13]

The question as to how and why good people become evil and do bad things—the human analog to Job's dilemma as to why God allows bad things to happen to good people—has been a fundamental paradox of human behavior since we evolved social consciousness. The question of man's place in relationship to himself—the continuing moral and ethical dilemma of good versus evil—can be viewed as a companion question to that of man's place in relationship to the universe. These are both central issues of human existence that have dominated human contemplation from the time *Homo sapiens* became conscious, spawning numerous religions, cultures, and philosophies. Only in very recent times has a perspective inspired by science and "rationality" been brought to bear on these profound questions in the hope of providing a complementary framework for understanding their origins and providing possibly new insights and answers. The provocative work of Milgram and Zimbardo strongly suggests that the conundrum as to why good people can do very bad things originates in peer pressure situations, fear of rejection, and a desire to be part of a group where power and control are conferred on individuals by authority. Zimbardo has become an articulate and vocal advocate for the recognition that this powerful dynamic, which seems to be built into our psyches independent of cultural origins and which has wreaked horrors over the centuries, be explicitly recognized and addressed rather than resorting to our instinctual tendency to put the blame simplistically on individual "bad apples," national characteristics, or cultural norms.

Urban Psychology: The Stresses and Strains of Life in the Big City

Sadly, Milgram died from a heart attack at the relatively young age of fifty-one. He was a major contributor to altering our commonly accepted views of human nature and, in particular, to showing that an individual's actions and behavior are strongly influenced by interaction with their community. His obedience experiments showed that one need not be evil or aberrant to act inhumanely. This relationship between an individual and his or her community naturally led him to the broader issue of the psychological dimension of urban life. In 1970 he published a provocative paper in *Science* titled "The Experience of Living in Cities," which laid the foundations for the newly developing field of urban psychology and which became the central focus of his ensuing interests.[14]

Milgram was very much struck by the psychological harshness of life in the big city. The general perception was that outside of the local environment an individual generally went about his or her business avoiding interaction and involvement, rarely acknowledging other people or events that might engender participation or commitment. So much so that most people are very loath to intervene or even call for help when they are witness to crime, violence, or some other crisis event. He devised a sequence of innovative experiments to investigate the apparent lack of trust, the greater sense of fear and anxiety, and the general lack of civility and politeness that seem to characterize life in large metropolises relative to that in small towns. For instance, he had individual investigators ring doorbells saying that they had misplaced the address of a nearby friend and would like to use the phone. He found that there was a large increase by factors of three to five in the proportion of entries allowed into homes in a small town relative to a large city. Furthermore, 75 percent of the city respondents answered the inquiry by shouting through closed doors or by peering out through peepholes, whereas in small towns 75 percent opened the door.

In a related experiment, Milgram's friend Zimbardo arranged for a car to be left for almost three days near the Bronx campus of New York University and for a similar car to be left for the same period near Stanford University in Palo Alto. For those not familiar, Palo Alto is a highly affluent small town south of San Francisco—a quintessential example of American suburbia, which I happened to be living in at the time of this experiment and so can attest to its rather low-key sleepy ambience. The license plates on both cars were removed and the hoods left open to "encourage" potential vandals. Within twenty-four hours the New York car was stripped of all movable parts and by the end of the three days was reduced to a metal skeleton. A big surprise was that most of the destruction occurred during daylight hours, fully observed by "disinterested" passersby. In contrast, the car in Palo Alto was left untouched.

In conceptualizing this dark psychosocial side of urban life, Milgram borrowed the term "overload" from electrical circuit and systems science theory.

In big cities we are continually bombarded with so many sights, so many sounds, so many "happenings," and so many other people at such a high rate that we are simply unable to process the entire barrage of sensory information. If we tried to respond to every stimulus, our cognitive and psychological circuitry would break down and, in a word, we would blow a fuse just like an overloaded electrical circuit. And sadly, some of us do. Milgram suggested that the kinds of "antisocial" behaviors we perceive and experience in large cities are in fact adaptive responses for coping with the sensory onslaught of city life, implying that without such adaptations we'd all blow our fuses.

I'm sure that the irony in Milgram's observations and speculations concerning the negative sociopsychological consequences of urban overload has not escaped you. The very aspect of city life that I have been extolling as the underlying driver for idea and wealth creation, for innovation, and for its attractiveness, namely, the increased connectivity between people and the concomitant urban buzz so celebrated by Jane Jacobs, is here revealed as one of the inevitable prices we have to pay for the benefits larger cities confer. This is another dimension of "the good, the bad, and the ugly" consequences of increased connectivity with its resulting superlinear scaling as city size increases. Systematically having more per capita not only means higher wages, more patents, and more restaurants, greater opportunity, more social activity, and a greater buzz, but also more crime and disease—and living with greater stress, anxiety, and fear, and with less trust and civility. As I will discuss shortly, much of this can be traced to the increasing pace of life in larger cities, which is a predictable consequence of the network theory.

6. HOW MANY CLOSE FRIENDS DO YOU REALLY HAVE? DUNBAR AND HIS NUMBERS

In the previous couple of sections I gave a broad overview of some of the general characteristics of social interactions in cities. This provides a natural segue to discussing how the systematic self-similarity and fractal-like geometry of urban infrastructural networks is mirrored in social networks. First, it is worth repeating that the six degrees of separation phenomenon tells us that despite appearances we are considerably more closely connected to one another than most of us are aware. Furthermore, small-world networks typically manifest power law scaling reflecting underlying self-similar characteristics and a preponderance of cliques of individuals. Such modular group structures are a central feature of our social life, whether they are our family, our circle of close friends, our department at work, our neighborhood, or our entire city.

Understanding and deconstructing the hierarchy of social group structures has been a major focus of attention in sociology and anthropology for more than fifty years, but it was only in the last twenty or so that some of their quantitative features have become apparent. Some of this has been driven by the work of the evolutionary psychologist Robin Dunbar and his collaborators, who proposed that an average individual's entire social network can be deconstructed into a hierarchical sequence of discrete nested clusters whose sizes follow a surprisingly regular pattern.[15] The size of the group at each level systematically increases as one progresses up the hierarchy from, say, family to city, while the strength of the bonding between people within the groups systematically decreases. So, for example, most people have a very strong connection with their immediate family members but only a very weak one with the bus driver or members of the city council.

Partly inspired by work on social primate communities and partly by anthropological studies of human societies ranging from hunter-gatherers to modern corporations, Dunbar discovered that this hierarchy appeared to have a surprisingly regular mathematical structure obeying very simple scaling rules reminiscent of self-similar fractal-like behavior. He and his collaborators found that at the lowest level of the hierarchy the number of people with whom the average individual has his or her strongest relationships is, at any one time, only about five. These are the people we are closest to and care most deeply about; they are usually family members—parents, children, or spouses—but they could be extremely close friends or partners. In surveys designed to measure the size of this core social group one of its defining characteristics was "the set of individuals from whom the respondent would seek personal advice or help in times of severe emotional and financial distress."

The next level up contains those you usually refer to as close friends with whom you enjoy spending meaningful time and might still turn to in time of need even if they are not on as intimate terms with you as your inner circle. This typically comprises around fifteen people. In the level above this are people you might still call friends though you would only rarely invite them to dinner but would likely invite them to a party or gathering. This might consist of coworkers, neighbors down the street, or relatives you don't see very often. There are typically about fifty such people in this group.

The next level pretty much defines the limit of your social horizon as far as personal interactions are concerned and consists of people you might refer to as "casual friends"—you know their names and remain in social contact with them. This group typically comprises about 150 people. It is this number that is usually referred to as the Dunbar number that has gained a certain degree of attention in the popular media.

You will notice that the sequence of numbers that quantify the magnitude of these successive levels of the group hierarchy—5, 15, 50, 150—are sequentially related to each other by an approximately constant scaling factor of about three. This regularity is the familiar fractal-like pattern we saw not only in the network hierarchy of our own circulatory and respiratory systems but also in transport patterns in cities. In addition to the actual flows in these networks, the major geometric difference between them is the value of the *branching*

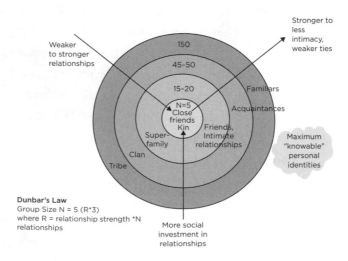

A schematic of the sequence of Dunbar's numbers reflecting a fractal-like hierarchy in the modular structure of social interactions: note that the interaction strength decreases inversely with the size of the modular group.

ratio—the number of units, people in this case, at one level of the hierarchy relative to the next. There is evidence that in social networks this pattern with a branching ratio of three persists beyond the 150 level to groups having sizes of approximately 500, 1,500, and so on. The precise value of these numbers should not be taken too seriously because there is significant variance in the data. The important point for our purposes is that viewed through a coarse-grained lens, social networks exhibit an approximate fractal-like pattern and that this seems to hold true across a broad spectrum of different social organizations. Even though this pattern remains approximately static, the individual members of the network may change with time or may move from one level to another as you become closer or more distant in your relationship with them. For instance, a parent may move out of your inner circle and be replaced by a spouse or a close friend, or you might meet someone casually at a party who subsequently becomes part of your 150. Regardless of such changes, the generic structure of the network with four to six people comprising your core group, and the nested group structure whose discrete size increases by factors of about three up to 150 or so, remains intact.

The number of around 150 represents the maximum number of individuals that a typical person can still keep track of and consider casual friends and therefore members of his or her ongoing social network. Consequently, this is the approximate size of a group in which all individuals still know one another sufficiently well for the group to remain coherent and maintain ongoing social relationships. Dunbar found many such examples of functioning social units whose size congregated around this magic number, ranging from bands of hunter-gatherers to army companies in the Roman Empire, in sixteenth-century Spain, and in the twentieth-century Soviet Union.

He speculated that this apparent universality has its origins in the evolution of the cognitive structure of the brain: we simply do not have the computational capacity to manage social relationships effectively beyond this size. This suggests that increasing the group size beyond this number will result in significantly less social stability, coherence, and connectivity, ultimately leading to its disintegration. For situations where group identity and cohesiveness are perceived as central for the group to function successfully, recognizing this limitation and the broader implications of social network structure is clearly of great importance. This is especially true in situations where stability, knowledge of other individuals, and social relationships are integral to performance. Companies, the military, government administrations and bureaucracies, universities, and research organizations are among the many institutions where this kind of information and this way of thinking could potentially be beneficial in improving performance, productivity, and the general well-being of all members of the organization.

Dunbar originally estimated this number using a simple scaling argument by extrapolating from the group size of primate communities to human society. He and his colleagues had discovered that the group size of social primates scales with the neocortex volume of their brains as a classic power law. The neocortex is the most sophisticated part of the brain, controlling and processing higher functions such as sensory perception, the generation of motor commands, spatial reasoning, conscious thought and language, and therefore the computational capacity to participate in complex social relationships. This presumed connection between brain size and the ability to form social groups is called the *social brain hypothesis*. Dunbar went much further by suggesting that this relationship is causal in that human intelligence evolved primarily as a response to the challenge of forming large and complex social groups rather than the usual explanation that it is a direct consequence of meeting ecological challenges.[16] Regardless of causation, he used the correlation with brain size to estimate the number 150 as the idealized size of analogous human social groups.

Because brain size scales almost linearly with metabolic rate, one could just as well have used metabolic rates of humans relative to primates, rather than the relative volumes of their neocortices, to determine the idealized size of human social groups. The same rough estimate of 150 would be obtained, leading one to argue, contrary to Dunbar, that evolutionarily this number is related to ecological challenges involving resources and metabolism rather than to cognitive challenges of group formation. One cannot distinguish between these two hypotheses—that is, whether group structure evolved in response to social rather than metabolic ecological pressures—without having an underlying theory to guide and strengthen the analysis and provide further testable predictions. This simply highlights the classic dilemma as to what extent, if any, causation can be inferred from correlation: just because two things are correlated does not mean that one caused the other.

Having said this, I must confess that I like the general idea that social network structure has its origins in evolutionary pressures whether social or environmental, because it implies that the self-similar fractal nature of social networks is encoded in our DNA and therefore in the neural system of our brains. Furthermore, because the geometry of white and gray matter in our brains, which forms the neural circuitry responsible for all of our cognitive functions, is itself a fractal-like hierarchical network, this suggests that the hidden fractal nature of social networks is actually a representation of the physical structure of our brains. This speculation can be taken one step further by invoking the idea that the structure and organization of cities are determined by the structure and dynamics of social networks, in which case the universal fractality of cities can be viewed as a projection of the universal fractality of social networks.

Putting all of this together we are led to the outrageous speculation that cities are effectively a scaled representation of the structure of the human brain. It's a pretty wild conjecture, but it graphically incorporates the idea that there is a universal character to cities. In a nutshell: cities are a representation of how people interact with one another and this is encoded in our neural networks and therefore in the structure and organization of our brains. In a curious way, this is perhaps more than just metaphor and would mean that a map of a city representing its physical and socioeconomic flows is a nonlinear representation of the geometry and flows in the neural network of our brains.

7. WORDS AND CITIES

Unlike in biology, surprisingly little attention had been paid to scaling laws for cities, urban systems, or companies prior to our investigations. This may be because few people suspected that such complex, historically contingent man-made systems as these would manifest any sort of systematic quantitative regularity. In addition, there is much less of a tradition of this kind of modeling and of confrontation of theory with data in urban studies than in either biology or physics. There was, however, one major exception to this and that is a famous scaling law known as Zipf's law for the ranking of cities in terms of their population size. This is shown graphically in Figure 39.

It's an intriguing observation: in its simplest form, it states that the rank order of a city is inversely proportional to its population size. Thus, the largest city in an urban system should be about twice the size of the second largest, three times the size of the third largest, four times the size of the fourth largest, and so on. So, for example, in the 2010 census, the biggest city in the United States was New York with a population of 8,491,079. According to Zipf's law, the second largest city, Los Angeles, should have a population of approximately one half of this, namely 4,245,539, the third largest, Chicago, should have a population of about one third, or 2,830,359, and Houston, the fourth largest, should have a population of about one quarter, namely 2,122,769, and so on. The actual numbers were: Los Angeles, 3,928,864; Chicago, 2,722,389; and Houston, 2,239,558, all in reasonably good agreement with Zipf's law to within less than 7 percent.

Zipf's law owes its name to the Harvard linguist George Kingsley Zipf, who popularized it through his fascinating book *Human Behavior and the Principle of Least Effort,* published in 1949.[17] He first enunciated his law in 1935, not for cities but for the frequency of word use in languages. As originally stated, the law says that the frequency of occurrence of any word in a corpus of written text such as all of Shakespeare's plays, the Bible, or even this book is inversely proportional to its rank in the frequency table. Thus, the most frequent word occurs approximately twice as often as the second most frequent, three times as often as the third, and so on, as shown in Figure 40. For example, analysis of English texts shows that the most frequent word is, not surprisingly, "the," which accounts for roughly 7 percent of all words used, while the second-place word "of" accounts for about half as many, namely 3.5 percent of all words, followed by "and" with about a third as many, namely 2.3 percent, and so on.

Even more mystifying is that this same law is valid across an astonishing array of examples, including the rank size distributions of ships, trees, sand particles, meteorites, oil fields, file sizes of Internet traffic, and many more. Figure 41 shows how the distribution of company sizes follows this law. Given

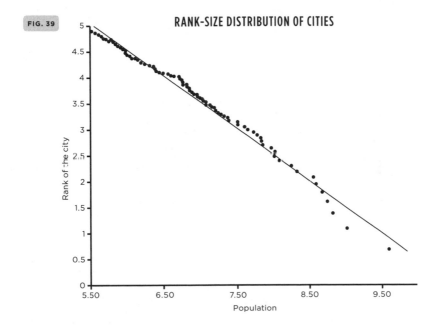

FIG. 39

RANK-SIZE DISTRIBUTION OF CITIES

Rank of the city

Population

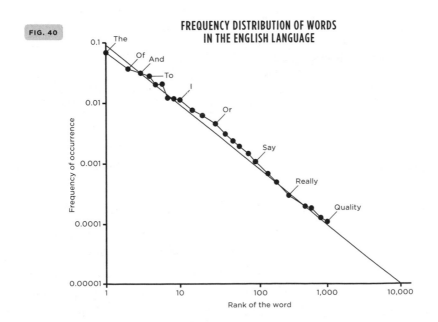

FIG. 40

**FREQUENCY DISTRIBUTION OF WORDS
IN THE ENGLISH LANGUAGE**

Frequency of occurrence

Rank of the word

The
Of And
To
I
Or
Say
Really
Quality

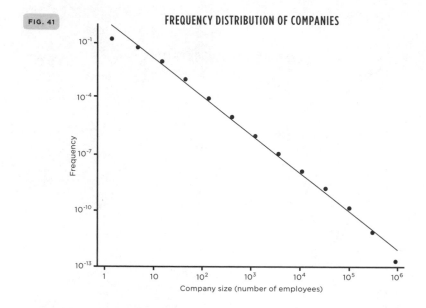

FIG. 41

FREQUENCY DISTRIBUTION OF COMPANIES

Frequency

Company size (number of employees)

Preceding and this page: (39) Rank-size distributions of cities in the United States: their rank is plotted on the vertical axis and their population plotted on the horizontal. Notice that in both these cases there are large deviations for the most frequent entities ("the" for words and New York for cities). (40) Zipf's law of rank-size distributions of words in the English language: the frequency of occurrence of words is plotted on the vertical axis and their rank plotted on the horizontal. (41) Rank-size distributions of companies in the United States: as in (39), their rank is plotted on the vertical axis and their size (the number of employees) plotted on the horizontal.

its amazing generality and some of its implications, Zipf's law has taken on a curious mystique among many researchers and writers whose imagination has been caught by its surprising simplicity. Zipf and many others following him have pondered its origins, but no generally agreed-upon explanation has yet emerged.

In economics, Zipf's law actually predates Zipf. Much earlier it had been discovered by the influential Italian economist Vilfredo Pareto, who expressed it as a frequency distribution of incomes in a population rather than in terms of their ranking. This distribution, which is valid for many other economic metrics like income, wealth, and the size of companies, follows a simple power law with an exponent of approximately -2. When expressed in terms of rankings, this exponent corresponds to Zipf's law. It quantifies the obvious economic fact that there are very few very wealthy people or large institutions but an enormous number of very poor or small ones. Pareto's law, or the Pareto principle, has often been loosely stated in the form of the so-called 80/20 rule, in which the richest 20 percent of a population controls 80 percent of the total income, which is approximately true for the entire globe. Similarly, roughly 80 percent of a company's profits come from 20 percent of its customers, as do 80 percent of its complaints. This asymmetry, in which there are only a very small number of very large entities but a huge number of very small ones, is characteristic of Zipf's law. For instance, you need only about 20 percent of the dictionary to understand 80 percent of literature and roughly 80 percent of the population lives in the top 20 percent of the largest cities. Everything in between approximately follows the inverse proportionately according to the power law.

Despite their generality, there are often large deviations from Zipf's and Pareto's "laws" and it would be naive to conclude that there is some fixed universal principle that determines the precise nature of these frequency distributions without considering them in the much broader context of many other dynamical processes. For instance, just knowing that the size of cities in urban systems follows a Zipf pattern is hardly sufficient input for developing a principled comprehensive science of cities. At a minimum, all the other scaling laws for the entire spectrum of urban activity that I have already alluded to encompassing flows of energy, resources, and information, beyond just knowing size frequency distribution, are needed. Although these distributions are indeed fascinating, I take a much more modest view of them as another phenomenological scaling law among many others but without special fundamental significance.

Nevertheless, the fact that these Zipf-like distributions are found across such a diverse set of phenomena suggests that they express some general systemic property that is independent of the character and detailed dynamics of the specific entities under consideration. This is reminiscent of the ubiquitous generality of the "bell-curve" distribution used for describing statistical variations around some average value. Technically this is called a *Gaussian* or *normal* distribution and arises mathematically whenever a sequence of events or entities, whatever they are, are randomly distributed, uncorrelated, and independent of one another. So, for example, the average height of men in the United States is about five feet ten inches (1.77 meters) and the frequency distribution of their heights around this mean value—that is, how many men there are of a given height—closely follows a classic Gaussian bell-curve distribution. This tells us what the likelihood is of someone being a specific height. Gaussian statistics are used across all of science, technology, economics, and finance to assign statistical probabilities of various events happening such as in weather forecasting or for drawing conclusions from polling surveys. However, it is sometimes forgotten that these probability estimates are based on the assumption that individual "events," whether comparing today's temperature with the historical record, or one person's height with another's, are independent of one another and can therefore be considered uncorrelated.

The canonical Gaussian bell curve is so pervasive and so taken for granted that it is generally assumed without much thought that that's how "everything" is distributed. Consequently, power law distributions like those of Zipf and Pareto barely saw the light of day. It was natural to assume that cities, incomes, and words would be randomly distributed following the classic bell curve. If that were so, it would predict that there are far fewer very large cities, very large companies, and very rich people, and far fewer common words than there actually are, because all of these follow power law distributions that have much longer tails, meaning that there are many more rare events than would be expected had they been random and obeyed Gaussian statistics. This difference is sometimes characterized by saying that power laws have "fat tails." Clearly, words in a book are correlated and not random because they have to form meaningful sentences, as are cities, because they are part of a unified urban system. It's therefore not so surprising that these distributions are not Gaussian.

Many of the most interesting phenomena that we have touched upon fall into this category, including the occurrence of disasters such as earthquakes, financial market crashes, and forest fires. All of these have fat-tail distributions with many more rare events, such as enormous earthquakes, large market crashes, and raging forest fires, than would have been predicted by assuming that they were random events following a classic Gaussian distribution. Furthermore, because these are approximately self-similar processes, the same dynamics occur at all scales. Thus the same generic mechanism that leads to a small adjustment in a financial market is at play when that market suffers a major crash. This is in marked contrast to the random nature implicit in Gaussian statistics, where events at different scales are presumed to be independent and uncorrelated. Ironically, economists and financial analysts traditionally use Gaussian statistics in their analysis, ignoring the preponderance of fat tails and therefore correlations. Caveat emptor!

Given the connection to the occurrence of rare events, it is not surprising that power law distributions and models based on fractal-like behavior have been gaining greater currency in the burgeoning field of *risk management*. A common metric used to address risk, whether in financial markets, industrial project failures, legal liabilities, credit loans, accidents, earthquakes, fire, terrorism, and so on, is the *composite risk index,* which is defined as the impact of the risk event multiplied by the probability of its occurrence. The impact is usually expressed in terms of the dollar cost of the estimated damage, and the probability by some version of a power law. As society becomes ever more complex and risk averse, developing a science of risk is becoming of increasing importance, so understanding fat tails and rare events is an area of increasing interest in both the academic and corporate communities.

8. THE FRACTAL CITY: INTEGRATING THE SOCIAL WITH THE PHYSICAL

The two dominant components that constitute a city, its physical infrastructure and its socioeconomic activity, can both be conceptualized as approximately self-similar fractal-like network structures. Fractals are often the result of an evolutionary process that tends toward optimizing specific features, such as ensuring that all cells in an organism or all people in a city are supplied by energy and information, or maximizing efficiency by minimizing transportation times or times for accomplishing tasks with minimal energy. Less obvious is what is being optimized in social networks. There is, for instance, no satisfactory explanation based on underlying principles for understanding the hierarchical structure observed by Dunbar, or for the origin of his sequence of numbers. Even if the social brain hypothesis is correct, it doesn't explain either the origin of the fractal nature of social groups or where the number 150 comes from. There are hints that such generic properties follow from the conjecture made earlier that self-interest—that is, the desire of all individuals and companies to maximize their assets and income—coupled with the concept of maximal filling of social space are the underlying driving forces. There is certainly a great deal yet to be done in constructing a quantitative theory of social networks, and many exciting challenges await future investigation.

All socioeconomic activity in cities involves the interaction between people. Employment, wealth creation, innovation and ideas, the spread of infectious diseases, health care, crime, policing, education, entertainment, and indeed, all of the pursuits that characterize modern *Homo sapiens* and are emblematic of urban life are sustained and generated by the continual exchange of information, goods, and money *between* people. The job of the city is to facilitate and enhance this process by providing the appropriate infrastructure such as parks, restaurants, cafés, sports stadiums, cinemas, theaters, public squares, plazas, office buildings, and meeting halls to encourage and increase social connectivity.

Consequently, all of the socioeconomic metrics that reflect such activity and which were discussed earlier when reviewing urban scaling laws are proportional to the number of links, or interactions that take place, between people within the city. If it were possible for everyone to interact with everyone else so that over the period of a year, for example, each person connects meaningfully with every other person in the city, then the total number of interactions between people could be easily calculated from a simple formula: it is the total number of people in the city multiplied by the total number of people that any individual can connect to in the city. This last number is just the total number of people minus one. For instance, if you are one of ten people in a group, you can connect to only nine others. In addition, you have to divide the answer by two because you don't count the link between you and another person as different from the link between that other person and you. They are symmetric and one and the same.

Thus the total possible number of pair-wise links between people in a city is given by the total number of people in the city multiplied by the total number minus one—all divided by two. This may seem a bit of a mouthful but it's actually quite simple, so let me explain with some examples.

If there are just two people, you and your partner for instance, then according to this formula the total number of links is $2 \times (2-1) \div 2 = 2 \times 1 \div 2 = 1$, which is obviously correct: there's just a single link connecting the two of you. Let's suppose another person is added to form a sort of ménage à trois, then with three people the formula says there are $3 \times (3-2) \div 2 = 3 \times 2 \div 2 = 3$ independent pair-wise interactions, which is easily seen to be correct: A with B, B with C, and C with A. Now increase the group size to four, then the number of links becomes $4 \times 3 \div 2 = 6$, which is double what it was when there were just three people even though only a single person has been added. Suppose we now double the group size from four to eight, then the number of links increases from six to $8 \times 7 \div 2 = 28$, which is more than a factor of four larger. If it is doubled again to sixteen, then the number of links likewise increases by another factor of approximately four from 28 to 120. In fact, each time the size is doubled, the number of links increases approximately fourfold. The lesson is clear: *the number of links between people increases much faster than the increase in the number of people in the group and, to a very good approximation, is given by just one half of the square of the number of people in the group.*

This simple nonlinear quadratic relationship between the maximal number of links between people and the size of the group has all sorts of very interesting social consequences. For instance, my wife, Jacqueline, particularly enjoys dinner parties if a single conversation can be sustained by the entire group, so she is reluctant to participate in dinner parties larger than six. With six people there are $6 \times 5 \div 2 = 15$ possible pair-wise independent conversations that have to be "suppressed" for a single collective one to emerge and be maintained. This is just about possible, and it's tempting to speculate that it's because the number of other guests, five, corresponds to Dunbar's number for the group size of an average individual's inner circle. With ten at the table, there are a whopping forty-five such dyadic possibilities, which inevitably leads to a balkanization of the group as it disintegrates into two, three, or more separate conversations. Of course, many people prefer this modality but it is worth remembering that if you want a certain kind of group intimacy, having more than about six people is going to make it quite a challenge.

In a similar vein, the size of my grandparents' family was relatively large as was typical of most families until quite recently. It consisted of ten people: eight children and two adults. Consequently, there were forty-five concurrent relationships across a spectrum of ages and personalities, creating a diversity of interactions. If these loosely followed a Dunbar pattern where each child was strongly connected to two or three of their siblings in addition to their parents, not everyone could love everyone else equally, which is usually the case. On the other hand, my immediate nuclear family consists of my wife and two children, a group of just four people with therefore only six separate relationships. Each of my children therefore had to deal with only five different relationships, compared with almost ten times as many, forty-four to be precise, that my dear old mum had to handle, even though her family was only two and a half times larger. Without casting judgment on the pros and cons of small versus large families, it's hard not to be struck by this enormous difference in family dynamics and speculate about the profound psychosocial consequences that such a change must have generated during the twentieth century as the size of families shrank.

Now let's go back and take a look at how this plays out in an entire city. If it were possible for everyone to interact meaningfully with everyone else as in one great big happy family, then the above argument would imply that all socioeconomic metrics should scale with the *square* of the population size. This would mean an exponent of 2, which is certainly superlinear (it's bigger than 1), but significantly larger than 1.15. However, this represents the extreme and totally unrealistic case where the entire population is in a frenzied state of continuous and complete interaction with itself, being churned about much like raisins or nuts in cake dough driven by a super-high-speed electric mixer. This is clearly impossible—and certainly not desirable. Even in a modest-size city of only 200,000 people there are roughly 20 billion possible relationships, and even if each person devoted just one minute a year to each relationship, they would have to spend their entire waking life relating to other people, leaving no time for anything else. Imagine extending that to a New York or Tokyo.

There is also the constraint of the Dunbar number, according to which we even have difficulty sustaining any sort of meaningful relationship with more than about 150 people, let alone a couple of hundred thousand or several million. It is this restriction to a relatively small number of interactions that drives the superlinear exponent to be significantly smaller than its maximum possible value of 2.

This exercise shows that there is a natural explanation for why social connectivity and therefore socioeconomic quantities scale *superlinearly* with population size. *Socioeconomic quantities are the sum of the interactions or links between people* and therefore depend on how correlated they are. In the extreme case when everyone is interacting with everyone else we saw that this leads to a superlinear power law whose exponent is 2. However, in reality there are significant constraints on the intensity and magnitude of how many people an individual can interact with, and this drastically reduces the value of the exponent to be less than 2.

The underlying reason we are limited in the number and rate of interactions we can sustain with other people in the city is rooted in the hidden constraints imposed by space and time. We simply can't be in all places at all times. An obvious though subtle fundamental constraint is that all of our interactions and relationships necessarily take place in a physical setting, whether in houses, offices, theaters, stores, or on the street. No matter how you communicate with other people, even if it's via satellite at the speed of light on your cell phone, or buying all of your goods and supplies on the Internet, you have to be *somewhere*. You may be sitting in a room inside a building, standing or walking in the street, or riding the subway or bus, but wherever you are, you are necessarily in some physical *place*. I emphasize this obvious fact because the development of the Internet and the rapidly evolving field of network science have spawned an unfortunate and misleading impression that social networks are somehow suspended in space as if no longer bound by the constraints of gravity and the nasty encumbrances of the physical world. This is exemplified by the conventional images of social networks as hubs and links that I introduced earlier and which is illustrated on page 298. These topological representations of social interactions are abstractions inspired by network theory and portray individuals as ephemeral beings suspended in hyperspace devoid of physicality, rather than as real people sitting in kitchens, coffeehouses, offices, or buses talking to one another. It is surprising that despite the enormous amount of recent research on the structure, organization, and mathematics of social networks, almost none acknowledge, let alone embrace, their direct and necessary coupling to the grungy reality of the physical world. And that physical world is primarily that of the urban environment.

And this is where the infrastructure of a city comes in: as I have previously emphasized, its role in a city is to enhance and facilitate social interaction. Which brings up another obvious point: not only must we be located somewhere in the city, but equally important for at least some of the time, we also have to be going from that someplace to some other place. People in cities cannot be static; their mobility is essential for its viability and vitality. We are persistently on the move from one place to another whether going to an office or factory for work, returning home to sleep and eat, going to a store to buy food, or going to a theater for entertainment. When viewed over timescales of days and weeks, people in cities are effectively in a state of continuous motion, which is inextricably intertwined with, and constrained by, its transport system. Mobility and social interaction, both so essential for the successful operation of a city, bring together the constraints of space and time—you cannot remain static and you do have to be somewhere—which interweave the structure, organization, and dynamics of social and infrastructural networks.

In chapters 3 and 4 the origin of universal scaling laws in biology was explained and a big-picture theory for understanding many aspects of living systems developed based on generic mathematical properties of networks. In a similar fashion, the ideas invoking general properties of social and infrastructural networks in cities have to be translated into mathematics in order to develop an analogous big-picture theory of cities from which urban scaling laws, for example, can be derived. In what follows, I will try to explain how this is accomplished without resorting to fancy technical details by focusing on the conceptual framework and the essential features of what is involved.

In this spirit, individuals are considered to be the "invariant terminal units" of social networks, meaning that on average each person operates in roughly the same amount of social and physical space in a city. This is in keeping with the implications of a "universal" Dunbar number and the space-time limitations on mobile activity in cities that we just discussed. Recall that the physical space in which we operate is spanned by space-filling fractal networks, such as roads and utility lines that service infrastructural terminal units such as houses, stores, and office buildings where we reside, work, and interact, and between which we also have to move. The integration of these two kinds of networks, namely, the requirement that socioeconomic interaction represented by space-filling fractal-like *social networks* must be anchored to the physicality of a city as represented by space-filling fractal-like *infrastructural networks*, determines the number of interactions an average urban dweller can sustain in a city. And as discussed earlier, it is this number that determines how socioeconomic activity scales with population size.

The biological metaphor of the city as a living organism derives primarily from its being perceived in terms of its physicality. This is most apparent in the networks that carry energy and resources in the form of electricity, gas, water, cars, trucks, and people, and it is this component of cities that is the close analog to the networks that proliferate in biology such as our cardiovascular and respiratory systems or the vasculature of plants and trees. Combining the ideas of space filling, invariant terminal units, and optimization (minimizing travel times and energy use, for example) results in these networks also being fractal-like with infrastructural metrics scaling as power laws with sublinear exponents indicative of economies of scale obeying the 15 percent rule.

When these constraints on the mobility and physical interaction space of people in cities are imposed on the structure of social networks, an important and far-reaching result emerges: the number of interactions with other people in a city that an average person maintains scales *inversely* to the way that the degree of infrastructure scales with city size. In other words, the degree to which the scaling of infrastructure and energy use is *sublinear* is predicted to be the same as the degree to which the scaling of the number of an average individual's social interactions is *superlinear*. Consequently, the exponent controlling social interactions, and therefore all socioeconomic metrics—the universal 15 percent rule for how the good, the bad, and the ugly scale with city size—is *bigger than 1* (1.15) to the same degree that the exponent controlling infrastructure and flows of energy and resources is *less than 1* (0.85), as observed in the data. Pictorially, the degree to which all of the slopes in Figures 34–38 exceed 1 is the same as the degree to which they are less than 1 in Figure 33.

In this network scaling sense, it is no accident therefore that the physical and the social mirror each other, so much so that we can think of the *physical city*—with its networks of buildings, roads, and electrical, gas, and water lines—as an inverse nonlinear representation of the *socioeconomic city*—with its networks of social interactions. The city is indeed the people.

The approximate 15 percent increase in social interactions and therefore in socioeconomic metrics such as income, patents, and crime generated with every doubling of city size can be interpreted as a bonus, or payoff, arising from the 15 percent savings in physical infrastructure and energy use. The systematic increase in social interaction is the essential driver of socioeconomic activity in cities: wealth creation, innovation, violent crime, and a greater sense of buzz and opportunity are all propagated and enhanced through social networks and greater interpersonal interaction.

But this can equally well be interpreted in a complementary way by viewing cities as catalytic facilitators or crucibles for social chemistry in which the increase in social interactions enhances creativity, innovation, and opportunity whose dividend is an increase in infrastructural economies of scale. Just as raising the temperature of a gas or liquid increases the rate in the number of collisions between molecules, so increasing the size of a city increases the rate and number of interactions between its citizens. Metaphorically speaking, increasing the size of a city can therefore be thought of as raising its temperature. In this sense, New York, London, Rio, and Shanghai are truly hot cities, especially compared with Santa Fe where I live, and the proverbial image of a "melting pot," originally applied to New York City, is an apt expression of this metaphor.

In this vein, the hallmark of a successful city, regardless of size, is that it provides the physical ambience, culture, and landscape for facilitating and enhancing *diverse* social interactivity by its attractive cityscapes and gathering places, user-friendly and accessible transport and communication systems, and a supportive sense of community, commerce, culture, commitment, and leadership. Cities are effectively machines for stimulating and integrating the continuous positive feedback dynamics between the physical and the social, each multiplicatively enhancing the other. Indeed, as will be explained in the next chapter, it is this multiplicative mechanism that is ultimately responsible for the open-ended exponential growth that is characteristic of economies and cities, and to which we have become addicted, if not enslaved.

It is perhaps not so surprising that there is a correlation between increased social interaction, socioeconomic activity, and greater economies of scale. What is surprising, however, is that this pivotal interrelationship follows such simple mathematical rules that can be expressed in an elegant universal form: *the sublinearity of infrastructure and energy use is the exact inverse of the super-linearity of socioeconomic activity.* Consequently, to the same 15 percent degree, the bigger the city the *more* each person earns, creates, innovates, and interacts—and the more each person experiences crime, disease, entertainment, and opportunity—and all of this at a cost that requires *less* infrastructure and energy for each of them. This is the genius of the city. No wonder so many people are drawn to them.

This tight coupling between enhanced socioeconomic activity and infra-structural economies of scale, encapsulated in the inverse relationship between the two, has its mechanistic origins in an analogous inverse relationship between their underlying network structures. Although social and physical networks share common generic features such as being fractal-like, space filling, and having invariant terminal units, there are some essential differences. A major one that has huge consequences is the way in which the sizes and flows *within* the networks scale as one progresses down through their fractal-like hierarchies.[18]

In infrastructural network systems, such as transport, water, gas, electrical, and sewer lines, the sizes and flows in the pipes, cables, roads, et cetera, systematically *increase* from terminal units that service individual houses and buildings up through the network to major conduits and arteries that connect to some central source, place, or repository, in much the same way that the sizes and flows in our cardiovascular system systematically *increase* from our capillaries up to our aorta and thence to our heart. This is the origin of *sublinear* scaling and economies of scale. In contrast, in socioeconomic networks—those responsible for wealth creation, innovation, crime, and so forth—the inverse behavior is at play as was explained when we discussed the hierarchy of Dunbar numbers. The strengths of social interaction and the flows of information exchange are greatest between terminal units (that is, between individuals) and systematically *decrease* up the hierarchy of group structures from families and other groups to increasingly larger clusters, leading to *superlinear* scaling, increasing returns, and an accelerating pace of life.

8

CONSEQUENCES AND PREDICTIONS

From Mobility and the Pace of Life to Social Connectivity,
Diversity, Metabolism, and Growth

I n this chapter I am going to explore some of the consequences of our big-picture theory of cities developed in the previous chapter. Even though the theory remains a work in progress, I will show from a few examples that much of what we experience in cities, and more generally in our daily participation in socioeconomic activity, is embodied in this quantitative framework. In this respect it should be viewed as complementary to traditional social science and economic theories whose character is typically more qualitative, more local-ized, more based on narrative, less analytic, and less mechanistic. Critical from the physics perspective is to make quantitative predictions that are sub-sequently confronted with data, and in some cases *big data*.

The theory already passes its first test, namely, to provide a natural expla-nation for the origin of the many scaling laws reviewed earlier. It also explains their universal character across diverse metrics and urban systems, as well as the self-similar and fractal nature of cities. Furthermore, the analysis con-denses and explains an enormous amount of data implicitly encapsulated in the many scaling laws that cover much of what we can measure about the structure and organization of cities, including the socioeconomic life of their citizens.

Although this represents a significant accomplishment, it is just a beginning. It provides a point of departure for extending the theory to a broad range of problems pertinent not just to cities and urbanization but also to economies and fundamental questions of growth, innovation, and sustainability. An important component of this is to test and confirm the theory by confronting new predictions with data, such as for metrics quantifying the social connectivity between people, their movement within cities, and the attractiveness of specific locations. For example, how many people visit a given place in a city? How often do they go there and from how far away? What is the diversity distribution of occupations and businesses? How many eye doctors, criminal lawyers, shop assistants, computer programmers, or beauticians can we expect to find in a city? Which of these grow and which contract? What is the origin of the accelerating pace of life and open-ended growth? And finally the key question addressed in chapter 10, can any of this be sustainable?

1. THE INCREASING PACE OF LIFE

I showed in the previous chapter that increasing a city's size generates more social interactions per capita while at the same time, and to the same degree, costing less. This dynamic is manifested in the extraordinary enhancement in innovation, creativity, and open-ended growth as city size increases. At the same time, it also leads to another profound feature of modern life, namely that its pace seems to be continually speeding up.

As was discussed earlier, if we think of social networks as layered hierarchies beginning with individuals as the "invariant terminal units" and progressing systematically up through modular groupings of *increasing* size from families, close friends, and colleagues to acquaintances, working clusters, and organizations, then the corresponding strengths of interaction and amounts of information exchanged at each level systematically *decrease*, resulting in *superlinear* scaling. A typical person is considerably more connected and spends much more time exchanging significantly more information with other individuals, whether they're family, close friends, or colleagues, than with much larger, more anonymous collectives such as the administrations of their city or workplace.

The opposite hierarchy pertains to infrastructural networks. Sizes and flows systematically *increase* from terminal units (houses and buildings) up through the network, leading to *sublinear* scaling and economies of scale I showed in chapter 3 when considering circulatory and respiratory systems of organisms that a further consequence of this kind of network architecture is the *systematic slowing down of the pace of life* as the size of organisms increases. Bigger animals live longer, have slower heart and respiratory rates, take longer to grow, mature, and produce offspring, and generally live life at a slower pace. Biological time systematically and predictably expands as size increases following the quarter-power scaling laws. A scurrying mouse is in many respects just a revved-up, scaled-down, ponderous elephant.

Knowing the inverse linkage between these two different kinds of networks, it should come as no great surprise that precisely the opposite behavior arises in social networks. Rather than the pace of life systematically decreasing with size, the superlinear dynamics of social networks leads to a systematic *increase in the pace of life:* diseases spread faster, businesses are born and die more often, commerce is transacted more rapidly, and people even walk faster, all following the 15 percent rule. This is the underlying scientific reason why we all sense that life is faster in a New York City than in a Santa Fe and that it has ubiquitously accelerated during our lifetimes as cities and their economies grew.

This effective speeding up of time is an emergent phenomenon generated by the continuous positive feedback mechanisms inherent in social networks in which social interactions beget ever more interactions, ideas stimulate yet more ideas, and wealth creates more wealth as size increases. It is a reflection of the incessant churning that is the very essence of city dynamics and leads to the multiplicative enhancement in social connectivity between people that is manifested as superlinear scaling and the systematic speeding up of socioeconomic time. Just as biological time systematically and predictably *expands* as size increases following quarter-power scaling laws, so socioeconomic time *contracts* following the 15 percent scaling laws, both following mathematical rules determined by underlying network geometry and dynamics.

2. LIFE ON AN ACCELERATING TREADMILL: THE CITY AS THE INCREDIBLE SHRINKING TIME MACHINE

Even if you are very young you probably don't need much convincing that almost all aspects of life have been speeding up during your lifetime. It's certainly true for me. Even though I am now in my mid-seventies with many of the big hurdles and challenges of life behind me, I still find myself struggling to keep up with the ever-present treadmill that seems to be getting progressively faster and faster. My inbox is always full no matter how many messages I delete and how many I answer, I am dangerously behind with completing not just this year's but last year's taxes, there are continuous seminars, meetings, and events that I would love and am supposed to attend, I struggle to remember the semi-infinite number of passwords that allow me to access myself through my various accounts and affiliations, and on and on. You surely have your own version of this, a similar litany of time pressures that never seem to abate no matter how hard you try to defeat them. And this is even worse if you live in a large city, have small children, or run a business.

This speeding up of socioeconomic time is integral to modern life in the Urbanocene. Nevertheless, like many of us, I harbor a romantic image that not so long ago life was less hectic, less pressured, and more relaxed and that there was actually time to think and contemplate. But read what the great German poet, writer, scientist, and statesman Johann Wolfgang von Goethe said on the subject almost two hundred years ago in 1825, soon after the beginning of the Industrial Revolution[1]:

> *Everything nowadays is ultra, everything is being transcended continually in thought as well as in action. No one knows himself any longer; no one can grasp the element in which he lives and works or the materials that he handles. Pure simplicity is out of the question; of simplifiers we have enough. Young people are stirred up much too early in life and then carried away in the whirl of the times. Wealth and rapidity are what the world admires. . . . Railways, quick mails, steamships, and every possible kind of rapid communication are what the educated world seeks but it*

only over-educates itself and thereby persists in its mediocrity. It is, moreover, the result of universalization that a mediocre culture [then] becomes [the] common [culture]. . . .

Although it's a curious mixture of comments on the accelerating pace of life and the resulting erosion of culture and values, expressed in a slightly archaic language, it sounds hauntingly familiar.

So it's hardly news that the pace of life has been accelerating, but what is surprising is that it has a universal character that can be quantified and verified by analyzing data. Furthermore, it can be understood scientifically using the mathematics of social networks by relating it to the positive feedback mechanisms that enhance creativity and innovation, and which are the source of the many benefits and costs of social interaction and urbanization. In this sense cities are time accelerator machines.

The contraction of socioeconomic time is one of the most remarkable and far-reaching features of modern existence. Despite pervading all of our lives, it has hardly received the attention it deserves. I'd like to give a personal anecdote that exemplifies the speeding up of time and the concomitant changes that accompany it.

I first came to the United States in September 1961 to attend graduate school in physics at Stanford University in California. I took a steam train from King's Cross Station in London up to Liverpool, where I boarded the Canadian steamship the *Empress of England* and sailed for almost ten days across the Atlantic, down the St. Lawrence River, eventually disembarking in Montreal. I stayed overnight before taking a Greyhound bus that deposited me in California four days later, having spent one night at the YMCA in Chicago, where I changed buses. The entire journey was an extraordinary experience that transported me across many dimensions, not least of which was an amazing introduction to the variety, diversity, and eccentricity of American life, including an appreciation of its immense geographical size. Fifty-five years later I am still trying to process everything I experienced on that road trip as I continue to grapple with the meaning and enigma of America and all that it stands for.

Although I came from a rather modest background, such a journey would have been typical of most students at that time. Overall, it took me well over two weeks to get from London to Los Angeles, where I stayed with a friend before being driven up to Palo Alto. Today, even the poorest student would complete the journey from London to L.A. in less than twenty-four hours, and the vast majority would do it in many hours less. A direct flight nowadays takes about eleven hours, and even by the late 1950s one could comfortably fly from London to Los Angeles in less than fifteen hours if one could afford it. However, had I undertaken this same journey a hundred or so years earlier it could have taken me many months.

This is just one graphic example of how the time required for travel has dramatically shrunk over the last couple of hundred years. It is often expressed with the platitude that the world has shrunk. Obviously it hasn't shrunk—the distance between London and Los Angeles is still 5,470 miles—what has shrunk is time, and this has had profound consequences for every aspect of life from the personal to the geopolitical. In 1914 the famous Scottish mapmaker John Bartholomew, cartographer royal to King George V, published *An Atlas of Economic Geography,* a wonderful collection of data and factoids about economic activity, resources, health and climatic conditions, and goodness knows what else concerning all known places across the globe.[2] One of his unique illustrations was a world map showing how long it took to get to any general area on the planet. It's quite illuminating. For instance, the boundaries of Europe were about five days' journey apart, whereas today they have shrunk to a mere few hours. Similarly, the boundaries of the British Empire extended over several weeks in 1914, but today its ghostly remains can be traversed in less than a day. Most of central Africa, South America, and Australia required in excess of forty days' travel, and even Sydney was over a month away.

But travel time is just one manifestation of the extraordinary acceleration of the pace of life that has been enabled by the dizzying proliferation of time-shrinking innovations. Just in my lifetime alone we have experienced transitions to jet airplanes and bullet trains in travel; personal computers, cell phones, and the Internet in communication; home shopping and fast-food drive-through restaurants in food and material supplies; microwaves, washing machines, and dishwashers in home aids; gas chambers, carpet bombing, and nuclear weapons in warfare; and so on. And before any of these came along, think of the revolutionary changes wrought by the steam engine, the telephone, photography, film, television, and radio.

One of the great ironies in all of these marvelous inventions (with the possible exception of the gruesome weapons of destruction) is that they all promised to make life easier and more manageable and therefore give us *more* time. Indeed, when I was a young man, pundits and futurists were speaking of the glorious future anticipated from such time-saving inventions, and a topic that was much discussed was what we would do with all free time that would now be at our disposal. With cheap energy available from nuclear sources and these fantastic machines doing all our manual and mental labor, the workweek would be short and we would have large swaths of time to really enjoy the good life with our families and friends, a little like the boring privileged lives of aristocratic ladies and gentlemen of previous centuries. In 1930 the great economist John Maynard Keynes wrote:

> For the first time since his creation man will be faced with his real, his permanent problem—how to use his freedom from pressing economic cares, how to occupy the leisure, which science and compound interest will have won for him, to live wisely and agreeably and well.

And in 1956, Sir Charles Darwin, grandson of *the* Charles Darwin, wrote an essay on the forthcoming Age of Leisure in the magazine *New Scientist* in which he argued:

> Take it that there are fifty hours a week of possible working time. The technologists, working for fifty hours a week, will be making inventions so the rest of the world need only work twenty-five hours a week. The more leisured members of the community will have to play games for the other twenty-five hours so they may be kept out of mischief. . . . Is the majority of mankind really able to face the choice of leisure enjoyments, or will it not be necessary to provide adults with something like the compulsory games of the schoolboy?

They could not have been more wrong. The main challenge they foresaw was how to keep people occupied so that they wouldn't become bored to death.

Instead of giving us *more* time, "science and compound interest" driven by "technologists working for fifty hours a week" have, in fact, given us *less* time. The multiplicative compounding of socioeconomic interactivity engendered by urbanization has inevitably led to the *contraction of time.* Rather than being bored to death, our actual challenge is to avoid anxiety attacks, psychotic breakdowns, heart attacks, and strokes resulting from being accelerated to death.

I suppose that in a rather different sense one could argue that unwittingly Sir Charles did in fact get part of it right. After all, one could argue that the greatest impact that television and the IT revolution have had on society has been to "provide adults with something like the compulsory games of the schoolboy." What else are Facebook, Twitter, Instagram, selfies, texts, and all of the other entertainment media that dominate our lives and fill up our time? Well, they do serve other purposes and can certainly increase the quality of life, but their addictive temptations have been hard to resist. It's tempting to perceive them as having evolved into "compulsory games" or as having replaced religion as the twenty-first-century version of Marx's "opiate of the people." In any case, these are prime examples of recent innovations that have contributed to the acceleration of social time.

Below, I will introduce a theory of growth inspired by the network scaling theory and show that innovations and paradigm shifts need to be made at a faster and faster pace in order to maintain continuous open-ended growth, thereby contributing even further to the acceleration of time. Before discussing this, however, I want to present some explicit examples, some using big data, to substantiate and test various *quantitative* predictions of the theory including the increasing pace of life.

3. COMMUTING TIME AND THE SIZE OF CITIES

In the 1970s an Israeli transportation engineer, Yacov Zahavi, wrote a series of fascinating reports on transportation in cities for the U.S. Department of Transportation and later for the World Bank to help address specific concerns about traffic and mobility as cities continued to grow and gridlock was becoming the norm. As expected, these reports were data rich, quite detailed, and aimed at providing solutions to specific urban transport problems. However, in addition to presenting standard analyses from a classic consulting engineer's perspective, Zahavi unexpectedly cast his results in a coarse-grained big-picture framework much as a theoretical physicist might have done. His model, which he grandiosely dubbed the "Unified Mechanism of Travel Model," takes no account of either the physical or social structure of cities, nor of the fractal nature of road networks, but is based almost entirely on optimizing the economic costs of travel for an average individual relative to his or her income (roughly speaking, travelers use the fastest travel mode they can afford). Although his model doesn't seem to have generated universal acclaim and wasn't published in academic journals, one of its many intriguing conclusions has entered into urban folklore and provides an interesting twist on the question of the increasing pace of life.

Using data from cities across several countries, including the United States, England, Germany, and some developing nations, Zahavi discovered the surprising result that the total amount of time an average individual spends on travel each day is approximately the same regardless of the city size or the mode of transportation. Apparently, we tend to spend about an hour each day traveling, whoever and wherever we are. Roughly speaking, the average commute time from home to work is about half an hour each way independent of the city or means of transportation.

Thus even though some people travel faster, some by car or train, some by bus or subway, and some much slower by bicycling or walking, on average all of us spend up to an hour or so traveling to and from work. So the increase in transportation speed resulting from the marvelous innovations of the past couple of hundred years has not been used to reduce commuting time but instead has been used to *increase commuting distances*. People have taken advantage of these advancements to live farther away and simply travel longer distances to work. The conclusion is clear: the size of cities has to some degree been determined by the efficiency of their transportation systems for delivering people to their workplaces in not much more than half an hour's time.

Zahavi's fascinating observations made a powerful impression on the Italian physicist Cesare Marchetti, who was a senior scientist at the International

Institute for Applied Systems Analysis (IIASA) in Vienna. IIASA has been a major player in questions of global climate change, environmental impact, and economic sustainability, and this is where Marchetti's interests and contributions have mostly been. He became intrigued by Zahavi's work and in 1994 published an extensive paper elaborating on the approximate invariance of daily commute times and promoted the idea that the true invariant is actually overall daily *travel* time, which he called *exposure time*.[3] So even if an individual's daily commute time is less than an hour, then he or she instinctively makes up for it by other activities such as a daily constitutional walk or jog. In support of this Marchetti wryly remarked, "Even people in prison for a life sentence, having nothing to do and nowhere to go, walk around for one hour a day, in the open."

Because walking speed is about 5 kilometers an hour, the typical extent of a "walking city" is about 5 kilometers across (about 3 miles), corresponding to an area of about 20 square kilometers (about 7 square miles). According to Marchetti, "There are no city walls of large, ancient cities (up to 1800), be it Rome or Persepolis, which have a diameter greater than 5km or a 2.5km radius. Even Venice *today*, still a pedestrian city, has exactly 5km as the maximum dimension of the connected *center*." With the introduction of horse tramways and buses, electric and steam trains, and ultimately automobiles, the size of cities could grow but, according to Marchetti, constrained by the one-hour rule. With cars able to travel at 40 kilometers an hour (25 mph), cities, and more generally metropolitan areas, could expand to as much as 40 kilometers or 25 miles across, which is typical of the catchment area for most large cities. This corresponds to an area of about 12 hectares or 450 square miles, more than fifty times the area of a walking city.

This surprising observation of the approximately one-hour invariant that communal human beings have spent traveling each day, whether they lived in ancient Rome, a medieval town, a Greek village, or twentieth-century New York, has become known as *Marchetti's constant*, even though it was originally discovered by Zahavi. As a rough guide it clearly has important implications for the design and structure of cities. As planners begin to design green carless communities and as more cities ban automobiles from their centers, understanding and implementing the implied constraints of Marchetti's constant becomes an important consideration for maintaining the functionality of the city.

4. THE INCREASING PACE OF WALKING

Zahavi and Marchetti presumed that for a given mode of transportation, such as walking or driving, travel speed did not change with city size. As we saw above, Marchetti estimated the approximate size of a city whose dominant form of mobility is walking by assuming that the average walking speed is 5 kilometers per hour (3.1 mph). But in large cities with diverse modes of transportation, walking occurs in busy areas where individuals effectively become part of a crowd and social network dynamics come into play. We are subliminally influenced by the presence of other individuals and infected by the increasing pace of life, unconsciously finding ourselves in a hurry to get to a store, the theater, or to meet a friend. In small towns and cities, pedestrian streets are rarely crowded and the general pace of life is much more leisurely. Consequently, one might expect walking speed to increase with city size, and it's tempting to speculate that it obeys the 15 percent rule because the mechanism underlying the increase is partially driven by social interaction.

Amusingly, data confirm that walking speeds do indeed increase with city size following an approximate power law, though its exponent is somewhat less than the canonical 0.15, being closer to 0.10 (see Figure 42). This is hardly surprising given the simplicity of the model and that social interactions are only partially responsible for this whimsical effect. It's interesting to note that, according to the data, average walking speeds increase by almost a factor of two from small towns with just a few thousand inhabitants to cities with populations of over a million, where the average walking speed is a brisk 6.5 kilometers per hour (4 mph). It's very likely that this saturates, and walking speeds don't appreciably increase in much larger cities because there are obvious biophysical limitations to how fast people can comfortably walk.

An unexpected expression of this hidden dynamic is the recent introduction of fast lanes for walking in the British city of Liverpool. Apparently, people were getting so frustrated at fellow pedestrians not walking fast enough

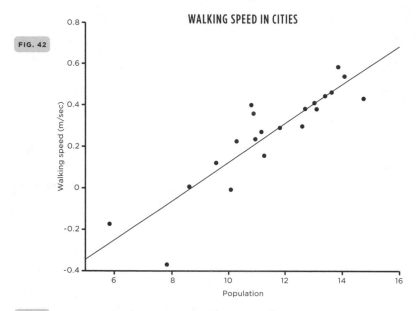

WALKING SPEED IN CITIES

FIG. 42

FIG. 43

(42) The average walking speed plotted logarithmically against population size showing how it systematically increases across various European cities. (43) Fast lanes for pedestrians in Liverpool, England.

that the city introduced a special fast lane for walkers (see Figure 43). The caption on the photograph is an amazing expression of the increasing pace of life: *half* of all people surveyed said they were inhibited from shopping downtown because of the *slow* pace of walking set by other pedestrians. This has spurred a burgeoning interest from cities around the world to follow Liverpool's lead, so I suspect we'll be seeing more of this curious phenomenon on streets in downtown areas of major cities.

5. YOU ARE NOT ALONE: MOBILE TELEPHONES
AS DETECTORS OF HUMAN BEHAVIOR

One of the most revolutionary manifestations of our highly connected twenty-first-century world is the extraordinary ubiquity of mobile (or cell) phones. The easy access to cheap, sophisticated smart phones coupled to the Internet has been a major contributor to ramping up the pace of life and shrinking time. Instantly transmitted sound bites in the form of tweets, text messages, and e-mails have swamped traditional telephone communication, not to speak of the carefully composed letters of yesteryear or, god forbid, intimate face-to-face conversation. I'll return to discuss some of the implications and unintended consequences of this amazing innovation later, but here I want to focus on a mostly unappreciated aspect of cell phones that has begun to revolutionize a small corner of science.

You probably know that your provider keeps track of every phone call or text message you make, when you made it, how long it lasted, to whom you were talking or texting, where you were and where they were... and very likely, in some cases, what was being said. This is an enormous amount of data and in principle provides unprecedented and highly detailed information on social interaction and mobility, especially because nowadays almost everyone uses such devices. There are now more cell phones in use than people on the planet. In the United States alone there are more than a trillion phone calls made each year, and on average a typical person spends more than three hours a day fixated by his or her cell phone. Almost twice as many people have access to cell phones in the world as they do to toilets—an interesting comment on our priorities.

This has been a huge boon to even the poorest countries, as they have been able to leapfrog traditional technology and jump immediately into twenty-first-century communication infrastructure at a fraction of the huge cost of installing and maintaining landlines, which most were unable to afford at the scale of coverage that mobile technology provides. It is no surprise that proportionally the greatest percentage of mobile phone usage occurs in developing countries.

So analyzing massive data sets of cell phone calls can potentially provide us with new and testable quantitative insight into the structure and dynamics of social networks and the spatial relationships between place and people and, by extension, into the structure and dynamics of cities. This unforeseen consequence of mobile phones and other IT devices has ushered in the era of big data and smart cities with the somewhat hyperbolic promise that these will provide the tools for solving all of our problems. And not just the infrastructural challenges in cities: the promise extends to all aspects of life from health and pollution to crime and entertainment. This is just one manifestation of the rapid emergence of "smart" industries based on the availability of enormous amounts of data that we ourselves unwittingly generate, whether through our mobile devices, our mobility, or our health records. This developing paradigm certainly provides us with new and powerful tools that, used sensibly, can undoubtedly have beneficial consequences in addition to providing new ways for companies and entrepreneurial individuals to create yet more wealth. Later, however, I will sound some strong words of caution aimed primarily at the naïveté and even danger inherent in this approach.[4]

Here I want to concentrate on how cell phone data can be used scientifically to test predictions and consequences of our emerging theoretical framework for understanding cities. Apart from the fact that cities, and social systems more generally, are complex adaptive systems, one of the traditional obstacles to developing quantitative testable theories in the social sciences has been the obvious difficulty of obtaining large amounts of credible data and performing controlled experiments. A major reason that the physical and biological sciences have made the tremendous progress they have is that the systems under study can be manipulated or contrived to test specific well-defined predictions and consequences derived from proposed hypotheses, theories, and models.

Giant particle accelerators, such as the Large Hadron Collider in Geneva, Switzerland, where the Higgs particle was recently discovered, are a quintessential example of such artificially controlled experimentation. By combining results from the analysis of many experiments involving ultrahigh energy collisions between particles with the development of a sophisticated mathematical theory, physicists have over many years discovered and determined the properties of the fundamental subatomic constituents of matter as well as the forces of interaction between them. This has led to one of the great achievements of twentieth-century science, the development of the *standard model of the elementary particles*. This incorporates, integrates, and explains a breathtaking spectrum of the world around us including electricity, magnetism, Newton's laws of motion, Einstein's theory of relativity, quantum mechanics, electrons, photons, quarks, gluons, protons, neutrons, Higgs particles, and much more, all in a unifying mathematical framework whose detailed predictions have been spectacularly verified by continuous and ongoing experimental tests.

Equally spectacular is that the energies and distances being probed in such experiments shed light on phenomena that determined the evolution of the universe after the Big Bang. In these experiments we are artificially re-creating events that literally last took place at the beginning of the universe. The resulting theoretical framework has given us a credible quantitative understanding of how galaxies formed and why the heavens look the way they do. Obviously we can't do experiments on the heavens themselves or on the universe—they are what they are and represent a singular unique event that, unlike experiments in the laboratory, cannot be repeated. We can only observe. Like geology and for that matter the social sciences, astronomy is a *historical* science in that we can test our theories only by making postdictions for what should have happened according to the equations and narratives of our theories, and then searching in the appropriate place for their verification. This is none other than Newton's strategy when he derived Kepler's laws of planetary motion from his fundamental laws of motion and the law of gravity, which had been developed to explain the mundane motion of objects moving in the immediate world around him. He couldn't do experiments directly on the planets themselves, but could compare his predictions for their motion with Kepler's observations and measurements and so verify that they were correct. Over the past hundred years this strategy has proven to be remarkably successful in both astrophysics and geology. Consequently, we are confident that we understand how both the universe and the Earth got to be the way they are. So in the case of these historical sciences success has been achieved by a shrewd integration of sophisticated observation coupled with appropriate traditional experimentation on proxy situations in the here and now.

Despite the obvious difficulties when it comes to social systems, social scientists have been very imaginative in devising analogous quantitative experiments to inspire and test hypotheses, and these have proven to give insight into social structure and dynamics. Many involve surveys and responses to various questionnaires and are subject to limitations that depend on the role of the experimental teams who have to interact with the subjects. Consequently it is very difficult to obtain large amounts of data for more than a relatively small sampling across a broad enough range of people and social situations, and this can lead to questions of credibility and generality of results and conclusions.

The beauty of mobile phone data, or data from social media such as Facebook or Twitter, for investigating social behavior is that these sorts of problems can be significantly mitigated. Not that using such data doesn't bring its own tricky issues. How representative of the entire population are cell phone users, and how representative of social interactions are cell phone calls? These are debatable questions, but what is clear is that this form of communication is now a dominant feature of social behavior, providing a quantitative window onto how, where, and when we interact.

6. TESTING AND VERIFYING THE THEORY: SOCIAL CONNECTIVITY IN CITIES

Carlo Ratti is an Italian architect/designer in the architecture department at MIT, where he runs an outfit with the catchy title Senseable City Lab. I first met Carlo when we were on the same program at the annual DLD conference in Munich. This aspires to be a sort of TED-like affair, though it's narrower in scope with more emphasis on art and design. Like TED and the Davos meeting of the World Economic Forum, it's basically a several-day networking cocktail party with a dense program of talks in an ambience of "this is where it's at," trying to project an image of futuristic culture, high-tech commerce, and "innovation." An eclectic mix of interesting and even influential people are in attendance, and there are occasional glimpses of substance and brilliant insight in the presentations, though these are modulated with a heavy dose of flaky, somewhat superficial bullshit wrapped up in superb PowerPoint presentations. This is the way of the world, and there are now many such high-profile events just like these. Despite their shortcomings they do serve an important purpose of creating cross-currents and exposing businesspeople, entrepreneurs, technologists, artists, writers, the media, politicians, and an occasional scientist to new and innovative, if occasionally crazy and provocative, ideas and, of course, to other such people—a little bit like a city does, though greatly compressed in space and time. By the way, like TED, you would be hard pressed to find anyone who any longer remembers what the acronym DLD stands for; I vaguely recall that the D in both of them stands for "design." Acronyms are also the way of the world and represent yet another subtle manifestation of the accelerating pace of life. 2M2H. LOL.

Although not a scientist, Carlo became enthusiastic about bringing a scientific perspective to understanding cities and tried to convince me that cell phone data was a great way of testing the theory and of investigating other aspects of urban dynamics. I was skeptical, mostly because I didn't think cell phone usage was sufficiently broad, diverse, and representative that it could be justified as a credible proxy for measuring social interaction and mobility. However, Carlo is not a man easily deterred. Slowly I began to pay attention to statistics on the extraordinary growth of cell phone usage, especially in developing nations where they were now being used by up to 90 percent of the population, and slowly I began to appreciate that Carlo and others like him were right. Many researchers were beginning to take advantage of this new source of data, mostly to investigate network structure and dynamics and provide insight into processes like the spread of disease and ideas.

Inspired by our scaling work, Carlo had hired several smart young physicist/engineers to pursue the use of cell phone data, and together with Luis Bettencourt and me in Santa Fe we started a collaboration to test one of the fundamental predictions of the theory. One of the most intriguing aspects of the scaling of cities is its universal character. As we've seen, seemingly unrelated socioeconomic quantities ranging from income and patent production to the incidence of crime and disease scale superlinearly with city size with a similar exponent of about 1.15. In the previous chapter I argued that this surprising commonality across different cities, urban systems, and metrics reflects the degree of interaction between people and has its origins in the universal structure of social networks. People across the globe behave in pretty much the same way regardless of history, culture, and geography. Thus without having to resort to any fancy mathematical theory, this idea predicts that the number of interactions between people in cities should scale with city size in the same way that all of the diverse socioeconomic quantities scale, namely, as a superlinear power law with an exponent of around 1.15 regardless of the urban system. In other words, the systematic 15 percent increase in socioeconomic activity with every doubling of city size, whether in wages and patent production or crime and disease, should track a predicted 15 percent increase in the interaction between people.

So how do you measure the number of interactions between people? Traditional methods have relied on written surveys, which are time consuming, labor intensive, and subject to sampling biases because they are necessarily limited in scope. Even if some of these challenges could be overcome, carrying out such surveys across an entire urban system consisting of hundreds of cities is daunting and probably not feasible. On the other hand, the recent availability of large-scale data sets automatically collected from mobile phone networks covering a large representative percentage of the population across the globe opens up unprecedented possibilities for systematically studying social dynamics and organization in *all* cities. Fortunately, our MIT colleagues had access to such large data sets consisting of billions of anonymized call records (meaning that we didn't know the names or numbers of the callers). Obviously, some of these were just one-off calls, so only those where there was a reciprocal call between two callers within a set period of time were counted. From this, the total number of interactions between callers, the total volume of calls, and the total time spent calling within each city was extracted.[5]

Our analysis was based on two independent data sets: mobile phone data from Portugal and landline data from the United Kingdom. The results are

Opposite page: (44) The scaling of four disparate urban metrics—income, GDP, crime, and patents—rescaled from Figures 34–38 to show how they all scale with a similar exponent of about 1.15. (45) The scaling of the connectivity between people as measured by the number of reciprocal phone calls between individuals across cities in Portugal and the United Kingdom showing a similar exponent confirming the prediction of the theory. (46) The size of modular groups of friends of individuals is approximately the same regardless of the size of the city.

FIG. 44

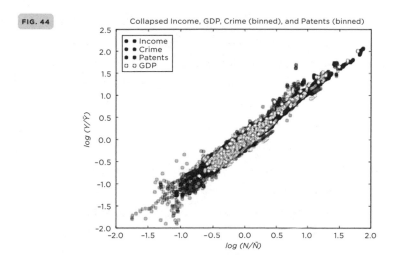

FIG. 45

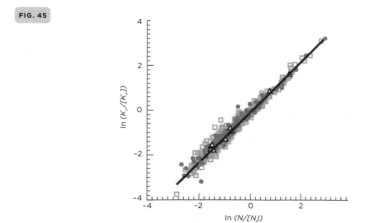

FIG. 46

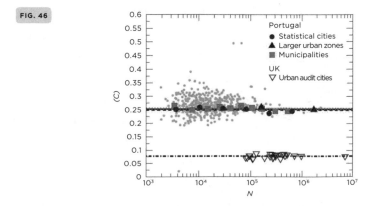

shown in Figures 44–46, where the total number of contacts between people in a city over an extended period of time is plotted logarithmically versus the population size of the city. As you can see, a classic straight line is revealed for both sets of data, indicating power law scaling with the exponent in both cases having the *same* value very close to the predicted 1.15, in spectacular agreement with the hypothesis. To illustrate this visually, I have placed this graph next to a composite plot derived from Figures 34–38 in chapter 7 showing the universality of socioeconomic urban scaling: a subset of these disparate indicators—GDP, income, patent production, and crime—has been plotted on the same graph to show how they all scale together when appropriately rescaled.

This result is a very satisfying confirmation of the hypothesis that social interactions do indeed underlie the universal scaling of urban characteristics. Further confirmation is provided by the observation that both the total time people spent on phone call interactions and the total volume of all of their calls systematically increase with city size in a similar fashion. These results also verify that the accelerating pace of life originates in the increasing connectivity and positive feedback enhancement in social networks as city size increases. For instance, over the fifteen-month period in which the Portuguese data were collected, an average resident of Lisbon, with a population of 560,000, spent about twice as much time on about twice as many reciprocated calls as an average resident of Lixa, a small rural town whose population is only about 4,200. In addition, if the one-off *non*reciprocal calls are included in the analysis, then the scaling exponent systematically increases, suggesting that the number of individual solicitations, such as commercial advertisements and political canvassing, is proportionately higher in larger cities. Life in the big city is faster and more intense in every way, including being increasingly bombarded by even more mindless nonsense than in a small town.

Actually, not quite in *every* way. When we investigated how many of an individual's contacts—his or her "friends"—are friends of one another, the answer was quite unexpected. In general, an individual's entire social network spans a highly diverse group of people ranging from close family, friends, and colleagues to relatively distant casual relationships, such as their car mechanic or handyman. Many of these know and interact with one another, but most do not. For example, your dear old mum may never talk to or barely even know your closest colleague at work, despite your having a close relationship with both of them. So how many members of your entire social network—the sum total of all your contacts—talk to one another? This subset defines your "extended family" and the size of your social module. Because there is significantly greater access to larger numbers of people in bigger cities, you might expect that your extended family would likewise be concomitantly bigger and that it would scale superlinearly in much the same way that other socioeconomic quantities do. I certainly did. But to our great surprise, the data revealed that quite to the contrary, it doesn't scale at all. The size of an average individual's modular cluster of acquaintances who interact with one another is an approximate invariant—it doesn't change with city size. For example, the size of the "extended family" of an average individual living in Lisbon, which has more than 500,000 inhabitants, is no larger than that of an average individual living in the small town of Lixa with less than 5,000 inhabitants. So even in large cities we live in groups that are as tightly knit as those in small towns or villages. This is a bit like the invariance of the Dunbar numbers I talked about in the previous chapter and, like those, probably reflects something fundamental about how our neurological structure has evolved to cope with processing social information in large groups.

There is, however, an important qualitative difference in the nature of these modular groups in villages relative to those in large cities. In a real village we are limited to a community that is imposed on us by sheer proximity resulting from its small size, whereas in a city we are freer to choose our own "village" by taking advantage of the much greater opportunity and diversity afforded by a greater population and to seek out people whose interests, profession, ethnicity, sexual orientation, and so on are similar to our own. This sense of freedom provided by a greater diversity across many aspects of life is one of the major attractions of urban life and a significant contributor to rapidly increasing global urbanization.

7. THE REMARKABLY REGULAR STRUCTURE
OF MOVEMENT IN CITIES

The extraordinary diversity and multidimensionality of cities has led to a host of images and metaphors that try to capture particular instantiations of what a city is. The walking city, the techno-city, the green city, the eco-city, the garden city, the postindustrial city, the sustainable city, the resilient city. . . and, of course, the smart city. The list goes on and on. Each of these expresses an important characteristic of cities, but none captures the essential feature encapsulated in the rhetorical Shakespearean question "What is the city but the people?" Most images and metaphors of cities conjure up their physical footprint and tend to ignore the central role played by social interaction. This critical component is captured in a different kind of metaphor such as the city as a cauldron, a crucible, a mixing bowl, or a reactor in which the churning of social interactions catalyzes social and economic activity: the people's city, the collective city, the anthro-city.

The image of the city as a large vat in which people are being continually churned, blended, and agitated can be viscerally felt in any of the world's great cities. It is most apparent in the continuous, and sometimes frenetic, movement of people in downtown and commercial areas in what often appears to be an almost random motion much like molecules in a gas or liquid. And in much the same way that bulk properties of gases or liquids, such as their temperature, pressure, color, and smell, result from molecular collisions and chemical reactions, so the properties of cities emerge from social collisions and the chemistry of and between people.

Metaphors can be useful but sometimes they can be misleading, and this is one of those instances. For despite appearances, the motion of people in cities is not at all like the random motion of molecules in a gas or particles in a reactor. Instead, it is overwhelmingly systematic and directed. Very few journeys are random. Almost all, regardless of the form of conveyance, involve willful travel from one specific place to another: mostly from home to work, to a store, to a school or cinema, and so forth . . . *and* back again. Furthermore, most travelers seek the fastest and shortest route, one that takes the least time and traverses the shortest distance. Ideally, this would mean that everyone would like to travel along straight lines but, given the obvious physical constraints of cities, this is impossible. There is no choice but to follow the meandering roads and rail lines, so, in general, any specific journey involves following a zigzagging route. However, when viewed at a larger scale through a coarse-grained lens by averaging over all journeys for all people over a long enough period of time, the preferred route between any two specific locations approximates a straight line. Loosely speaking this means that on average people effectively travel radially along the spokes of circles whose center is their specific destination, which acts as a hub.

With this assumption it is possible to derive an extremely simple but very powerful mathematical result for the movement of people in cities. Here's what it says: Consider any location in a city; this could be a "central place" such as a downtown area or street, a shopping mall or district, but it could just as well be some arbitrary residential area such as where you live. The mathematical theorem predicts how many people visit this location from any distance away and how often they do it. More specifically, it states that the *number of visitors should scale inversely as the square of both the distance traveled and the frequency of visitation.*

Mathematically, an inverse square law is just a simple version of the kinds of power law scaling we've been discussing throughout the book. In that language the prediction of movement in cities can be restated as saying that the number of people traveling to a specific location scales with *both* the distance traveled and the visitation frequency as a power law whose exponent is -2. Thus if the number of travelers is plotted *logarithmically* against either the distance traveled keeping the visitation frequency fixed, or vice versa, against the visitation frequency keeping the travel distance fixed, a straight line should result in *both* cases with the *same* slope of -2 (recall that the minus sign simply means that the line slopes downward). I should emphasize that as with all of the scaling laws, an average over a long enough period of time, such as six months or a year, is presumed, in order to smooth over daily fluctuations, or the differences between weekdays and weekends.

As can be readily seen from Figure 47, these predictions are spectacularly confirmed by the data. Indeed, the observed scaling is remarkably tight, with slopes in excellent agreement with the prediction of –2. Particularly satisfying is that the same predicted inverse square law is observed across the globe in diverse cities with very different cultures, geographies, and degrees of development: we see an identical behavior in North America (Boston), Asia (Singapore), Europe (Lisbon), and Africa (Doha). Furthermore, when each of these metropolitan areas is deconstructed into specific locations, the same inverse square law is manifested at each of these within the city, as shown, for example, in Figures 48 and 49 for a sampling of locations in both Boston and Singapore.

Let me give a simple example to illustrate how the theorem works. Suppose that on average 1,600 people visit the area around Park Street, Boston, from 4 kilometers away once a month. How many people visit there from *twice* as far away (8 km) with the same frequency of once a month? The inverse square law tells us that ¼ (= ½²) as many make the visit, so only 400 people (¼ × 1,600) visit Park Street from 8 kilometers away once a month. How about from five times as far away, 20 kilometers? The answer is ¹⁄₂₅ (= ⅕²) as many, which is just 64 people (¹⁄₂₅ × 1,600) visiting once a month. You get the idea. But there's more: you can likewise ask what happens if you change the frequency of visitation. For instance, suppose we ask how many people visit Park Street from 4 kilometers away but now with a greater frequency of *twice* a month. This also obeys the inverse square law so the number is ¼ (= ½²) as many, namely 400 people. And similarly, if you ask how many people visit there from the same distance of 4 kilometers away but *five times* a month, the answer is 64 people (¹⁄₂₅ × 1,600).

Notice that this is the same number that visit Park Street from five times as far away (20 km) with a frequency of just once a month. Thus the number of people visiting from 4 kilometers away *five* times a month is the same as the number visiting from *five* times farther away (20 km) once a month (64 in both cases of our specific example). This result does not depend on the specific numbers I chose for the illustration. It is an example of an amazing general symmetry of mobility: *if the distance traveled multiplied by the frequency of visits to any specific location is kept the same, then the number of people visiting also remains the same.* In our example we had 4 kilometers × 5 times a month = 20 in the first case and 20 kilometers × once a month = 20 in the second. This invariance is valid for any visiting distance and for any frequency of visitation to any area in any city. These predictions are verified by the data and manifested in the various graphs of Figures 48 and 49 where you can

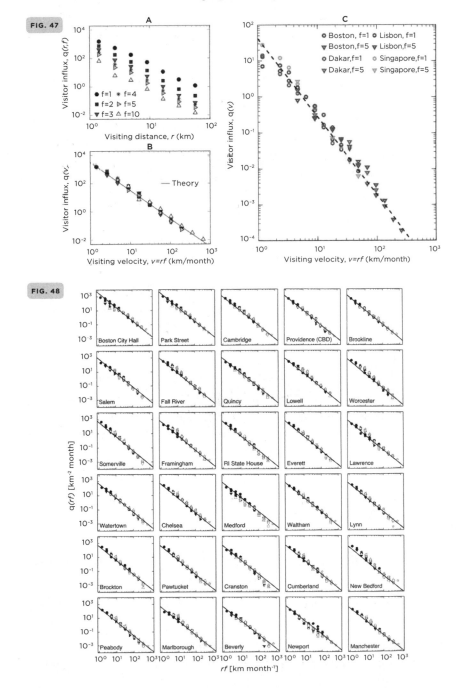

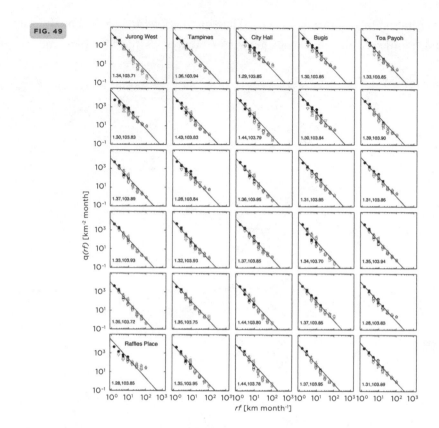

FIG. 49

Opposite and this page: (47) (a) Flux of people visiting from varying distances to a specific location in Boston with different but fixed frequencies (*f* times a month) showing agreement with the inverse square law. (b) Same data as in (a) but showing how all of the different frequencies and distances collapse to a single line when plotted against the single variable frequency × distance. (c) Similar plot to (b) showing how the visitor flux obeys the same predicted inverse square law in very different cities around the globe. (48) Similar plot to (c) but at multiple locations in Boston and (49) Singapore; the solid lines are the predictions from the theory.

explicitly see that the pattern of visitation remains unchanged when the product of distance times frequency has the same value.

I want to emphasize just how remarkable and unexpected this prediction is, given the extraordinary complexity and diversity of movement and transport in cities. When one thinks of the seemingly chaotic random and diverse movement and mobility of people in cities like New York, London, Delhi, or

São Paulo, it's hard not to view this extremely simple picture of a hidden order and regularity as being most unlikely and even absurd. Each individual's random decision to travel to and from a specific location via an optimal route regardless of whether they walk, take the subway, ride the bus, drive their car, or even do all of the above, is predicted to result in a coherent collective flow just as the random motion of trillions of individual water molecules results in a smooth, coherent flow when you turn on the tap in your kitchen.

As I explained above, cell phone data provide detailed information not only about who you talked to and for how long, but also where you were and when you did it. In effect, each of us is carrying a device that is keeping track of where we are at any time. It's as if we were able to put a tag on every molecule in a room and thereby know its location, how fast it's moving, who it bumps into, and so forth. Because there are more than ten thousand trillion trillion (10^{28}) molecules in a modest-size room, this would represent the mother of all big data. However, this information is actually not very useful, especially for gases in equilibrium—it's massive overkill. The powerful techniques of statistical physics and thermodynamics have been successfully developed to understand and describe macroscopic properties of gases, such as their temperature, pressure, phase transitions, and so on, without needing to know the gory details of the motion of all of their constituent molecules. In cities, on the other hand, such information is extremely valuable not only because we ourselves are the molecules, but also because, unlike gases, cities are complex adaptive systems with intricate network structures exchanging both energy and information. Mobile phone data provide us with a powerful tool for determining the structure and dynamics of these networks and, consequently, for quantitatively testing theoretical predictions.

This is potentially a powerful tool for planning and development because it provides a framework for estimating the flux of people to and from specific areas of a city. Building a new mall or developing a new housing project requires accurate, or at least credible, estimates of the flow of traffic and people to ensure sufficient and efficient transport needs. Much of this is done using computer models, which are certainly very useful, but the simulations tend to be locally focused, ignoring the relationship to a bigger, integrated systemic dynamic of the city, and are rarely based on fundamental principles.

The analysis of these massive data sets of mobile phone calls used to test the theory was brilliantly carried out by a Swiss engineer, Markus Schläpfer, working with the Hungarian physicist Michael Szell, two of the bright young postdocs hired by Carlo Ratti at MIT. Markus later joined us at the Santa Fe Institute in 2013, where we began this particular collaboration. Of the many projects he worked on, a particularly interesting one was with Luis on analyzing how the heights and volumes of buildings relate to city size. Markus has since moved on to the prestigious ETH in Zurich, his hometown, where he is engaged with a large collaborative program called the Future Cities Lab, which is based in Singapore and supported by their government.

8. OVERPERFORMERS AND UNDERPERFORMERS

Most of us find rankings intriguing whether they're for cities, schools, universities, corporations, states, or countries, not to mention football teams and tennis players. Basic to this is, of course, the choice of metrics and methodologies used to make the rankings. How questions are posed and what segment of the population is sampled can strongly determine the outcome of surveys and polls, which can have significant consequences in politics and commerce. Such rankings play an increasingly influential role in decision making by individuals, planners, and policy makers in government and industry. How a city or state ranks on the world or national stage in health, education, taxes, employment, and crime can have a powerful influence on how it is perceived by investors, corporations, and vacationers.

In sports, a question that's perennially debated is who was the greatest player or team of all time, a question that is obviously unanswerable from any objective viewpoint. The problem is not just what metrics are reasonable proxies for "greatness," but that it typically involves comparing apples with oranges, often over different periods of time. In this context let's briefly revisit the discussion on weight lifting introduced in chapter 2 and recall Galileo's seminal insight that the strength of the limbs of animals should scale sublinearly as the ⅔ power of their body weight. This prediction is confirmed by data on champion weight lifters, as shown in Figure 7. I introduced the idea that this scaling curve should be viewed as the baseline about which performance should be measured: it tells us how much an idealized champion weight lifter of a given body weight should lift. In a similar fashion, the ¾ power scaling law for metabolic rates shown in Figure 1 tells us what the metabolic rate of an idealized organism of a given size "should" be. "Idealized" in that case is understood to mean a system that is "optimized" in terms of the energy use, dynamics, and geometry of network structures, as was explained in chapter 3.

This framework was used as a point of departure to give a science-based measure of performance. Four of the six champions lifted loads they should have, given their body weights and the expectations from the scaling law. On the other hand, the middleweight *over*performed relative to the expectations for his size, whereas the heavyweight *under*performed. So despite the heavyweight lifting a greater load than anyone else, from a scientific perspective he was actually the *weakest* of all the champions while the middleweight was the *strongest*.

In situations like this where we are able to develop a quantitative scientific framework for performance based on scaling, we can also construct meaningful metrics for rankings and for comparison across different competitions. More mechanical sports such as weight lifting, rowing, and even running potentially lend themselves to this methodology, whereas team sports like football and basketball are considerably more challenging. Deviations from scaling therefore provide a principled metric for individual performance and a quantitative point of departure for investigating why, as in the weight lifting case, the middleweight *over*performed and the heavyweight *under*performed relative to their size.

This strategy of ranking performance effectively levels the playing field by removing the dominant variation arising simply from differences in size to reveal the essential individual skill of each competitor. Below I will apply this idea to cities, but before doing so I want to use it in conjunction with the mobility analysis in cities to indicate how it can potentially be used as an important planning and development tool.

As exemplified in Figures 48 and 49, the data for travel to specific locations in cities fits the theoretical prediction extremely well. However, if you look closely at the graphs for Boston you can see that there are two particular loca-

tions, the airport and the football stadium, where there are significant deviations and the fits aren't quite as good. Given the special role of both places, this
is perhaps not so surprising precisely because they draw on a relatively narrow
subset of people who use them for very specific reasons: either for taking a trip
or for going to a football game.

Although the data for the airport still cluster reasonably well around the
prediction, the greatest deviations occur in the number of people traveling either from short distances away or relatively infrequently. This subset actually
constitutes the majority of people using the airport. In contrast, those coming
from farther away or who use the airport most frequently conform very well to
the predicted scaling curve, though they constitute the minority of users.
Knowing and understanding these patterns of usage both in the general trend
and in their variance is clearly important for planning and managing transport flows to and within the airport and how these relate to transport in the
metropolitan area as a whole.

In Singapore there is only one such major outlier, namely, Raffles Place,
which is the central core of the city-state's financial district. In addition, it is a
major transportation hub and a gateway to a large tourist area. The data on the
number of visitors actually scale reasonably well, but the exponent is significantly smaller than in the rest of Singapore's locations, all of which are in excellent agreement with the predicted value of −2. In addition, it exhibits much
larger fluctuations around scaling than in the rest of Singapore. This translates
into fewer people coming either from close by or less often than expected, and
more people coming from farther away or more often than expected. This may
well be because of the special nature of Singapore as a small island nation
whose core district, Raffles Place, is not near its geographical center but rather
borders on its boundary with the ocean.

As in the case of Boston's airport and stadium, it's important for planning,
designing, and controlling transport and movement for both this specific special location as well as across the entire city to recognize that Raffles Place is
an outlier from the dominant mobility pattern observed in every other location in the city. Equally important is that this can be quantified and understood within the context of the entire urban system.

9. THE STRUCTURE OF WEALTH, INNOVATION, CRIME, AND RESILIENCE: THE INDIVIDUALITY AND RANKING OF CITIES

How rich, creative, or safe can we expect a city to be? How can we establish which cities are the most innovative, the most violent, or the most effective at generating wealth? How do they rank according to economic activity, the cost of living, the crime rate, the number of AIDS cases, or the happiness of their populations?

The conventional answer is to use simple per capita measures as performance indices and rank order of cities accordingly. Almost all official statistics and policy documents on wages, income, gross domestic product (GDP), crime, unemployment rates, innovation rates, cost of living indices, morbidity and mortality rates, and poverty rates are compiled by governmental agencies and international bodies worldwide in terms of both total aggregate and per capita metrics. Furthermore, well-known composite indices of urban performance and the quality of life, such as those assembled by the World Economic Forum and magazines like *Fortune*, *Forbes*, and *The Economist*, primarily rely on naive linear combinations of such measures.[6]

Because we have quantitative scaling curves for many of these urban characteristics and a theoretical framework for their underlying dynamics we can do much better in devising a scientific basis for assessing performance and ranking cities.

The ubiquitous use of per capita indicators for ranking and comparing cities is particularly egregious because it implicitly assumes that the baseline, or null hypothesis, for any urban characteristic is that it scales *linearly* with population size. In other words, it presumes that an idealized city is just the linear sum of the activities of all of its citizens, thereby ignoring its most essential feature and the very point of its existence, namely, that it is a collective emergent agglomeration resulting from *nonlinear* social and organizational interactions. Cities are quintessentially *complex adaptive systems* and, as such, are significantly more than just the simple linear sum of their individual components and constituents, whether buildings, roads, people, or money. This is expressed by the superlinear scaling laws whose exponents are 1.15 rather than

1.00. This approximately 15 percent increase in all socioeconomic activity with every doubling of the population size happens almost independently of administrators, politicians, planners, history, geographical location, and culture.

In assessing the performance of a particular city, we therefore need to determine how well it performs *relative* to what it has accomplished just because of its population size. By analogy with the discussion on determining the strongest champion weight lifter by measuring how much each deviated from his expected performance relative to the idealized scaling of body strength, one can quantify an individual city's performance by how much its various metrics deviate from their expected values relative to the idealized scaling laws. This strategy separates the truly local nature of a city's organization and dynamics from the general dynamics and structure common to all cities. As a result, several fundamental questions about any individual city can be addressed, such as how exceptional it is relative to its peers, what timescales are relevant for local policy to take effect, what are the local relationships between economic development, crime, and innovation, to what extent is it unique, and to what extent can it be considered a member of a family of like cities.

My colleagues Luis, José, and Debbie carried out such an analysis for the entire U.S. urban system consisting of 360 Metropolitan Statistical Areas (MSAs) for a suite of metrics.[7] A sample of the results is presented in Figure 50), where the deviations from scaling for personal income and patent production for cities in the United States in 2003 are plotted logarithmically on the vertical axis against the rank order of each city. We called these deviations *Scale-Adjusted Metropolitan Indicators* (SAMIs). The horizontal axis across the center of these graphs is the line along which the SAMI is zero and there is no deviation from what is predicted from the size of the city. As can be seen, *every* city deviates to some extent from its expected values. Those to the left denote *above*-average performance, whereas those to the right denote *below*-average performance. This provides a meaningful ranking of a city's individuality and uniqueness beyond what is effectively guaranteed just because it's a city of a certain size. Without delving into details of this analysis, I want to make a few salient points about some of the results.

First, compared with conventional per capita indicators, which place seven of the largest twenty cities in the top twenty in terms of their GDPs, our

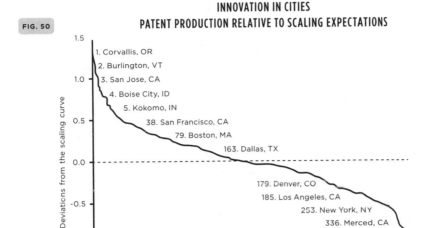

INNOVATION IN CITIES
PATENT PRODUCTION RELATIVE TO SCALING EXPECTATIONS

FIG. 50

1. Corvallis, OR
2. Burlington, VT
3. San Jose, CA
4. Boise City, ID
5. Kokomo, IN
38. San Francisco, CA
79. Boston, MA
163. Dallas, TX

179. Denver, CO
185. Los Angeles, CA
253. New York, NY
336. Merced, CA
337. Yuma, AZ
338. Visalia, CA
339. Shreveport, LA
340. McAllen, TX

Deviations from the scaling curve

Rank of city

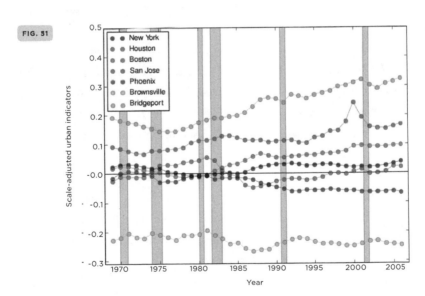

FIG. 51

- New York
- Houston
- Boston
- San Jose
- Phoenix
- Brownsville
- Bridgeport

Scale-adjusted urban indicators

Year

(50) Deviations in the number of patents produced in U.S. cities in 2003 relative to the expected number from scaling laws given the size of the city, plotted against their rank. Those to the left are overperformers with Corvallis being ranked first, whereas those to the right are underperformers with McAllen being ranked last. (51) The time evolution of these deviations (SAMIs) showing their long persistence time.

science-based metrics rank *none* of these cities in the top twenty. In other words, once the data are adjusted for the generic superlinear effects of population size, these cities don't fare so well. Mayors of these cities who take credit and boast that their policies have led to economic success as evidenced by their city's being near the top of the per capita GDP rankings are therefore giving a misleading impression.

Amusingly from this point of view, New York City as a whole turns out to be quite an average city, marginally richer than its size might predict (rank 88th in income, 184th in GDP), not very inventive (178th in patents), but surprisingly safe (267th in violent crime). On the other hand, San Francisco is the most exceptional large city, being rich (11th in income), creative (19th in patents), and fairly safe (181st in violent crime). The truly exceptional cities are typically smaller, such as Bridgeport for income (due to all those bankers and hedge fund managers from New York City living in the suburbs), Corvallis (home of Hewlett-Packard research lab and a respectable university, Oregon State), and San Jose (which encompasses Silicon Valley—what more need be said?) for patents, and Logan (the Mormon culture?) and Bangor (who knows?) for safety.

This is just for a single year (2003), and it is natural to ask how any of this changes with time. Unfortunately, readily accessible data on all of these metrics are hard to come by prior to about 1960. However, an analysis covering data over the last forty to fifty years reveals some very intriguing results, as illustrated in Figure 51, where the temporal evolution of deviations for personal income in a few typical cities is shown. Perhaps the most salient feature is how relatively slowly fundamental change actually occurs. Cities that were overperforming in the 1960s, such as Bridgeport and San Jose, tend to remain rich and innovative today, whereas cities that were underperforming in the 1960s, such as Brownsville, are still near the bottom of the rankings. So even as the population has increased and the overall GDP and standard of living have risen across the entire urban system, *relative* individual performance hasn't changed very much. Roughly speaking, all cities rise and fall together, or to put it bluntly: if a city was doing well in 1960 it's likely to be doing well now, and if it was crappy then, it's likely to be crappy still.

Once a city has gained an advantage, or disadvantage, relative to its scaling expectation, this tends to be preserved over decades. In this sense, either for good or for bad, cities are remarkably robust and resilient—they are hard to change and almost impossible to kill. Think of Detroit and New Orleans, and more drastically of Dresden, Hiroshima, and Nagasaki, all of which have to varying degrees survived what were perceived as major threats to their very existence. All are actually doing fine and will be around for a very long time.

A fascinating example of persistent advantage is San Jose, which includes Silicon Valley, the place everyone wants to be. It's hardly a surprise that this is a major overperformer in terms of wealth creation and innovation. But what is a surprise is that San Jose was already overperforming in the 1960s, and almost to the same degree as it is now, as graphically illustrated in Figure 51. This also demonstrates that this overperformance has been sustained and even reinforced for more than forty years despite the short-term boom and bust technological and economic cycle in 1999–2000, at the end of which the city relaxed back to its long-term basal trend. Put slightly differently: apart from a relatively small bump in the late 1990s, the continued success of San Jose was already set well *before* the birth of Silicon Valley. So rather than seeing Silicon Valley as generating the success of San Jose and lifting it up in the conventional socioeconomic rankings, this suggests that it was the other way around and that it was some intangible in the culture and DNA of San Jose that helped nurture the extraordinary success of Silicon Valley.[8]

It takes decades for significant change to be realized. This has serious implications for urban policy and leadership because the timescale of political processes by which decisions about a city's future are made is at best just a few years, and for most politicians two years is infinity. Nowadays, their success depends on rapid returns and instant gratification in order to conform to political pressures and the demands of the electoral process. Very few mayors can afford to think in a time frame of twenty to fifty years and put their major efforts toward promoting strategies that will leave a truly long-term legacy of significant achievement.

10. PRELUDE TO SUSTAINABILITY:
A SHORT DIGRESSION ON WATER

In the developed world we take much of our infrastructure for granted and rarely appreciate the scale and cost that goes into providing amenities like clean, safe drinking water each time we turn on the faucet in our kitchens. This is a huge privilege and, as I discussed in an earlier chapter, is a prime reason why our longevity made such a huge leap beginning in the late nineteenth century. Providing such basic services for all people across the globe is an enormous challenge as we urbanize the planet. Safe water is progressively becoming a source of increased social friction, especially as the climate changes and produces unpredictable periods of severe drought or massive floods, both of which compromise supply and delivery systems. This is already a major issue in many developing countries and hints of it have begun to be seen even in the United States with serious problems arising in supply systems, as in Flint, Michigan, and severe water shortages occurring throughout many of the western states.

I live in the small city of Santa Fe, New Mexico, whose population is about 100,000, in what is usually identified as a semiarid high desert climate where only about 14 inches of rain falls each year. Appropriately, water is very expensive and there are high penalties for overuse. The city boasts one of the highest costs for water of any city in the United States—about two and a half times above the national average and about 50 percent higher than the next most expensive city, which, surprisingly, is drizzly Seattle with almost 40 inches of rain a year. But equally surprising is that it is *six* times higher than one of the cheapest cities, Salt Lake City, which gets only 16.5 inches of rain a year. Even more bizarre is that the cost of water in the desert cities of Phoenix, whose population is 4.5 million, and Las Vegas, whose population is almost 2 million, is barely higher than in Salt Lake City, even though Phoenix gets only 8 inches of rain a year and glitzy Las Vegas only a measly 4 inches. Go figure!

This kind of profligacy is everywhere. For instance, most people don't realize that even California megalopolises like Los Angeles and San Francisco with almost one hundred times the population of Santa Fe, or fancy leafy-green towns with luxurious vegetation like Palo Alto, the home of Stanford University, or Mountain View, the home of Google, get pretty much the same amount of rainfall as Santa Fe, namely, about 14 inches a year. And much of their water use goes to keeping their lawns and gardens looking as if they were growing in Singapore, whose annual rainfall is 92 inches.

The good news is that most urban communities in the United States and across the globe are becoming much more sensitive to these issues, recognizing that clean water is a precious commodity that is being depleted at an alarming rate and that it cannot be taken for granted. Most are beginning to implement policies to significantly reduce water consumption but, like many "green" conservation measures, these may be too little too late.

The point is that in all of these communities enormous resources have been put into the engineering infrastructure to artificially provide them with abundant water transported over great distances and/or from very deep aquifers, with the implicit assumption that these sources are inexhaustible and forever cheap—questionable, at best. As urbanization and questions of sustainability become ever more pressing, the politics and economics of water will progressively become ever more contentious, much like that of oil and other sources of energy in the twentieth century. And, as in the case of oil, major conflicts may eventually be fought over its access and ownership.

Curiously, however, it is worth bearing in mind that in contrast to oil, the planet has more than enough water, just as it has more than enough solar energy to support the entire mass of humanity essentially forever. Adapting our technological and socioeconomic strategies to this simple fact is critical to our long-term survival—a mind-set that we should have adopted long ago by promoting renewable solar energy and desalinization. Have we been so shortsighted and parochial that we are doomed like the sailors in Samuel Taylor Coleridge's famous poem "The Rime of the Ancient Mariner" to a collective nightmare of thirst?

> Water, water, every where,
> And all the boards did shrink;
> Water, water, every where,
> Nor any drop to drink.

Before returning to science and the city I want to give a sense of the *scale* of a major city's water system and what it entails. New York City is rightly famous for being a trendsetter in almost everything, but one of its achievements that is not often appreciated is its water system. Its quality and taste are often considered to be superior not just to water from other municipalities, but to fancy bottled water and at a fraction of the cost without the absurd waste of throwaway plastic containers. Next time you're in New York you can save yourself a few bucks and at the same time get a superior product by simply filling your water bottle from the tap.

Water is supplied to the city from watersheds that are up to one hundred miles north of the city and primarily fed by gravity flow, thereby saving huge energy costs in pumping. The water system has a storage capacity of 550 billion gallons (about 2 billion cubic meters), which provides more than 1.2 billion gallons a day (about 4.5 million cubic meters) of clean drinking water to more than 9 million residents. These are huge numbers, and to appreciate their size note that they are equivalent to delivering about 100,000 of those silly little 16-ounce or 500-millileter plastic bottles of water *every second*. To accomplish this astonishing feat, water leaves the reservoirs through massive "pipes" that are actually deep underground concrete tunnels. A new $5 billion project is presently under way to build a new tunnel to augment two older ones. It will eventually span sixty miles in length at a depth of up to eight hundred feet, and its diameter when leaving the reservoir is twenty-four feet (bigger even than Godzilla's aorta). This steps down in progressive sections as water is distributed through the network hierarchy to the entire New York metropolitan area, eventually feeding the main pipes buried beneath streets whose diameters are anywhere from four to twelve inches depending on the building density (obviously the largest pipes are in downtown Manhattan). Water is distributed from these mains to individual homes through one-inch pipes that are then further downsized to half an inch before they reach kitchen sinks and toilets.

This hierarchical geometry of New York's water system is typical of all urban water systems across the globe, except, of course, that the overall scale varies depending on the size of the city. This template is quite similar to our own circulatory system, even to the extent that both networks are space filling and the terminal units in both systems are approximately invariant. Santa Fe's water system is very much smaller than New York's, but the size of the one-inch pipes delivering water to my house and the half-inch ones delivering water to my toilet are the same as those in New York in much the same way that the size of our capillaries is pretty much the same as those of mice or blue whales. This fractal-like behavior is reflected in the total length of all of the pipes in New York's water network system: added together from the reservoir down to the street mains they stretch for approximately 6,500 miles (10,500 km). In other words, put end to end the total length of all of the pipes in the system would run from New York to Los Angeles and back again. This is pretty impressive, but it pales in comparison to the length of all of the vessels in your circulatory system: if you put all of them end to end it, too, would also stretch from New York to Los Angeles and back, yet all of it fits inside your body.

11. THE SOCIOECONOMIC DIVERSITY OF BUSINESS ACTIVITY IN CITIES

Like resilience and innovation, *diversity* has become a much overused buzzword commonly employed to characterize a successful city. Indeed, the ever-changing admixture of individuals, ethnicities, cultural activities, businesses, services, and social interactions is a defining characteristic of urban life. A major socioeconomic component of this is the abundance of different types of businesses in cities. While all cities necessarily have a similar essential core of businesses—lawyers, doctors, restaurants, garbage collectors, teachers, administrators, and so on—only a very few have specialized subcategories such as maritime lawyers, tropical disease doctors, blacksmiths, chess store proprietors, nuclear physicists, and hedge fund managers.

Consequently, quantifying the diversity of business types is potentially problematic because any systematic classification scheme is subject to an arbitrary designation of specific categories; any business type can be further subdivided as long as a defining distinction can be made. Restaurants, for example, can be decomposed into fine dining, fast food, et cetera, as well as into multiple levels of cuisine, price, quality, and so on. There are broad categories such as Asian, European, and American, but Asian, for instance, can be divided into Chinese, Indian, Thai, Indonesian, Vietnamese, et cetera, and Chinese itself can be further subdivided into Cantonese, Szechuan, dim sum, and so on. The lesson is clear: urban diversity is scale-dependent in the sense that it depends on the resolution with which it is perceived. This is not so different in spirit from the problem first recognized by Lewis Fry Richardson when he tried to measure the length of various coastlines and boundaries, and which led Benoit Mandelbrot to invent the concept of fractals.

Luckily, the challenge of formally categorizing businesses has been addressed, at least in North America, by the compilation of an extraordinary data set that includes the records of almost all business establishments in the United States (more than 20 million). This is the result of an impressive collaboration between the United States, Canada, and Mexico, and is called the North American Industry Classification System (NAICS).[9] An *establishment* is just any single physical location where business is conducted. Consequently, individual businesses that are part of a national chain, such as a Walmart store or a McDonald's franchise, would be counted as separate establishments. Establishments are often viewed as the fundamental units of economic analysis because innovation, wealth generation, entrepreneurship, and job creation all manifest themselves through the formation and growth of such workplaces. The NAICS classification scheme employs a six-digit code at the most detailed industry level. The first two digits designate the largest business sector, the third digit designates the subsector, and so on, thereby capturing economic life at an extraordinarily fine-grained level.

My colleagues Luis, José, and Debbie carried out an analysis of these data with our postdoc Hyejin Youn taking the lead role. Hyejin was trained in statistical physics in Seoul, South Korea, and had joined SFI to finish her doctorate. She first worked on the origin and structure of languages before joining our collaboration. She has now established herself as an expert on innovation in technology and is currently a fellow of the Institute of New Economic Thinking (INET) at Oxford University—a new program funded by the financier George Soros.

As we saw in the analysis of other urban metrics, the data reveal surprisingly simple and unexpected regularities. For instance, the total number of establishments in each city regardless of what business they conduct turns out to be linearly proportional to its population size. Double the size of a city and on average you'll find twice as many businesses. The proportionality constant is 21.6, meaning that there is approximately one establishment for about every 22 people in a city, *regardless of the city size*. Or to put it slightly differently, on average a new workplace is created each time the population of a city increases by just 22 people, whether in a small town or a large metropolis. This is an unexpectedly small number and usually comes as quite a surprise to most people, even those dealing in business and commerce. Similarly, the data also show that the total number of employees working in these establishments also scales approximately linearly with population size: on average, there are only about 8 employees for every establishment, again regardless of the size of the city. This remarkable constancy of the average number of employees and the average number of establishments across cities of vastly different sizes and characters is not only contrary to previous wisdom, but also rather puzzling when viewed in light of the pervasive superlinear agglomeration effects that underlie all socioeconomic activity including per capita increases in productivity, wages, GDP, and patent production.[10]

To get a deeper understanding of this and to reveal the business character of a city, it's illuminating to ask how many different *types* of businesses there are in a city. This is like asking how many different species of animals there are in an ecosystem. The simplest coarse-grained measure of economic diversity of a city is obtained by just counting the number of distinct types of establishments it has as a function of its population size. The data confirm that diversity systematically increases with population size at all levels of resolution, as defined by the NAICS data set. Unfortunately, the classification scheme is unable to capture the full extent of economic diversity in the largest cities because it is unable to differentiate among very closely related types of establishments such as, for instance, between a northern and southern Italian restaurant. However, an extrapolation of the data strongly suggests that if we could measure diversity to the finest possible resolution it would scale logarithmically with city size.

Compared with the usual canonical power law behavior that most metrics obey, this represents an extremely slow increase with population size. For instance, an increase of the population by a factor of one hundred from, say,

100,000 to 10 million results in the addition of one hundred times as many businesses but an increase of only a factor of two in their diversity. To put it slightly differently: doubling the size of a city results in doubling the total number of establishments, but only a meager 5 percent increase in new *kinds* of businesses. Almost all of this increase in diversity is reflected in a greater degree of specialization and interdependence involving larger numbers of people, both as workers and as clients. This is an important observation because it shows that increasing diversity is closely linked to increasing specialization, and this acts as a major driver of higher productivity following the 15 percent rule.

A more detailed way of assessing economic diversity is to dig deeper and examine the specific component types of establishments that constitute the business landscape of individual cities. How many lawyers, doctors, restaurants, or contractors are there in each city, and how many of these are corporate lawyers, orthopedic surgeons, Indonesian restaurants, or plumbing contractors? As an example of such an analysis, Figure 52 shows the abundances of the one hundred leading business types for a selection of cities in the United States. These are plotted in the classic rank-size format that was used when discussing Zipf's law for the frequency distribution of words in a language and cities in an urban system. When giving a talk, I preface the showing of this graph by asking the audience what they think is the most abundant business type in New York City. So far, no one has yet come up with the correct answer, including audiences of business and corporate gurus operating in New York City itself. It's amazing what you can learn from taking a simple analytic, principled approach to problems.

Well, in New York the most abundant business type is offices of physicians. Weird, especially because physicians are ranked only fifth in Phoenix, which has a huge retirement community of decrepit old farts like me, and seventh in San Jose, which is possibly less surprising given all those young obsessive-compulsive California joggers and health enthusiasts. Less surprising is that after physicians, the next highest ranked in New York are offices of lawyers, followed by restaurants. In fact, restaurants rank high in all cities, being first, for example, in Chicago, Phoenix, and San Jose. Eating out, whether at a fancy Four Seasons restaurant or at a McDonald's fast-food stop,

FIG. 52

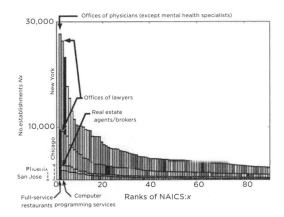

(52) The number of establishments ranked in descending order of their frequencies (from common to rare) for New York City, Chicago, Phoenix, and San Jose. Establishment types are referred to by their NAICS classification. (53) Universal per capita rank-abundance curve of establishment types for *all* 366 metropolitan statistical areas of the United States; shown explicitly are New York City, Chicago, Phoenix, Detroit, San Jose, Champaign-Urbana, and Danville. The inset shows the first 200 types plotted logarithmically, obeying an approximate Zipf-like power law behavior.

FIG. 53

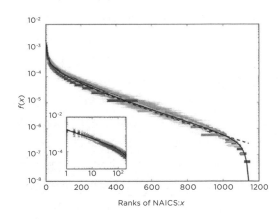

is clearly a major component of the socioeconomic activity of Americans. More generally, it's interesting to speculate what these sorts of rankings say about a city. For instance, after restaurants Phoenix ranks real estate second, which is perhaps not surprising in a rapidly growing city, whereas San Jose, the home of Silicon Valley, predictably ranks computer programming second. It's obvious why lawyers and restaurants would rank high in New York, but what is it about New York that there should be such a large abundance of doctors? Is life in the Big Apple that stressful and unhealthy? If you find this intriguing, you can see the analogous decomposition of economic activities in your favorite city in the supplementary material to our paper posted on the

Web. Clearly, for those running a city or thinking about its future or investing in its development, knowing these sorts of details about the composition of their business landscape provides an important input.

In contrast to the inherent universal properties of cities that are manifested in the scaling laws, these rank-size distributions of business types reflect the individuality and distinctive characteristics of each specific city as exhibited by the composition of its economic activities. They are the hallmark of each city and obviously do depend on its history, geography, and culture. It is therefore all the more remarkable that, despite the unique admixture of business types for each individual city, the *shape and form* of their distribution is mathematically the same for all of them. So much so, in fact, that with a simple scale transformation inspired by theory their rank abundances collapse onto a single unique universal curve common to *all* cities, as shown explicitly in Figure 53. When one considers the vast range of income, density, and population levels, let alone the uniqueness and diverse cultures that vary so widely in cities across the United States, this universality is quite surprising.

What is particularly satisfying is that this unexpected universality, as well as the actual form of the universal curve and the logarithmic scaling of diversity, can all be derived from theory. The universality is driven by the constraint that the sum total of all the different businesses in a city scales linearly with population size, regardless of the detailed composition of business types or of the city. The snakelike mathematical form of the actual distribution function in Figure 53 can be understood from a variant of a very general dynamical mechanism that has been successfully used for understanding rank-size distributions in many different areas, ranging from words and genes to species and cities. It goes under many different names including *preferential attachment, cumulative advantage, the rich get richer,* or *the Yule-Simon process.* It is based on a positive feedback mechanism in which new elements of the system (business types in this case) are added with a probability proportional to the abundances of how many are already there. The more there are, the more of that type are going to be added, so more frequent types get even more abundant with increasingly higher probability than less frequent types.[11]

A couple of colloquial examples may help: successful companies and universities attract the smartest people, resulting in their becoming more successful, thereby attracting even smarter people, leading to even greater success and so on, just as wealthy people attract favorable investment opportunities that generate even more wealth which they further invest to get even more wealthy. Hence the catchphrase *the rich get richer* and its implied, but usually unstated, corollary *the poor get poorer* to characterize this process. Or, as so articulately put by Jesus according to the Gospel of Matthew in the New Testament:

> *For everyone who has will be given more, and he will have an abun-*
> *dance. Whoever does not have, even what he has will be taken from him.*

This surprising declaration has been used by some fundamentalist Christians and others as justification for rampant capitalism—a sort of anti–Robin Hood slogan supporting the idea of taking from the poor to give to the rich. But while Jesus's remarks are a good example of *preferential attachment*, the quote is, not surprisingly, taken out of context. It is often conveniently forgotten that Jesus was actually referring to knowledge of the mysteries of the kingdom of heaven and not to material wealth. He was expressing a spiritual version of the very essence of diligent study, knowledge accumulation, and research and education as expressed by the ancient rabbis: *He who does not increase his knowledge decreases it.*

The first serious mathematical consideration of preferential attachment was made by Udny Yule, a Scottish statistician who in 1925 used it to explain the power-law distribution for the number of species per genus of flowering plants. The modern socioeconomic version of preferential attachment, or cumulative advantage, is due to Herbert Simon and is consequently nowadays referred to as the Yule-Simon process. Simon, by the way, was an extraordinary polymath and one of the most influential social scientists of the twentieth century. His research ranged across the fields of cognitive psychology, computer science, economics, management, philosophy of science, sociology, and political science. He was a founding father of several important scientific subdisciplines whose influence has gained great prominence in recent years, including artificial intelligence, information processing, decision making, problem solving, organization theory . . . and complex systems. He spent almost his entire academic career at Carnegie Mellon University in Pittsburgh and received the

Nobel Memorial Prize in Economics for his seminal work on decision-making processes in economic organizations.

The empirical and theoretical analyses of business diversity presented above show that all cities exhibit the same underlying dynamics in the development of their business ecology as they grow. Initially, small cities with a limited portfolio of economic activities need to create new businesses and functionalities at a fast rate. These basic activities constitute the economic core of every city, big and small; every city needs lawyers, doctors, shopkeepers, tradesmen, administrators, builders, et cetera. As cities grow and these basic core activities become saturated, the pace at which new functionalities are introduced slows down dramatically but never completely ceases. Once the set of individual building blocks is large enough, the resulting combination of talents and functions is sufficient to generate novel variations that expand the business landscape, giving rise to specialized establishments such as exotic restaurants, professional sports teams, and luxury stores, leading to greater economic productivity.

Although the theory cannot predict how specific business types rank in individual cities (why, for instance, doctors are first in New York but seventh in San Jose), it does predict how their rankings change as cities grow. The general rule is that business types whose abundances scale superlinearly with population size systematically rise in their rankings, whereas those that scale sublinearly systematically decrease. For instance, at the coarsest level of the NAICS classification scheme traditional sectors such as agriculture, mining, and utilities scale sublinearly; the theory predicts that the rankings and relative abundances of these industries decrease as cities get larger. On the other hand, informational and service businesses such as professional, scientific, and technical services, and management of companies and enterprises, scale superlinearly and are consequently predicted to increase disproportionally with city size, as observed. As a concrete example, consider the number of lawyers' offices. This scales superlinearly with an exponent close to the canonical 1.15, meaning that there are systematically more lawyers per capita in larger cities. The preferential attachment model predicts that as a city grows the ranking of lawyers should increase as an approximate power law with an exponent of about 0.4, as observed.[12] Such predictions can be made for any business type at any level of granularity.

Thus the scaling exponents that relate how the abundances of each business category scale with city size capture the disproportionate growth of different business sectors and parameterize them in a more systematic way than simple counts of businesses, or "expert" judgments on the nature of sectors, which are very often quite subjective. A critical aspect of this approach is that cities and businesses are complex adaptive systems and should consequently be viewed as an integrated system and not as isolated individual agents. By looking at all cities and the complete set of business sectors that constitute the entire urban economy in a country, this analysis couples the economic fabric of each city with that of the entire system of cities.

12. GROWTH AND THE METABOLISM OF CITIES

A major theme running throughout this book is that nothing grows without the input and transformation of energy and resources. This was the fundamental basis for the comprehensive theory that I presented in chapter 4 for quantitatively understanding the growth of biological systems, whether of individual organisms or communities. Recall the basic idea: Food is consumed, then digested and metabolized into a usable form that is transported through networks to supply cells, where some is allocated to the repair and maintenance of existing cells, some to replace those that have died, and some to create new ones that add to the overall biomass. This sequence is the basic template for how all growth occurs whether for organisms, communities, cities, companies, or even economies. Roughly speaking, incoming metabolized energy and resources are apportioned between general maintenance and repair, including replacement of what's already there and has decayed and the creation of new entities, whether cells, people, or infrastructure, that add to the system and increase its size. Thus the energy available for growth is just the difference between the rate at which energy can be supplied and the rate that is needed for maintenance.

On the supply side, metabolic rate in organisms scales *sublinearly* with the number of cells (following the generic ¾ power exponent derived from network constraints) while the demand increases approximately *linearly*. So as the organism increases in size, demand eventually outstrips supply because linear scaling grows faster than sublinear, with the consequence that the amount of energy available for growth continuously decreases, eventually going to zero resulting in the cessation of growth. In other words, growth stops because of the mismatch between the way maintenance and supply scale as size increases. The sublinear scaling of metabolic rate and the associated economies of scale arising from optimizing network performance are therefore responsible for why growth stops and why biological systems exhibit the bounded sigmoidal growth curves that were shown in Figures 15–18 of chapter 4. The same network mechanism that underlies sublinear scaling, economies of scale, and the cessation of growth is also responsible for the systematic slowing down of the pace of biological life as size increases—and for eventual death.

I now want to apply this framework to the growth of social organizations, beginning with cities. Because of its generality it can be readily extended to companies and entire economies, which I shall discuss in the next chapter. As was explained in chapter 7 cities are comprised of two generic components: their physical infrastructure, manifested as buildings, roads, et cetera, and their socioeconomic dynamics, manifested as ideas, innovation, wealth creation, and social capital. Both are networked systems, and their tight interconnectivity and interdependence result in the approximate complementarity between the corresponding sub- and superlinear scaling laws, namely, that the 15 percent savings with every doubling of size in the former are approximately equal to the 15 percent gain in the latter.

It is the first of these, the physical infrastructural component, that has close analogies to biology and provides the metaphor of the city as an organism. But as I have persistently emphasized, the city is much more than its physicality. Consequently, the concept of metabolic rate as the supply-side input that fuels growth and sustains a city has to be expanded to include socioeconomic activity. In addition to the electricity, gas, oil, water, materials, products, artifacts, and so on that are used and generated in a city, we have to add wealth, information, ideas, and social capital. At a more fundamental level all of these, whether physical or socioeconomic, are driven and sustained by the supply of energy. In addition to heating buildings, transporting materials and people, manufacturing goods, and providing gas, water, and electricity, every transaction, every dollar gained or lost, every conversation and meeting, every phone call and text message, every idea and every thought has to be fueled by energy. Furthermore, just as food must be metabolized into a form that is useful for supplying cells and sustaining life, so the incoming energy and resources digested by a city must be transformed into a form that can be used to supply, sustain, and grow socioeconomic activities such as wealth creation, innovation, and the quality of life. No one has articulated this more eloquently than the great urbanist Lewis Mumford[13]:

> *The chief function of the city is to convert power into form, energy into culture, dead matter into the living symbols of art, biological reproduction into social creativity.*

This extraordinary process, which can be thought of as the *social metabolism* of a city, is responsible for increasing our conventional *biological metabolic rate* derived from the food we eat from just 2,000 food calories a day or 100 watts to about 11,000 watts, the equivalent of 2 million food calories a day. Thus the actual energy content of the food input to the total energy budget of a city is a tiny portion of its overall consumption—less than 1 percent—and that's why I didn't include it in the above discussion even though it is obviously a critical component of urban life. This may seem paradoxical since we saw in the last section that food establishments are the most abundant business type in most cities, even exceeding lawyers. The point is that the overwhelming energy cost associated with food is *not* in the food itself (the 2,000 food calories a day per person) but in its production, transportation, distribution, and marketing through the supply chain from farms to stores to your house and ultimately to your mouth.

When one contemplates the huge number of different contributions to the total metabolism of a city it becomes clear that determining its value, whether in dollars or watts, becomes a major challenge that as far as I know has never been attempted in detail.[14] This is quite surprising given that this is fundamental to how cities and more generally economies function and grow. In addition to the need for collecting and analyzing huge amounts of data across a broad spectrum of diverse activities, there is the issue as to what should actually be counted as part of the social metabolism of a city. What are the independent contributions? Should we include, for instance, the energy costs of crime, police, patents, construction, investments, and research as independent contributions, or will we be double counting if we do as there are clear overlaps and interconnections among and between them?

For the purposes of understanding growth, however, this challenge can be elegantly finessed using the conceptual framework of our scaling theory. The critical point is that *all* socioeconomic contributions to social metabolism that underlie growth, including wealth creation and innovation, scale in approximately the same way following the classic superlinear power law with a common exponent of approximately 1.15. Because all of its subcomponents scale this way, the total social metabolic rate of a city must likewise scale superlinearly with an exponent of 1.15. This is the beauty of the scaling perspective— we don't need to know the details of what the individual contributions to the metabolism of a city are in order to determine its growth trajectory. This is because they are all interconnected and interrelated through the same integrated unifying dynamics of the social and infrastructural networks that constitute urban life.

The superlinear scaling of metabolism has profound consequences for growth. In contrast to the situation in biology, the supply of metabolic energy generated by cities as they grow increases *faster* than the needs and demands for its maintenance. Consequently, the amount available for growth, which is just the difference between its social metabolic rate and the requirements for maintenance, continues to increase as the city gets larger. The bigger the city gets, the faster it grows—a classic signal of open-ended exponential growth. A mathematical analysis indeed confirms that growth driven by *superlinear scaling* is actually faster than exponential: in fact, it's *superexponential.*

Even though the conceptual and mathematical structure of the growth equation is the same for organisms, social insect communities, and cities, the consequences are quite different: *sublinear scaling and economies of scale that dominate biology lead to stable bounded growth and the slowing down of the*

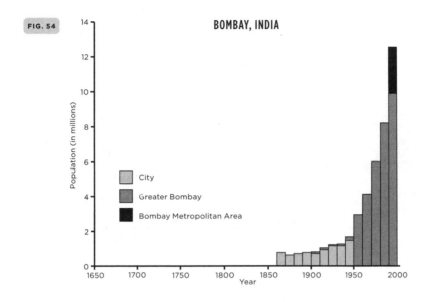

FIG. 54

BOMBAY, INDIA

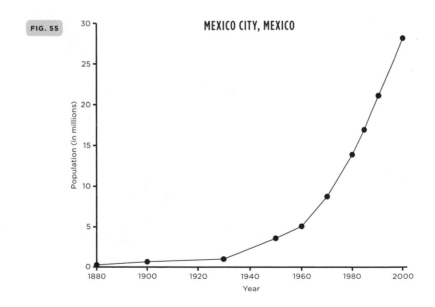

FIG. 55

MEXICO CITY, MEXICO

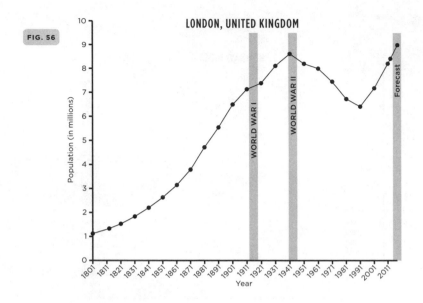

FIG. 56

LONDON, UNITED KINGDOM

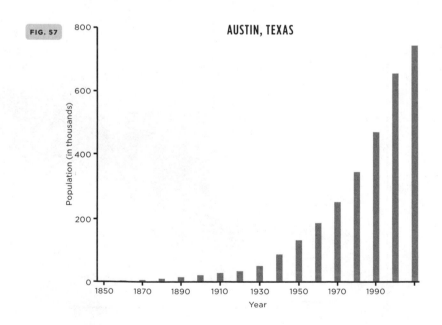

FIG. 57

AUSTIN, TEXAS

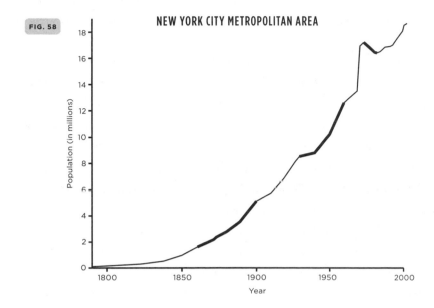

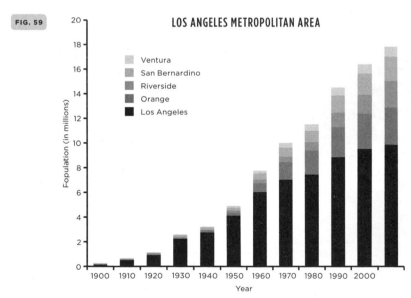

Pages 375–77: Growth curves for various cities across the globe illustrating the ubiquity of open-ended superexponential growth; in order, they are Bombay, Mexico City, London, Austin, the New York City metropolitan area, and the Los Angeles metropolitan area. Credible data do not exist before about 1850.

pace of life, whereas *superlinear scaling and increasing returns to scale that dominate socioeconomic activity lead to unbounded growth and to an accelerating pace of life.*

The continuous positive feedback mechanisms inherent in social networks that are responsible for the multiplicative enhancement of social connectivity and superlinear scaling lead naturally to open-ended superexponential growth and concomitantly to an increase in the pace of life. This is what has evolved over the past couple of hundred years as cities have exploded on the planet. Some examples from across the globe are shown in Figures 54–59; they include Old World cities (London), New World cities (New York, Austin, several California cities, and Mexico City), and Asian cities (Bombay). The major point I want to stress here is that the growth equation driven by superlinear scaling leads to a mathematical formula whose predictions are consistent with the generic superexponential growth seen in these graphs.

Notice, however, that both London and New York show periods of contraction and stagnation. I will return to discuss these effects in chapter 10, where I'll pick up the story on open-ended growth in a larger context by relating it to the role of innovation cycles and the accelerating pace of life and how these affect the critical question of sustainability.

9

TOWARD A SCIENCE OF COMPANIES

Companies, like people and households, are fundamental elements of the socioeconomic life of cities and states. Innovation, wealth generation, entrepreneurship, and job creation are all manifested through the formation and growth of businesses, firms, and corporations, all of which I shall refer to generically as *companies*. Companies dominate the economy. For example, the total worth of all publicly traded companies in the United States—technically referred to as their *total market capitalization*—is more than $21 trillion, which is more than 15 percent larger than the entire GDP. The value and annual sales of any of the very largest companies, such as Walmart, Shell, Exxon, Amazon, Google, and Microsoft, approach half a trillion dollars ($500 billion), implying that a relatively small number of them have the lion's share of the total market.

Given our previous revelations concerning the rank and size frequency distributions of personal incomes (Pareto's law) and cities (Zipf's law), it will come as no great surprise to learn that this lopsidedness is a reflection of a similar power law distribution for the ranking of companies in terms of their market capitalization or annual sales.[1] This was already shown in Figure 41 in chapter 8. Thus there are very few extremely large companies, but an enormous number of very small ones, with all of those in between following a simple systematic power law distribution. So while there are almost 30 million independent businesses in the United States, most of which are private with very few employees, there are only about four thousand publicly traded companies, and these constitute the bulk of economic activity.

Given this observation it is natural to ask, as we did for cities and organisms, whether companies scale in terms of their measurable metrics such as their sales, assets, expenses, and profits. Do companies manifest systematic regularities that transcend their size, individuality, and business sector? And if so, could there possibly be a quantitative, predictive *science of companies* paralleling the *science of cities* that was developed in previous chapters? Is it possible to understand the general quantitative features of their life histories, how they grow, mature, and eventually die?

As with cities there is an extensive body of literature on companies stretching back to Adam Smith and the founding of modern economics. Much of the work is qualitative, often gathered through case studies of specific companies or business sectors, from which the general dynamical and organizational features of companies are intuited. Historically, companies have been viewed as the necessary agents that organize people to work collaboratively to take advantage of economies of scale, thereby reducing the transaction costs of production or services between the manufacturer or provider and the consumer. The drive to minimize costs so as to maximize profits and gain greater market share has been extraordinarily successful in creating the modern market economy by providing goods and services at affordable prices to vast numbers of people. Despite all of its pitfalls, abuses, and negative unintended consequences, this free market credo has been instrumental in creating an unprecedented standard of living across the globe. It is potentially a harsh and simplistic vision that often ignores quality and, more important, the role of explicit corporate social responsibility as a basic complementary component for why companies exist beyond the primitive drive to maximize profits and compensation.

Most of the literature on companies has been developed from the vantage points of economics, finance, law, and organizational studies, though in more recent years ideas from ecology and evolutionary biology have begun to gain prominence. There is also a large popular literature written by successful investors and CEOs revealing the secrets of their own success, which they tend to extrapolate to explain and prescribe what makes some companies successful and others failures. To varying degrees, all of these provide insights into the nature, dynamics, and structure of companies, but none brings a broad scientific perspective to the problem in the sense I have been using in this book.[2]

The mechanisms that have traditionally been suggested for understanding companies can be divided into three broad categories: transaction costs, organizational structure, and competition in the marketplace. Although these are interrelated they have very often been treated separately. In the language of the framework developed in previous chapters these can be expressed as follows: (1) Minimizing *transaction costs* reflects economies of scale driven by an optimization principle, such as maximizing profits. (2) *Organizational structure* is the network system within a company that conveys information, resources, and capital to support, sustain, and grow the enterprise. (3) *Competition* results in the evolutionary pressures and selection processes inherent in the ecology of the marketplace.

Automobiles, computers, ballpoint pens, and insurance portfolios cannot be produced on a grand scale without creating a complex organizational structure, which must be adaptive if it is going to survive in a competitive market. As in cities, this necessitates the integration of energy, resources, and capital—the metabolism of the company—with the exchange of information in order to fuel innovation and creativity. In this sense, companies at all scales are classic complex adaptive systems and it is this framework, rooted in the scaling paradigm, that I want to explore. To what extent can a quantitative, mechanistic theory for understanding their growth, longevity, and organization be developed that is complementary to traditional ways of viewing them?

There have been surprisingly few studies on the nature of companies using large data sets spanning the entire spectrum of economic activities and company histories. Most of these have been carried out by researchers inspired by ideas from complex systems, a good example of which is the discovery that the size distribution of companies follows a systematic Zipf-like power law (shown in Figure 41). This insight was made by the computational social scientist Robert Axtell, who was trained in public policy and computer science at Carnegie

Mellon University, where he came under the influence of the great polymath Herbert Simon, whom I mentioned earlier.

Axtell, who is at George Mason University in Virginia and is also on the external faculty of the Santa Fe Institute, is a leading expert in *agent-based modeling*, which is a computational technique used for simulating systems composed of huge numbers of components.[3] Basically, the strategy involves postulating simple rules governing the interactions between individual constituent agents, which could be companies, cities, or people, coupled with an algorithm that specifies how they evolve in time and letting the resulting system run on a computer. More sophisticated versions include rules for learning, adaptation, and even reproduction so as to model more realistic evolutionary processes.

With the development of powerful computers, agent-based modeling has become a standard tool for gaining insight into many problems in ecological and social systems such as modeling the structure of terrorist organizations, the Internet, traffic patterns, stock-market behavior, pandemics, ecosystem dynamics, and business strategies. Over the past few years Axtell has used agent-based modeling to try to simulate the entire ecosystem of companies in the United States, encompassing more than six million companies and 120 million workers. This ambitious project relies heavily on census data both as input to constrain the simulation and to test its outcomes.

More recently, he has teamed up with other prominent members of the SFI community, including Doyne Farmer, now a professor at Oxford, and John Geanakoplos, a well-known economist at Yale, to extend this project to try to simulate the entire economy. This is truly ambitious, requiring enormous amounts of input data on everything from financial transactions and industrial production to real estate, government spending, taxes, business investments, foreign trade and investments, and even consumer behavior. The hope is that such an integrated simulation of the whole economy could provide a realistic test bed for evaluating different strategies for economic stimulus, such as whether to reduce taxes or increase public spending; and perhaps most important, to be able to predict tipping points or forecast imminent crises so as to avoid potential recessions or even eventual collapse.[4]

It is sobering that no such detailed model for how the economy actually works exists and that policy is typically determined by relatively localized, sometimes intuitive, ideas of how it should work. Very little of the thinking explicitly acknowledges that the economy is an ever-evolving complex adaptive system and that deconstructing its multitudinous interdependent components into finer and finer semiautonomous subsystems can lead to misleading, even dangerous, conclusions as testified by the history of economic forecasting. Like long-term weather forecasting, this is a notoriously difficult challenge, and to be fair to economists we should recognize that they are pretty good at forecasting the relatively short term, provided the system remains stable. Traditional economic theory relies heavily on the economy remaining in an approximately equilibrium state. The serious challenge is to be able to predict outlying events, major transitions, critical points, and devastating economic hurricanes and tornadoes where their record has mostly been pretty dismal.

Nassim Taleb, author of the best-selling, highly influential book *The Black Swan,* has been particularly harsh on economists despite, or maybe because of, having been trained in business and finance.[5] He has held positions at several distinguished universities including New York University and Oxford and has focused on the importance of coming to terms with outlying events and developing a deeper understanding of risk. He has been brutally outspoken in his condemnation of classical economic thinking with hyperbolic comments such as: "Years ago, I noticed one thing about economics, and that is that economists didn't get anything right." He has even called for the Nobel Memorial Prize in economics to be withdrawn, saying that the damage from economic theories can be devastating. I may disagree with some of Taleb's ideas and polemics but it's important and healthy to have such outspoken mavericks challenging the orthodoxy, especially when it's had such a poor record and its proclamations have major implications for our lives.

The great virtue of agent-based modeling is its potential for providing an alternative framework for addressing some of these big issues by treating the entire system as an integrated entity rather than as a sum of its idealized bits and pieces. It recognizes up front that the economy is typically not in equilibrium but is an evolving system with emergent properties that result from the underlying interactions between its multiple constituent parts.

It does, however, have some serious shortcomings. To begin with, a crucial input is the specification of the rules for how agents behave, interact, and make decisions, and in many cases this has to be based on intuition rather than on fundamental knowledge or principles. Furthermore, it is often very difficult to interpret the results of a detailed simulation and determine the causal relationships between different components and subunits of the system. It can therefore be unclear what the important drivers are that determine specific outputs versus those that are consequences of general principles common to all such systems. In its extreme version the underlying philosophy of agent-based modeling is antithetical to the traditional scientific framework, where the primary challenge is to reduce huge numbers of seemingly disparate and disconnected observations down to a few basic general principles or laws: as in biology, where the principle of natural selection applies to *all* organisms from cells to whales, or in physics, where Newton's laws apply to *all* motion from automobiles to planets. In contrast, the aim of agent-based modeling is to construct an almost one-to-one mapping of each specific system. General laws and principles that constrain its structure and dynamics play a secondary role. For example, in simulating a specific company, every individual worker, administrator, transaction, sale, cost, et cetera is in effect included and each company consequently treated as a separate, almost unique entity, typically without explicit regard either to its systematic behavior or its relationship to the bigger picture.

Clearly, both approaches are needed: the generality and parsimony of "universal" laws and systematic behavior reflecting the big picture and dominant forces shaping general behavior, coupled with and informed by detailed modeling reflecting the individuality and uniqueness of each company. In the case of cities, scaling laws revealed that 80 to 90 percent of their measurable characteristics are determined from just knowing their population size, with the remaining 10 to 20 percent being a measure of their individuality and uniqueness, which can be understood only from detailed studies that incorporate local historical, geographical, and cultural characteristics. It is in this spirit that I now want to explore to what extent this framework can be used to reveal emergent laws obeyed by companies.

1. IS WALMART A SCALED-UP BIG JOE'S LUMBER
AND GOOGLE A GREAT BIG BEAR?

The financial services company Standard & Poor's, best known for its stock market index of U.S.-based companies, the S&P 500, provides a valuable database of all publicly traded companies going back to 1950 with summaries of their financial statements and balance sheets. It's called Compustat. Unlike analogous databases for organisms and cities, this database does not come for free. S&P wants about $50,000 for access to it. That may be chicken feed to most investors, corporations, and business schools for whom it is intended, but for us mere academics it's a lot of money, equivalent to the annual salary of a postdoc. Unfortunately, when we organized the ISCOM project to study companies from a scaling perspective we didn't have that kind of money available, so the study of companies had to go on the back burner and the focus of the project was instead directed to cities, where data came for free.

The city work proved to be much more exciting and productive than I for one had foreseen, and it took much longer than anticipated to return to thinking about companies with the kind of focus the subject deserved, even after we had eventually gained access to the Compustat database via explorative funding from the National Science Foundation. Partly because of this, the analysis and the theoretical framework are less developed than for cities. Nevertheless, significant headway has been made and a coherent picture has emerged to provide the foundations of a coarse-grained science of companies.

The modern concept of a company and the kind of rapid market turnover we see today in which most companies don't survive for very long has been around for only the last couple of hundred years at most. This is a much shorter time period than the many hundreds or even thousands of years that cities and urban systems have been evolving and in marked contrast to the billions of years that biological life has been thriving. Consequently, there has been much less time for the market forces that act on companies to reach the kind of metastable configuration manifested in the systematic scaling laws obeyed by cities and organisms.

As explained in earlier chapters, scaling laws are a consequence of the optimization of the network structures that sustain these various systems result-ing from the continuous feedback mechanisms inherent in natural selection and the "survival of the fittest." In the case of cities, we would therefore expect the emergent scaling laws to exhibit much greater variance around idealized power laws than organisms do, because the time over which evolutionary forces have acted is so much shorter. Comparing fits to scaling in the two cases, such as Figure 1 for animal metabolic rates versus Figure 3 for the patent production in cities, confirms this prediction: there is a consistently larger spread around the fits for cities than for organisms. Extrapolating this to com-panies where "evolutionary" timescales are even shorter suggests that if they do indeed scale, there should be an even greater spread in the data around idealized scaling curves than for cities and organisms.

The Compustat data set used for the analysis consists of all 28,853 compa-nies that were traded on U.S. markets in the sixty years between 1950 and 2009. The database includes standard accounting measures such as the num-ber of employees, total sales, assets, expenses, and liabilities, each broken down into subcategories that include interest expenses, investments, inventory, de-preciation, and so on. The flow diagram shown below indicates how all of these are interrelated.

It was constructed by Marcus Hamilton, a young anthropologist whom we had hired as a postdoc to help lead this effort. Even as a student Marcus had a mission in life: to make anthropology and archaeology more quantitative, computational, and mechanistic. For good reasons these fields have been among the least of the social sciences in appreciating this perspective, so Mar-cus has had a tough journey. But for us he was perfect. After getting his doc-torate he worked with Jim Brown on global sustainability issues from an ecological and anthropological viewpoint before joining us at SFI. He has pio-neered some fascinating work on trying to understand hunter-gatherer socie-ties from our scaling perspective and together with José Lobo and me has been developing a theory for how and why our hunter-gatherer ancestors made the crucial transition to sedentary communities that eventually led to city forma-tion. Together with José and Marcus, I recently coauthored a paper that was published in a leading anthropology journal—one of the crowning achieve-ments of my career!

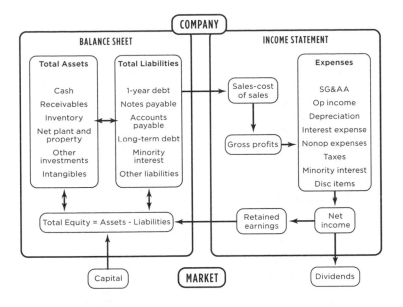

The initial results and conclusions of our investigation into the scaling of companies are very compelling. They provide a powerful basis for developing an understanding of their generic structure and life histories. Figures 60–63 show the sales, incomes, and assets of all 28,853 companies plotted logarithmically against their number of employees. These are the dominant financial characteristics of any company and are standard measures of their fiscal health and dynamics. As these graphs clearly demonstrate, companies do indeed scale following simple power laws and as anticipated they do so with a much greater spread around their average behavior than for either cities or organisms. So in this statistical sense, companies are approximately scaled, self-similar versions of one another: Walmart is an approximately scaled-up version of a much smaller, modest-size company. Even after taking this greater variance into account, this scaling result reveals remarkable regularities in the size and dynamics of companies and is quite surprising given the tremendous variety of different business sectors, locations, and age.

Before expounding further on this, it's instructive to examine how scaling regularities are extracted from big data sets having a large variance such as these. A standard strategy is to bin the data into a series of equal intervals

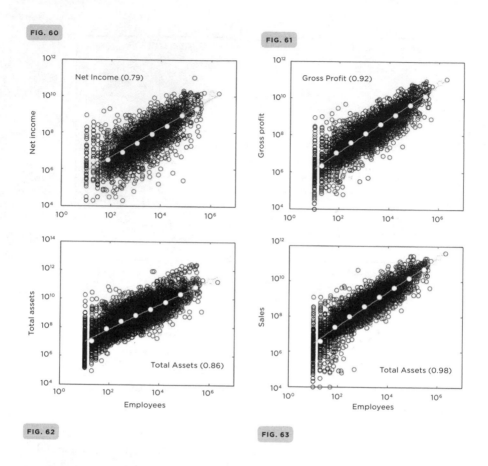

Income, profit, assets, and sales for all 28,853 publicly traded companies in the United States from 1950 to 2009 plotted logarithmically against their number of employees showing sublinear scaling with a substantial variance. The dotted line represents the result of the binning procedure explained in the text.

much like a histogram, and then take the average within each interval. This effectively averages over the fluctuations and reduces the large number of data points to a relatively small number, which is just the number of bins used to divide up the entire interval. The number of employees ranges by a factor of more than a million from the very smallest, mostly young companies with few employees up to giants like Walmart with more than a million. To illustrate the procedure, the data in Figures 60–63 have been binned into eight equal intervals each covering a single order of magnitude. Thus, the first bin includes all companies with fewer than 10 employees, the second all those with 10 to 100, the third all those with 100 to 1,000, and so on, the last bin containing all those with more than a million employees.

The six points resulting from averaging over each bin are shown as gray dots in the graph. They represent a highly coarse-grained reduction of the data, and as you can see follow a very good straight line supporting the idea that underlying the statistical spread is an idealized power law. Because the size and number of bins used is arbitrary, we could just as well have divided up the entire interval into ten, fifty, or one hundred bins rather than just eight, and test whether the straight line remains robust against increasingly finer resolutions of the data. It does. Although binning is not a rigorous mathematical procedure, the stability of obtaining approximately the same straight-line fit using different resolutions lends strong support to the hypothesis that on average companies are self-similar and satisfy power law scaling. The graph in Figure 4 at the opening of the book is in fact the result of this binning procedure, as is the graph in Figure 41 taken from Axtell's work on showing that companies follow Zipf's law. These results strongly suggest that companies, like cities and organisms, obey universal dynamics that transcend their individuality and uniqueness and that a coarse-grained science of companies is conceivable.

Additional evidence supporting this discovery came from an unlikely source, namely the Chinese stock market. In 2012 Zhang Jiang, a young faculty member in the School of Systems Science at Beijing Normal University, joined our collaboration. Jake, as he is known to most of us, visited SFI in 2010 and became enthusiastic about getting involved in the company project. He had access to a database similar to Compustat covering all Chinese companies participating in their emerging stock market. Following the collapse of the Cultural Revolution and the rise to power of Deng Xiaoping, economic reform led to the reestablishment of a securities market in China, and at the end of 1991, the Shanghai Stock Exchange opened for business.

When Jake analyzed the data, he found to our great satisfaction that Chinese companies scaled in a similar fashion to U.S. companies, as can be seen in Figures 64–67. This, however, came as something of a surprise considering that the Chinese market had been in operation for less than fifteen years. Ap-

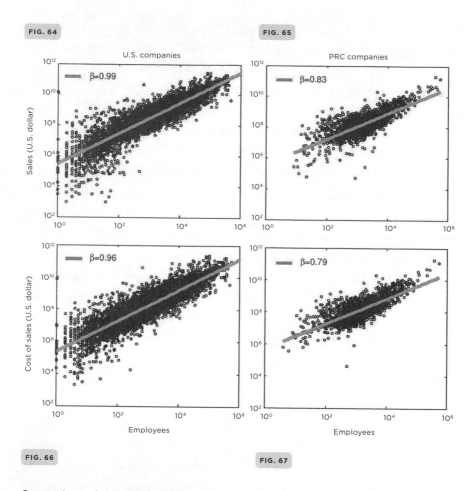

Comparison of the scaling of U.S. companies with those of China showing a similar behavior.

parently, in a vigorous fast-track setting, competitive "free" market dynamics are sufficiently potent for systematic trends to begin to emerge relatively quickly. This is no doubt related to the extraordinarily rapid pace at which the Chinese stock market and its overall economy have grown in such a short time. The Shanghai exchange is already the fifth largest in the world and in Asia is second only to Hong Kong's. Its total market capitalization is $3.5 trillion, compared with more than $21 trillion for the New York Stock Exchange and $7 trillion for Hong Kong's.

2. THE MYTH OF OPEN-ENDED GROWTH

A crucial aspect of the scaling of companies is that many of their key metrics scale *sublinearly* like organisms rather than *superlinearly* like cities. This suggests that companies are more like organisms than cities and are dominated by a version of economies of scale rather than by increasing returns and innovation. This has profound implications for their life history and in particular for their growth and mortality. As we saw in chapter 4, sublinear scaling in biology leads to bounded growth and a finite life span, whereas in chapter 8 we saw that the superlinear scaling of cities (and of economies) leads to open-ended growth.

Their sublinear scaling therefore suggests that companies also eventually stop growing and ultimately die, hardly the image that many CEOs would cherish. It's actually not quite as simple as that because the prediction for the growth of companies is more subtle than just a simple extrapolation from biology. To explain this I am going to present a simplified version of how the general theory applies to companies focusing on the essential features that determine their growth and mortality.

The sustained growth of a company is ultimately fueled by its profits (or net revenue), where these are defined as the difference between sales (or total income) and total expenses; expenses include salaries, costs, interest payments, and so on. To continue growth over a prolonged period, companies must eventually return a profit, part of which is sometimes used to pay dividends to shareholders. Together with other investors, they in turn may buy additional stocks and bonds to help support the future health and growth of the company. However, to understand their generic behavior it is more transparent to ignore dividends and investments, which are primarily important for smaller, younger companies, and concentrate on profits, which are the dominant driver of growth for larger ones.

As we've seen, growth in both organisms and cities is fueled by the difference between metabolism and maintenance. Using that language, the total income (or sales) of a company can be thought of as its "metabolism" while expenses can be thought of as its "maintenance" costs. In biology, metabolic rate scales sublinearly with size, so as organisms increase in size the supply of energy cannot keep up with the maintenance demands of cells, leading to the eventual cessation of growth. On the other hand, the social metabolic rate in cities scales superlinearly, so as cities grow the creation of social capital increasingly outpaces the demands of maintenance, leading to faster and faster open-ended growth.

So how does this dynamic play out in companies? Intriguingly, companies manifest yet another variation on this general theme by following a path that sits at the cusp between organisms and cities. Their effective metabolic rate is neither sub- nor superlinear but falls right in the middle by being linear. This is illustrated in Figures 63 and 64, where sales are plotted logarithmically against the number of employees showing a best fit with a slope very close to one. Expenses, on the other hand, scale in a more complicated fashion: they start out sublinearly but, as companies become larger, eventually transition to becoming approximately linear. Consequently, the difference between sales and expenses, which is the driver of growth, also eventually scales approximately linearly.

This is good news because, mathematically, linear scaling leads to exponential growth and this is what all companies strive for. Furthermore, this also shows why, on average, the economy continues to expand at an exponential rate because the overall performance of the market is effectively an average over the growth performances of all its individual participating companies. Although this may be good news for the overall economy, it sets a major challenge for each individual company because each one has to keep up with an exponentially expanding market. So even if a company is growing exponentially (the good news), this may not be sufficient for it to survive unless its expansion rate is at least that of the market (the bad news). This primitive version of the "survival of the fittest" for companies is the essence of the free market economy.

More good news is that the nonlinear scaling of maintenance expenses in younger companies, buoyed by investments and the ability to borrow large amounts relative to their size, fuels their rapid growth. Consequently, the idealized growth curve of companies has characteristics in common with classic sigmoidal growth in biology in that it starts out relatively rapidly but slows down as companies become larger and maintenance expenses transition to becoming linear. However, unlike biology, whose maintenance costs do not transition to linearity, companies do not cease growing but continue to grow exponentially, though at a more modest rate.

Let's see how this scenario compares with data. Figure 68 is a wonderful graph showing the growth of sales for all 28,853 companies in the Compustat data set plotted together in real calendar time, adjusted for inflation. To get all of them onto a single manageable graph, the vertical axis representing sales is logarithmic. Despite being a "spaghetti" plot, the graph is surprisingly illuminating. The overall trend is clear: as predicted, many young companies shoot out of the starter's block and grow rapidly before slowing down, while older, more mature ones that have survived continue growing but at a much slower rate. Furthermore, the upward trends of these older, slower-growing companies all follow an approximate straight line with similar shallow slopes. On this semilogarithmic plot, where the vertical axis (sales) is logarithmic but the horizontal one (time) is linear, a straight line means mathematically that sales are growing exponentially with time. Thus, on average, all surviving companies eventually settle down to a steady but slow exponential growth, as predicted.

This is very encouraging, but there's a potential pitfall that becomes apparent when the growth of each company is measured *relative* to the growth of the overall market. In that case, as can be clearly seen in Figure 70 where the overall growth of the market has been factored out, *all large mature companies have stopped growing.* Their growth curves when corrected for both inflation and the expansion of the market now look just like typical sigmoidal growth curves of organisms in which growth ceases at maturity, as illustrated in Figures 15–18 of chapter 4. This close similarity with the growth of organisms when viewed in this way provides a natural segue into whether this similarity extends to mortality and whether, like us, all companies are destined to die.

3. THE SURPRISING SIMPLICITY OF COMPANY MORTALITY

After growing rapidly in their youth, almost all companies with sales over about $10 million end up floating on top of the ripples of the stock market. Of these, many operate with their metaphorical noses just above the surface. This

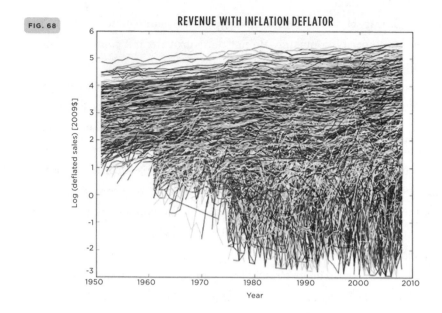

FIG. 68

REVENUE WITH INFLATION DEFLATOR

Log (deflated sales) [2009$]

Year

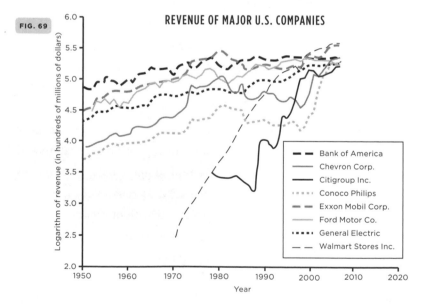

FIG. 69

REVENUE OF MAJOR U.S. COMPANIES

Logarithm of revenue (in hundreds of millions of dollars)

- – – Bank of America
- —— Chevron Corp.
- —— Citigroup Inc.
- ····· Conoco Philips
- – – Exxon Mobil Corp.
- —— Ford Motor Co.
- ····· General Electric
- – – Walmart Stores Inc.

Year

(68) "Spaghetti" graph showing the growth of sales for all 28,853 publicly traded companies plotted together in real time and adjusted for inflation; notice the rapid "hockey stick" rise of small, younger companies versus the relative slow growth of larger, mature ones. (69) Growth curves of a sampling of the oldest and largest companies showing their relatively slow growth; also shown is Walmart, which is a much younger company but whose sales leveled off to similar values after a rapid rise.

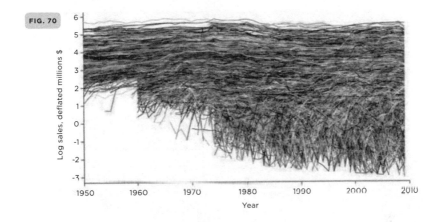

"Spaghetti" graph showing the growth of sales for all 28,853 publicly traded companies relative to the expansion of the overall market plotted together in real time, corrected for inflation. When adjusted for the expansion of the market, the largest companies cease growing.

is a precarious situation because if a big wave comes along they may well drown. Even with profits growing exponentially, let alone if they are suffering losses, companies become vulnerable if they are unable to keep up with the growth of the market. This is greatly exacerbated if a company is not sufficiently robust to withstand the continual ups and downs inherent in the market as well as in its own finances. A sizable fluctuation in the market or some unexpected external perturbation or shock at the wrong time can be devastating to a company whose sales and expenses are finely balanced. This may lead to contraction and decline from which a company might recover, but in severe cases it can be devastating and lead to its demise.

This sequence of events probably sounds familiar because it's not so very different from the process that leads to our own death. We, too, are finely balanced between metabolism and maintenance costs, a condition biologists refer to as *homeostasis*. The gradual buildup of unrepaired damage resulting from the wear and tear inherent in the process of living makes us less resilient and increasingly vulnerable to fluctuations and perturbations as we age. A case of the flu or pneumonia or a heart attack or stroke that might have been handled in youth and throughout midlife is often fatal once we have reached "old age."

Ultimately, we reach a stage where even a small perturbation such as a minor cold or heart flutter can lead to death.

While this image provides a useful metaphor for the mortality of companies, it represents only part of the picture. To dig a little deeper we must first define what we mean by death of a company because many of them disappear by mergers or acquisitions rather than by liquidating or going bankrupt. A useful definition is to use sales as the indicator of a company's viability, the idea being that if it's metabolizing then it's alive. Thus *birth* is defined as the time when a company first reports sales and *death* as the time when it ceases to do so. With this definition companies may die through a variety of processes: they may split, merge, or liquidate as economic and technological conditions change. While liquidation is often responsible for the death of companies, a much more common cause is their disappearance through mergers and acquisitions.

Of the 28,853 companies that have traded on U.S. markets since 1950, 22,469 (78 percent) had died by 2009. Of these 45 percent were acquired by or merged with other companies, while only about 9 percent went bankrupt or were liquidated; 3 percent privatized, 0.5 percent underwent leveraged buyouts, 0.5 percent went through reverse acquisitions, and the remainder disappeared for "other reasons."

Figures 71–74 show the survivorship and mortality curves for companies that were born and died within the period covered by the data set (1950–2009) as a function of how long they lived.[6] The curves have been separated into bankruptcies and liquidations on the one hand and into acquisitions and mergers on the other, and then further deconstructed according to the size of their sales. As can clearly be seen, the general structure of these curves is almost the same regardless of how the data are sliced, even when companies are separated into individual business sectors. In all cases the number of survivors falls rapidly immediately following their initial public offering, with fewer than 5 percent remaining alive after thirty years. Similarly, the mortality curves show that the number that have died reaches almost 100 percent within fifty years, with almost half of them having already disappeared in less than ten. It's tough being a company! The survival curves are well approximated by a simple exponential as shown in Figure 75, where the number of companies that have survived is plotted logarithmically versus their age; plotted this way, exponentials appear as straight lines.

You might have thought that these results would depend sensitively on whether death occurred via mergers and acquisitions rather than from bankruptcies and liquidations. However, as you can see, they both follow very similar exponential survival curves with only slightly different values for their mortality. One might also have expected the results to depend on which business sector a company is in. The dynamics and competitive market forces would seem to be quite different, for instance, in the energy sector compared with IT, transportation, or finance. Surprisingly, however, all business sectors show similar characteristic exponential survival curves with similar timescales: no matter which sector or what the stated cause is, only about half of the companies survive for more than ten years.

This is consistent with an analysis showing that companies scale in approximately the same way when broken down into separate business categories. Within each sector, power laws are obtained having exponents close to those found for the entire cohort of companies—those shown in Figure 75. In other words, the general dynamics and overall life history of companies are

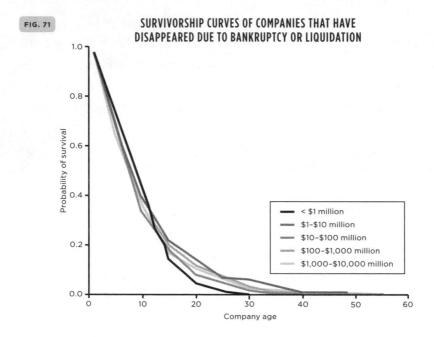

FIG. 71

**SURVIVORSHIP CURVES OF COMPANIES THAT HAVE
DISAPPEARED DUE TO BANKRUPTCY OR LIQUIDATION**

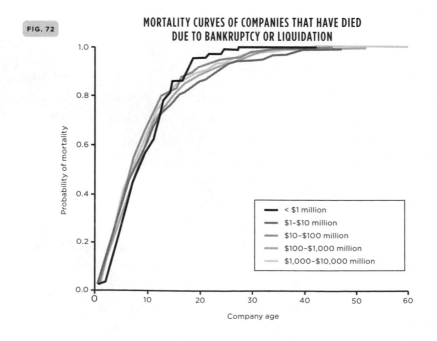

FIG. 72

**MORTALITY CURVES OF COMPANIES THAT HAVE DIED
DUE TO BANKRUPTCY OR LIQUIDATION**

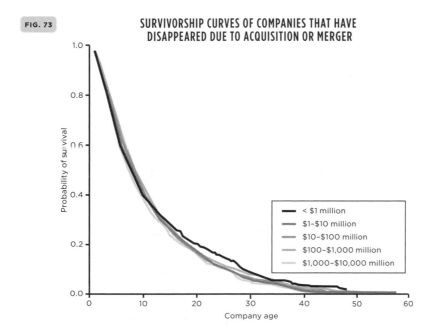

FIG. 73

SURVIVORSHIP CURVES OF COMPANIES THAT HAVE
DISAPPEARED DUE TO ACQUISITION OR MERGER

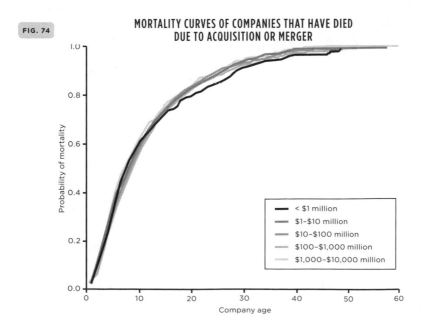

FIG. 74

MORTALITY CURVES OF COMPANIES THAT HAVE DIED
DUE TO ACQUISITION OR MERGER

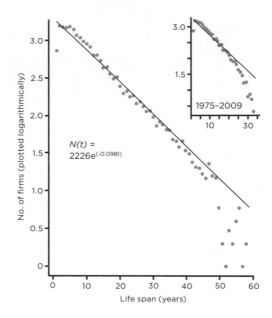

Pages 398–99: (71–74) Survivorship and mortality curves for U.S. publicly traded companies between 1950 and 2009 separated into bankruptcies and liquidations and acquisitions and mergers, and further deconstructed into different-size classes according to sales. Note how little variation there is among them. (75) Number of companies (N) plotted logarithmically versus their life span (t) showing classic exponential decay and therefore a constant mortality rate, as indicated by the straight line.

effectively independent of the business sector in which they operate. This strongly suggests that there is indeed a universal dynamic at play that determines their coarse-grained behavior, independent of their commercial activity or whether they are eventually going to go bankrupt or merge with or be bought by another company. In a word, this strongly supports the idea of a quantitative science of companies.

This is really quite amazing. After all, when we think of the birth, death, and general life history of companies as they struggle to establish and maintain themselves in the marketplace dealing with the vagaries, uncertainties, and unpredictability of economic life and the myriad specific decisions and accidents that led to the successes and failures that preceded their death, it's hard to believe that collectively they were following such simple general rules. This revelation echoes the surprise that organisms, ecosystems, and cities are likewise subject to generic constraints, despite the apparent uniqueness and individuality of their life histories.

Exponential survival curves similar to those exhibited by companies are manifested in many other collective systems such as bacterial colonies, animals and plants, and even the decay of radioactive materials. It is also believed that the mortality of prehistoric humans followed these curves before they became sedentary social creatures reaping the benefits derived from community structures and social organization. Our modern survival curve has evolved from being a classic exponential to developing a long plateau stretching over fifty years, as illustrated in Figure 25 in chapter 4, showing that we now live much longer on average than our hunter-gatherer forebears even though our maximum life span has remained pretty much what it always was.

What is the special property of exponentials that they describe the decay of so many disparate systems? It's simply that they arise whenever the death rate at any given time is directly proportional to the number that are still alive. This is equivalent to saying that the percentage of survivors that die within equal slices of time at any age remains the same. A simple example will make this clear: taking one year as the time slice, this says that the percentage of five-year-old companies that die before they reach six years old is the same as the percentage of fifty-year-old companies that die before they reach fifty-one. In other words: *the risk of a company's dying does not depend on its age or size.*

A lingering issue of possible concern is that the data cover only sixty years, so companies older than this are automatically excluded. Actually, it's worse than this because the analysis includes only those companies that were born and died in the time window between 1950 and 2009, thereby excluding all those that were born before 1950 and/or were still alive in 2009. This could clearly lead to a systematic bias in the estimates of life expectancy. A more complete analysis therefore needs to include these so-called censored companies, whose life spans are at least as long as and likely longer than the period over which they appear in the data set. This actually involves a sizable number of companies: in the sixty years covered, 6,873 firms were still alive at the end of the window in 2009. Fortunately, there is a well-established sophisticated methodology, called *survival analysis*, that has been developed precisely for addressing this issue.

Survival analysis was developed in medicine in order to estimate survival probabilities for patients who have undergone therapeutic interventions under test conditions. These tests have necessarily to be conducted over a limited time period, leading to the problem we face here, namely, that many subjects die after the test period has ended. The technique commonly used, called the Kaplan-Meier estimator, employs the entire data set and optimizes probabilities assuming that each death event is statistically independent of every other death.[7]

A detailed analysis using this technique was carried out for the complete cohort of companies in the Compustat data set, including all those that were previously censored, with the result that there was only a modest change from the previous censored estimates. The *half-life* of U.S. publicly traded companies was found to be close to 10.5 years, meaning that half of all companies that began trading in any given year have disappeared in 10.5 years.

Most of the hard work on this was done by an undergraduate intern, Madeleine Daepp, who joined us under the aegis of a wonderful program called Research Experience for Undergraduates (REU) funded primarily by the NSF. This provides support for undergraduates to spend a summer working in real-life research at institutions across the entire spectrum of scientific activity. At SFI we usually have about ten such bright young minds on-site, all of whom are treated as equal members of the institute and who work closely with individual researchers. It's a great experience for both us and them. Madeleine was a junior in mathematics at Washington University in St. Louis when she joined us and worked under the direct supervision of Marcus Hamilton. It's hard to complete such a project from scratch in just ten weeks, so Madeleine returned several times over the ensuing three years before the work was eventually completed and a successful paper published. I was delighted to learn just recently that she has been accepted to the PhD program in urban planning at MIT, which is one of the best in the world. I expect to hear great things of her in the future.

The survival analysis technique used for dealing with "incomplete observations" such as we have here was invented in 1958 by two statisticians, Edward Kaplan and Paul Meier. It has since been extended to areas outside of medicine and used to estimate, for example, how long people can expect to remain unemployed after a job loss or how long it takes for machine parts to fail. Amusingly, Kaplan and Meier each submitted similar but independent papers to the prestigious *Journal of the American Statistical Association* for publication, and a wise editor persuaded them to combine them into a single paper. This has since been cited in other scholarly papers more than 34,000 times, which is an extremely large number for an academic paper. For instance, Stephen Hawking's most famous paper, "Particle Creation by Black Holes," has been cited less than 5,000 times. Depending on the field, most papers are fortunate to receive even 25 citations. Several of my own that I thought were pretty damn good have been cited less than 10 times, which is pretty discouraging, although I am a coauthor of two of the most cited papers in ecology, each having more than 3,000 citations.

4. REQUIESCANT IN PACE

Although there are significant differences, it's hard not to be struck by how similar the growth and death of companies and organisms are when viewed through the lens of scaling—and how dissimilar they both are to cities. Companies are surprisingly biological and from an evolutionary perspective their mortality is an important ingredient for generating innovative vitality resulting from "creative destruction" and "the survival of the fittest." Just as all organisms must die in order that the new and novel may blossom, so it is that all companies disappear or morph to allow new innovative variations to flourish: better to have the excitement and innovation of a Google or Tesla than the stagnation of a geriatric IBM or General Motors. This is the underlying culture of the free market system.

The great turnover of companies and especially the continual churning of mergers and acquisitions are integral to the market process. And of course this means that the Googles and Teslas, which may seem invincible now, will themselves eventually fade away and disappear. From this point of view we should not lament the passing of any company—it's an essential component of economic life—we should only mourn and be concerned about the fate of the people who often suffer when companies disappear, whether they are the workers, management, or even the owners. If only we could tame the potential brutality and greed of the survival of the fittest and soften some its more egregious consequences by formulating a magic algorithm for how to balance the classic tension between regulation, government intervention, and uncontrolled rampant capitalism. This struggle painfully played itself out as we witnessed the struggle between the death throes of corporations that probably should have died and the desire to save jobs and protect the lives of workers because certain incompetent if not duplicitous corporations were deemed "too big to fail" during the 2008 financial crisis.

It may be a platitude, but it's nevertheless true that nothing stays the same. Both Standard & Poor's and the business magazine *Fortune* construct ongoing lists of the five hundred most successful companies, and there is a certain prestige associated with being named on both of these lists. Richard Foster, who was for twenty-two years a director and senior partner of the well-known business consultants McKinsey & Company, analyzed the tenure of companies on these lists and discovered that it has been regularly decreasing over the past sixty years. For example, he found that in 1958 a company could expect to stay on the S&P 500 for about sixty-one years, whereas today it's more like eighteen. Of the Fortune 500 companies in 1955, only sixty-one were still on the list in 2014. That's only a 12 percent survival rate, the other 88 percent having gone bankrupt, merged, or fallen from the list because of underperformance. More poignant, perhaps, is that most of those on the list in 1955 are unrecognizable and completely forgotten today; how many remember Armstrong Rubber or Pacific Vegetable Oil?

In 2000 Foster wrote an influential best-seller on business, aptly titled *Creative Destruction*.[8] He was much taken by the ideas on complexity being developed at the Santa Fe Institute, so much so that he joined the board of trustees and persuaded McKinsey to fund a professorship in finance which was held by Doyne Farmer. I got to know him when I first engaged with SFI in the late 1990s. He was convinced that the scaling and network ideas that we had been developing in biology could provide major insights into how companies function. He pointed out that there were no quantitative, mechanistic theories of companies, and that because they are very often compared to organisms, this approach might provide a novel way of developing such a theory. He generously offered to give me access to the large McKinsey database on companies and to help support a postdoc to carry out the research. I was still at Los Alamos at the time, engaged with running high energy physics and knew even less about companies than I do now. Furthermore, the research in biology was still at quite an early stage and I was not convinced that it could readily be extended to companies, so despite being flattered by the offer, I did not pursue it. In retrospect that was probably the right decision at the time, but it speaks volumes about Dick Foster that he had the foresight to see that the scaling approach might provide a useful basis for understanding companies. It took more than a decade of extensive work on organisms, ecosystems, and cities before we were in a position to address Dick's challenge.

Unfortunately, it's not a straightforward exercise to relate the observations regarding the tenure of companies on the S&P and Fortune 500 lists to their actual life spans without a detailed analysis that takes into account their age and whether or not they died. Nevertheless, the findings dramatically illustrate the fragility of seemingly powerful companies and are also a striking example of the speeding up of socioeconomic life.

The survival analyses tell us that there should be very few very old companies. An extrapolation of the theory and data predicts that the probability of a company's lasting for one hundred years is only about forty-five in a million, and for it to last two hundred years it's a minuscule one in a billion. These numbers should not be taken too seriously, but they do give us a sense of the scale of long-term survivability and provide an interesting insight into the characteristics of companies that have remained viable for hundreds of years. There are at least 100 million companies in the world, so if they all obey similar dynamics, then one would expect only about 4,500 to survive for a hundred years, but none for two hundred. However, it is well known that there are lots of companies, especially in Japan and Europe, that have lived for hundreds of years. Unfortunately, there are no comprehensible data sets or any systematic statistical analyses of these remarkable outliers, though there are plenty of anecdotal stories. Nevertheless, we can learn something instructive about the aging of companies from the general characteristics of these very long-lived outliers.

Most of them are of relatively modest size, operating in highly specialized niche markets, such as ancient inns, wineries, breweries, confectioners, restaurants, and the like. These have quite a different character from the kinds of companies we have been considering in the Compustat data set and the S&P and Fortune 500 lists. In contrast to most of these, these outliers have survived not by diversifying or innovating but by continuing to produce a perceived high-quality product for a small, dedicated clientele. Many have maintained their viability through reputation and consistency and have barely grown. Interestingly, most of them are Japanese. According to the Bank of Korea, of the 5,586 companies that were more than two hundred years old in 2008, over half (3,146 to be precise) were Japanese, 837 German, 222 Dutch, and 196 French. Furthermore, 90 percent of those that were more than one hundred years old had fewer than three hundred employees.

There are some wonderful examples of these geriatric survivors. For instance, the oldest shoemaker in Germany is the Eduard Meier company, founded in Munich in 1596, which became purveyors to the Bavarian aristocracy. It still has only a single store that sells, though no longer makes, quality upscale shoes. The oldest hotel in the world according to *Guinness World Records* is Nishiyama Onsen Keiunkan in Hayakawa, Japan, which was founded in 705. It has been in the same family for fifty-two generations and even in its modern incarnation has only thirty-seven rooms. Its main attraction seems to be its hot springs. The world's oldest company was purported to be Kongo Gumi, founded in Osaka, Japan, in 578. It was also a family business going back many generations, but after almost 1,500 years of continuously being in business it went into liquidation in 2006 and was purchased by the Takamatsu Corporation. And what was the niche market that Kongo Gumi cornered for 1,429 years? Building beautiful Buddhist temples. But sadly, with the changes in Japanese culture following the Second World War the demand for temples dried up and Kongo Gumi was unable to adapt fast enough.

5. WHY COMPANIES DIE, BUT CITIES DON'T

The power of scaling is that it can potentially reveal the underlying principles that determine the dominant behavior of highly complex systems. For organisms and cities this fruitfully led to a network-based theory for quantitatively understanding the principal features of their dynamics and structure, from which many of their salient properties could be understood. In both cases we know quite a bit about their network structures, whether they are circulatory systems, road networks, or social systems. On the other hand, despite an extensive literature on the subject, we know much less about the network structures of companies other than that they are for the most part hierarchical. Standard company organizational charts are typically top-down, having a treelike structure superficially suggestive of a classic self-similar fractal. That would explain why companies manifest power law scaling.

Unfortunately, however, we don't have extensive quantitative data on these organizational networks comparable to what we have for cities and organisms. For instance, we don't typically know how many people are functioning at each level, how much of the company's finances and resources are flowing between them, and how much information they are exchanging. And even if some of this were available, we need to have it across the entire size range of companies. Furthermore, it is not at all clear that the "official" organizational charts of companies represents the actuality of what the real operational network structures are. Who is really communicating with whom, how often are they doing it, how much do they exchange, and so on? What is really needed is access to all of the company's communication channels, such as the phone calls, the e-mails, the meetings, et cetera, quantified analogously to the cell phone data we used for helping to develop a science of cities. It's unlikely such comprehensive data exist and even less likely that we would ever gain ready access to them. Companies are quite wary of exposing themselves to outside investigators unless they are paying them exorbitant consultant fees, presumably so that they can maintain control. But if you want to understand how a company really functions, or want to develop a serious science of companies, then this is the kind of data that is ultimately needed.

Consequently, we don't have a well-developed mechanistic framework analogous to the network-based theory for organisms and to a lesser extent cities for analytically understanding the dynamics and structure of companies and, in particular, for calculating the values of their exponents. Nevertheless, just as we have been able to construct a theory of their growth trajectories, we can address the question of their mortality by extrapolating from what we have already learned.

I emphasized earlier that most companies operate close to a critical point where sales and expenses are finely balanced, making them potentially vulnerable to fluctuations and perturbations. A major shock at the wrong time can lead to their demise. Younger companies, which are buffered against this by an initial capital endowment, become particularly vulnerable once this initial infusion is expended if they are unable to turn a significant profit. This is sometimes referred to as the liability of adolescence.

The fact that companies scale sublinearly, rather than superlinearly like cities, suggests that they epitomize the triumph of economies of scale over innovation and idea creation. Companies typically operate as highly constrained top-down organizations that strive to increase efficiency of production and minimize operational costs so as to maximize profits. In contrast, cities embody the triumph of innovation over the hegemony of economies of scale. Cities aren't, of course, driven by a profit motive and have the luxury of being able to balance their books by raising taxes. They operate in a much more distributed fashion, with power spread across multiple organizational structures from mayors and councils to businesses and citizen action groups. No single group has absolute control. As such, they exude an almost laissez-faire, free-wheeling ambience relative to companies, taking advantage of the innovative benefits of social interactions whether good, bad, or ugly. Despite their apparent bumbling inefficiencies, cities are places of action and agents of change relative to companies, which by and large usually project an image of stasis unless they are young.

To achieve greater efficiency in the pursuit of greater market share and increased profits, companies stereotypically add more rules, regulations, protocols, and procedures at increasingly finer levels of organization, resulting in the increased bureaucratic control that is typically needed to administer, manage, and oversee their execution. This is often accomplished at the expense of innovation and R&D (research and development), which should be major components of a company's insurance policy for its long-term future and survivability. It's difficult to obtain meaningful data on "innovation" in companies because it's not straightforward to quantify. Innovation is not necessarily synonymous with R&D, especially as there are significant tax advantages in labeling all sorts of extraneous activities as R&D expenses. Nevertheless, from analyzing the Compustat data set we found that the relative amount allocated to R&D systematically *decreases* as company size increases, suggesting that support for innovation does not keep up with bureaucratic and administrative expenses as companies expand.

The increasing accumulation of rules and constraints is often accompanied by stagnating relationships with consumers and suppliers that lead companies to become less agile and more rigid and therefore less able to respond to significant change. In cities we saw that one very important hallmark is that they become ever more diverse as they grow. Their spectrum of business and economic activity is incessantly expanding as new sectors develop and new opportunities present themselves. In this sense cities are prototypically *multidimensional,* and this is strongly correlated with their superlinear scaling, open-ended growth, and expanding social networks—and a crucial component of their resilience, sustainability, and seeming immortality.

While the dimensionality of cities is continually expanding, the dimensionality of companies typically contracts from birth through adolescence, eventually stagnating or even further contracting as they mature and move into old age. When still young and competing for a place in the market, there is a youthful excitement and enthusiasm as new products are developed and ideas bubble up, some may be crazy and unrealistic and some grandiose and visionary. But market forces are at work so that only a few of these are successful as the company gains a foothold and an identity. As it grows, the feedback mechanisms inherent in the market lead to a narrowing of its product space and inevitably to greater specialization. The great challenge for companies is how to balance the positive feedback from market forces, which strongly encourage staying with "tried and true" products versus the long-term strategic need to develop new areas and commodities that may be risky and won't give immediate return.

Most companies tend to be shortsighted, conservative, and not very supportive of innovative or risky ideas, happy to stay almost entirely with their major successes while the going is good because these "guarantee" short-term returns. Consequently, they tend toward becoming more and more *unidimensional*. This reduction in diversity coupled with the predicament described earlier in which companies sit near a critical point is a classic indicator of reduced resilience and a recipe for eventual disaster. By the time a company realizes its condition it is often too late. Reconfiguring and reinventing become increasingly difficult and expensive. So when a large enough unanticipated fluctuation, perturbation, or shock comes along the company becomes seriously at risk and ripe for a takeover, buyout, or simply going belly-up. In a word, it is, as the Mafiosi put it, *il bacio della morte*—the kiss of death.[9]

10

THE VISION OF A GRAND UNIFIED
THEORY OF SUSTAINABILITY

I n this final chapter I want to bring together some of the threads that have
been developed throughout the book and weave them into a tapestry that I
hope will stimulate deeper thinking and speculation about the future of the
extraordinary exponentially expanding socioeconomic universe that we have
created.

One of the major challenges of the twenty-first century that will have to be
faced is the fundamental question as to whether human-engineered social systems, from economies to cities, which have only existed for the past five thousand years or so, can continue to coexist with the "natural" biological world
from which they emerged and which has been around for several billion years.
To sustain more than 10 billion people living in harmony with the biosphere at
a standard of living and quality of life comparable to what we now have requires that we develop a deep understanding of the principles and underlying
system dynamics of this social-environmental coupling. I have argued that a
critical component of this is to develop a deeper understanding of cities and
urbanization. Continuing to pursue limited and single-system approaches to
the many problems we face without developing a unifying framework risks the
possibility that we will squander huge financial and social capital and fail mis-
erably in addressing the really big question, resulting in dire consequences.

Existing strategies have, to a large extent, failed to come to terms with an essential feature of the long-term sustainability challenge embodied in the paradigm of complex adaptive systems; namely, *the pervasive interconnectedness and interdependency of energy, resources, and environmental, ecological, economic, social, and political systems.* One of the most important results to emerge from the work I discussed in chapters 7 and 8 is that all socioeconomic activity, from innovation and wealth creation to crime and disease—whether good, bad, or ugly—is quantitatively interrelated, as manifested in the universality of scaling laws. Almost all existing approaches to the challenge of global sustainability focus on relatively specific issues, such as the environmental consequences of future energy sources, the economic consequences of climate change, and the social impact of future energy and environmental choices. While such focused studies are of obvious importance and where most of our research efforts should be directed, they are not sufficient. They focus primarily on the trees and risk missing the forest.

It's time to recognize that a broad, multidisciplinary, multi-institutional, multinational initiative, guided by a broader, more integrated and unified perspective, should be playing a central role in guiding our scientific agenda in addressing this issue and informing policy. We need a broad and more integrated scientific framework that encompasses a quantitative, predictive, mechanistic theory for understanding the relationship between human-engineered systems, both social and physical, and the "natural" environment—a framework I call a *grand unified theory of sustainability.* It's time to initiate a massive international Manhattan-style project or Apollo-style program dedicated to addressing global sustainability in an integrated, systemic sense.[1]

ACCELERATING TREADMILLS, CYCLES OF INNOVATION, AND FINITE TIME SINGULARITIES

In biology, the network principles underlying economies of scale and sublinear scaling have two profound consequences. They constrain the pace of life—big animals live longer, evolve more slowly, and have slower heart rates, all to the same degree—and limit growth. In contrast, cities and economies are driven by social interactions whose feedback mechanisms lead to the opposite behavior. The pace of life systematically increases with population size: diseases spread faster, businesses are born and die more often, and people even walk faster in larger cities, all by approximately the same 15 percent rule. Moreover, the social network dynamic underlying superlinear scaling leads to open-ended growth, which is the primary assumption upon which modern cities and economies are based. Continuous adaptation, not equilibrium, is the rule.

This is a wonderfully consistent picture: the same conceptual framework based on underlying network dynamics and geometry with the same mathematical structure leads to quite different outcomes in these two very different cases, and both are strongly supported by a plethora of diverse data and observations. However, there is a big catch with potentially huge consequences. Even though the growth of organisms, cities, and economies follows essentially identical mathematical equations, their resulting solutions have subtle but crucial differences arising from one being driven by sublinear scaling (the economies of scale of organisms) and the other by superlinear scaling (the increasing returns to scale of cities and economies): in the superlinear case, the general solution exhibits an unexpectedly curious property technically known as a *finite time singularity,* which is a signal of inevitable change, and possibly of potential trouble ahead.

A finite time singularity simply means that the mathematical solution to the growth equation governing whatever is being considered—the population, the GDP, the number of patents, et cetera—*becomes infinitely large at some finite time,* as illustrated in Figure 76. This is obviously impossible, and that's why something has to change.

Before addressing some of the consequences of this phenomenon, let me first elaborate on some of its salient features. Simple power laws and exponentials are continuously increasing functions that also eventually become infinitely large, but they take an *infinite* time to do so. Another way of saying this is that in these cases the "singularity" has been pushed off to an infinite time into the future, thereby rendering it "harmless" relative to the potential impact of a *finite* time singularity. In the case of growth driven by superlinear scaling, the approach to the finite time singularity, represented by the solid line in Figure 76, is *faster* than exponential. This is often referred to as *super-*

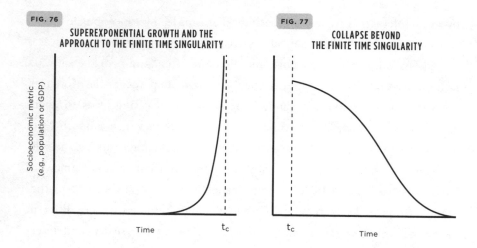

(76) Graph illustrating a *finite time singularity:* the quantity being plotted is growing superexponentially and becomes infinite at a finite time, t_c, denoted by the dotted vertical line. (77) Illustration of stagnation and collapse beyond the singularity.

exponential, a term I've already used earlier when discussing the growth of cities.

This kind of growth behavior is clearly unsustainable because it requires an unlimited, ever-increasing, and eventually infinite supply of energy and resources at some finite time in the future in order to maintain it. Left unchecked, the theory predicts that it triggers a transition to a phase that leads to stagnation and eventual collapse, as illustrated in Figure 77. This scenario sounds just like a rehash of the standard Malthusian argument that has been summarily dismissed by generations of economists: namely, that we won't be able to keep up with demand and that open-ended growth will eventually lead to catastrophe.

Which brings us to the crux of the matter. Because of the presence of a finite time singularity resulting from superlinear scaling, this scenario is categorically different from that of Malthus. If growth were purely exponential as assumed by Malthusians, neo-Malthusians, their followers, and critics, then the production of energy, resources, and food could at least in principle keep up with exponential expansion because all of the relevant characteristics of the economy or city remain finite, even if they continue to increase in size and become very large.

This cannot be achieved if you are growing superexponentially and approaching a finite time singularity. In this scenario demand gets progressively larger and larger, eventually becoming *infinite* within a *finite* period of time. It is simply not possible to supply an infinite amount of energy, resources, and food in a finite time. So if nothing else changes, this inextricably leads to stagnation and collapse, as illustrated in Figure 77. An extensive analysis carried out in 2001 by Didier Sornette and Anders Johansen, then at UCLA, showed that data on population growth and the growth of financial and economic indicators strongly support the theoretical predictions that we have been growing superexponentially and are indeed headed toward such a singularity.[2]

I want to emphasize that this situation is qualitatively quite different from classic Malthusian dynamics, where this is no such singularity. The existence of a singularity signifies that there has to be a transition from one phase of the system to another having very different characteristics, analogous to the way the condensation of steam to water which subsequently freezes to ice epitomizes transitions between different phases of the same system, each having quite different physical properties. And indeed, underlying such familiar phase transitions are singularities in the thermodynamic variables characterizing the system (water) but in terms of temperature rather than time (0°C for freezing and 100°C for boiling). Unfortunately, for cities and socioeconomic systems the phase transition stimulated by the finite time singularity is from superexponential growth to stagnation and collapse, and this could lead to potentially devastating consequences.

So how can such a collapse be avoided, and can it be achieved while still ensuring open-ended growth? The first point to appreciate is that these predictions assume that the parameters of the growth equation do not change. So one clear strategy for forestalling a potential catastrophe is to intervene before reaching the singularity by "resetting" the parameters. Moreover, to maintain open-ended growth with these new settings requires that the driving term in the equation—the "social metabolism"—needs to remain superlinear, meaning that the new dynamic must still be driven by the positive feedback forces of social interaction responsible for innovation, and for wealth and knowledge creation. Such an "intervention" is none other than what is usually referred to as an *innovation*. A major innovation effectively resets the clock by changing the conditions under which the system has been operating and growth occurring. Thus, *to avoid collapse a new innovation must be initiated that resets the clock, allowing growth to continue and the impending singularity to be avoided.*

Major innovations can therefore be viewed as mechanisms for ensuring a soft transition to a new phase by circumnavigating the potentially disastrous discontinuity inherent in the black hole of a finite time singularity. Having made the transition and "reset the clock" to avoid stagnation and collapse, the process begins all over again with the continuation of superexponential growth, eventually leading to a new finite time singularity which likewise has to be circumvented. The entire sequence is continually repeated, thereby pushing potential collapse as far into the future as the creativity, inventiveness, and resourcefulness of human beings allow. This can be restated as a sort of "theorem": *to sustain open-ended growth in light of resource limitation requires continuous cycles of paradigm-shifting innovations*, as illustrated in Figure 78.

In actuality the breaks between consecutive phases are not sharp and discontinuous as drawn in the figure but are smeared out over relatively short periods of time around each transition. After all, the Industrial Revolution didn't begin on a specific day, or even in a specific year, but emerged over a period of a few years around 1800, which is relatively small compared with the timescales of its influence.[3]

This result shouldn't come as any great surprise because this is how continued open-ended growth in both the population and socioeconomic activity has been sustained. On the grand scale the discoveries of iron, steam, coal, computation, and, most recently, digital information technology are among the major innovations that have fueled our continued growth and expansion. Indeed, the litany of such discoveries is a testament to our extraordinary ingenuity.

It was this essential feature that was missing in Malthus's original argument as well as in the subsequent arguments of most of its modern and more recent protagonists, beginning with Paul Ehrlich and the Club of Rome in the 1970s. Their admonitions were dismissed by most economists primarily be-

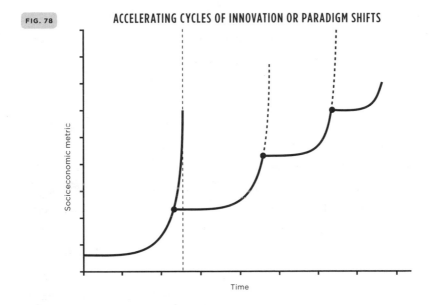

FIG. 78

ACCELERATING CYCLES OF INNOVATION OR PARADIGM SHIFTS

Socieconomic metric

Time

Graph of successive superexponential growth trajectories, each potentially leading to a finite time singularity (denoted by the vertical dotted line) and subsequent collapse, as in Figure 77, unless an innovation is made at a time prior to the singularity (denoted by the black dots), which resets the clock to start the entire cycle over again. For ease of presentation dotted lines indicating later finite time singularities associated with the black dots are suppressed. The image at the left represents the myth of Sisyphus.

cause they ignored the crucial role of innovation. The concept of business and economic cycles, and of implied cycles of innovation, has been around for a long time and is now standard rhetoric in economics and the business community, even though it is primarily based on broad phenomenological deductions with little fundamental theory or mechanistic understanding. It is implicitly taken for granted, and often taken as unquestioned dogma, that as long as human beings remain inventive we will stay ahead of any impending threat by continuous and ever more ingenious innovations.

Unfortunately, however, it's not quite as simple as that. There's yet another major catch, and it's a big one. The theory dictates that to sustain continuous growth *the time between successive innovations has to get shorter and shorter.* Thus paradigm-shifting discoveries, adaptations, and innovations must occur at an increasingly accelerated pace. Not only does the general pace of life inevitably quicken, but we must innovate at a faster and faster rate!

This is illustrated in Figure 78, where the black dots, which indicate the beginning of each new cycle of innovation, become increasingly closer together with time: in addition to the pace of life accelerating as we climb up each growth curve, we have to make major innovations and transition to a new state at an increasingly accelerated rate. The treadmill metaphor I used earlier in both chapters 1 and 8 when explaining the contraction of socioeconomic time and the increasing pace of life told only part of the story, and it's worth expanding upon it here. We're not only living on an accelerating treadmill that's always getting faster and faster, but at some stage we have to jump onto another treadmill that is accelerating even faster and sooner or later have to jump from that one onto yet another one that's going even faster. And this entire process has to be continually repeated into the future at a faster and faster rate.

This is an extraordinary image and sounds like some bizarre psychotic behavior. It's hard to believe that it can be sustained without our having a collective heart attack! It almost makes Sisyphus's task seem pretty mild. You may recall that the gods had condemned Sisyphus to ceaselessly push a huge rock up to the top of a mountain, from which it would immediately roll back down under its own weight only for him to have to start all over again from the bottom. Many reasons have been advanced for why Sisyphus was being so severely punished, but the two I like best that relate to the Sisyphean task we have created for ourselves are that he had stolen the secrets of the gods and had put Death in chains. No more need be said other than to recognize that our task is actually much harder than Sisyphus's because we have to push the rock back up to the top of the mountain at a faster rate each time.

These successive accelerating cycles of faster than exponential growth predicted by the theory are consistent with observations for cities, waves of technological change, and the world population (I have already referred to the work of Sornette and Johansen). As a specific example, consider the growth curve of New York City from 1790 to the present, shown in Figure 58 of chap-

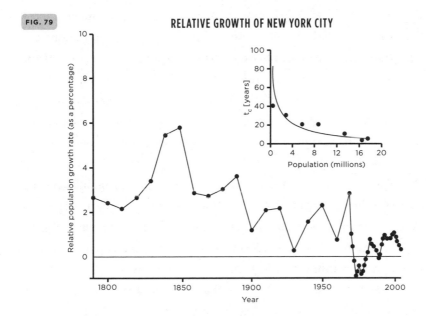

FIG. 79

RELATIVE GROWTH OF NEW YORK CITY

Growth of New York City from 1790 relative to the smooth superexponential dominant background showing successive cycles whose frequency systematically decreases in quantitative agreement with theoretical predictions (the curve of the inset).

ter 8. The various successive phases of its growth are highlighted in the enhanced solid black lines. By looking at how these deviate relative to the smooth dominant background of "pure" superexponential growth, the sequence of changes reflecting cyclic dynamics at the urban scale is clearly revealed. As you can see from Figure 79, the data support the idea of cycles whose frequency systematically increases with time. The inset shows how the time between these successive "innovations" gets progressively shorter consistent with the *quantitative* prediction from theory.

On a grander scale the prediction of accelerating cycles of major innovations is also strongly supported by data. One of the issues here is deciding which of the enormous number of possible innovations constitute major paradigm shifts. To some extent this is in the eyes of the beholder, but it's likely that most of us would agree that certain discoveries and innovations such as printing, coal, the telephone, and computers constitute a major "paradigm shift," whereas railways and cell phones may be more debatable. Unfortunately, there is no established quantitative "science of innovation" and therefore no universally agreed-upon criteria or data relating directly to major innovations and paradigm shifts, let alone to finite time singularities. So in order to confront theory with data we have to rely on informal studies and a certain degree of intuition. This situation may well change as innovation becomes an increasingly active area of investigation, with researchers beginning to grapple with questions such as what is innovation, how do we measure it, how does it happen, and how can it be facilitated?[4]

The well-known inventor and futurist Ray Kurzweil has compiled and analyzed a list of his candidates for major innovations in a form that is highly suitable for comparison with our predictions.[5] His results are shown in Figures 80 and 81, where the time between successive innovations is plotted against how long ago each innovation happened. Two versions are presented: one semilogarithmic (meaning that the vertical axis is logarithmic while the horizontal one is linear), and the other logarithmic for both axes. To get oriented, note that the first data point in the top left-hand corner of both graphs tells us that life began about 4×10^9 (four billion) years ago—measured along the horizontal axis—and that it took almost another two billion years before the next major innovation happened—measured along the vertical axis. It's interesting to note that when viewed on the linear timescale (Figure 80), everything appears to happen at the same time following the appearance of our earliest ancestors a million or so years ago. The curve falls precipitously, dramatically illustrating the acceleration of time. This graph also shows again why plotting such data logarithmically (Figure 81) is so much more illuminating in separating different events taking place over such vast timescales: for instance, plotted this way the appearance of the telephone just a hundred years ago can be temporally separated from the appearance of agriculture ten thousand years ago.

The theory explains and predicts this inverse relationship between the successive shortening of the time between innovations and how long ago they happened and is *quantitatively* consistent with the lines drawn on both graphs. I hasten to add, however, that the predictions pertain only to the part of these graphs where innovation is due to socioeconomic dynamics that engender human ingenuity; the theory makes no prediction about the rate of biological innovations. This leaves open the intriguing question as to whether an analo-

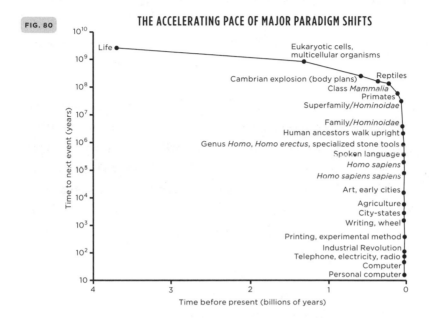

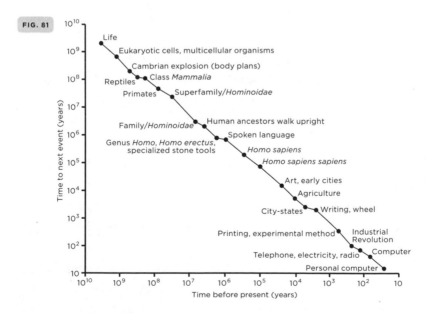

The time between major innovations versus how long ago the innovation happened plotted both semilogarithmically (80) and logarithmically (81).

gous singularity dynamic has played a similar role in driving biological innovation or whether the agreement and the simple power law relationship that
extends into this very early prehuman time period is just an accident or a result of Kurzweil's judicious choice of paradigm shifts. In any case, the agreement with theory where it applies is compelling and is supported by more
detailed analysis confined to just the past few hundred years.

The general concept of a *singularity* plays an important role in mathematics and theoretical physics. A singularity is a point at which a mathematical
function is no longer "well behaved" in some specific way, such as becoming
infinite in the manner I have been discussing. Defining how to tame such singularities stimulated enormous progress in nineteenth-century mathematics,
and this subsequently had a huge impact on theoretical physics. Its most famous popular consequence was the concept of black holes, which arose from
trying to understand the singularity structure of Einstein's theory of general
relativity.

Until it was popularized by Kurzweil in his 2005 book *The Singularity Is
Near: When Humans Transcend Biology,* "singularity" was a term that was
hardly used colloquially. Building on an earlier idea of a "technological singularity" introduced in 1993 by the science-fiction writer and computer scientist
Vernor Vinge, Kurzweil proposed that we are approaching a singularity in
which our bodies and brains will be augmented by genetic alterations, nano-
technology, and artificial intelligence to become hybrid cyborgs no longer
bound by the constraints of biology. It was suggested that this would result in
a collective intelligence that will be enormously more powerful than all present human intelligence combined. Or as Vinge so succinctly put it: "Within
thirty years, we will have the technological means to create superhuman intelligence. Shortly after, the human era will be ended."[6] This was written in 1993,
so the prediction was that this will come to pass by 2023, which is now just
seven years hence. I don't think so.

These are fascinating speculations which may well eventually come to pass,
but are presently in the realm of science fiction. This Pollyanna vision for the
future is almost diametrically opposite to the dire predictions of the neo-
Malthusians even though the conclusions in both cases are ironically based on
the same premise that exponential growth is not sustainable as is and that

something dramatic has to happen. Just as the Malthusians ignored the crucial role of innovation, the singularity enthusiasts ignore the crucial role of the entire socioeconomic dynamic of the planet, which in fact is the prime driver of the impending singularity. Neither case is anchored in a broader framework that embraces a quantitative mechanistic theory, so whatever their predictions are it's very hard to evaluate them scientifically. Perhaps the greatest conceptual irony, especially of the singularists, is that their conclusions and speculations are based on exponential growth, which doesn't actually lead to a singularity, at least not in a *finite time.*

Nevertheless, exponential growth may very well be unsustainable for the very kind of concrete reasons originally advanced by Malthus, namely, that we will be unable to produce enough food or sufficient energy, or that we will run out of essential resources such as phosphorus, oil, or titanium, and that at the same time, we will have failed to develop the appropriate technology to address these issues. In addition, we may be producing so much entropy that the resulting pollution, environmental damage, and other induced changes, particularly to the climate, become insurmountable, leading to untold unintended devastating consequences. However, I remind you that if this is the result of exponential growth, then there's nothing *in principle* hindering us from accomplishing what the optimists proclaim and continuing to grow by dint of innovating ourselves out of all of these multiple problems and threats. But *in practice* this could be very a different story, and I am not at all optimistic that it could be achieved.

But this is not the situation. The actuality is qualitatively quite different. As I emphasized earlier, we have been expanding at a *super*exponential rate rather than "just" exponentially, and this has been driven by the superlinear scaling of socioeconomic activity as a result of the multiplicative enhancement inherent in our social dynamics. This quintessential modern human dynamic has led to the speeding up of the pace of life and of the rate at which we have to make major innovations in order to combat the imminent threat of what finite time singularities portend. The image of an accelerating Sisyphus haunts us.

The time between the "Computer Age" and the "Information and Digital Age" was no more than about thirty years—to be compared with the thousands of years between the Stone, Bronze, and Iron ages. The clock by which we measure time on our watches and digital devices is very misleading; it is determined by the daily rotation of the Earth around its axis and its annual rotation around the sun. This astronomical time is linear and regular. But the actual clock by which we live our socioeconomic lives is an emergent phenomenon determined by the collective forces of social interaction: it is continually and systematically speeding up relative to objective astronomical time. We live our lives on the metaphorical accelerating socioeconomic treadmill. A major innovation that might have taken hundreds of years to evolve a thousand or more years ago may now take only thirty years. Soon it will have to take twenty-five, then twenty, then seventeen, and so on, and like Sisyphus we are destined to go on doing it, *if we insist on continually growing and expanding.* The resulting sequence of singularities, each of which threatens stagnation and collapse, will continue to pile up, leading to what mathematicians call an *essential singularity*—a sort of mother of all singularities.

The great John von Neumann, mathematician, physicist, computer scientist, and polymath, a man whose ideas and accomplishments have had a huge influence on your life, made the following remarkably prescient observation more than seventy years ago: "The ever accelerating progress of technology and changes in the mode of human life . . . gives the appearance of approaching some essential singularity in the history of the race beyond which human affairs, as we know them, could not continue."[7] Among von Neumann's many accomplishments before he died at the relatively young age of fifty-three in 1957 are his seminal role in the early development of quantum mechanics, his invention of game theory, which is a major tool in economic modeling, and the conceptual design of modern computers universally referred to as the von Neumann architecture.

So can we imagine making an innovation as powerful and influential as the invention of the Internet every fifteen, ten, or even five years? This is a classic reductio ad absurdum argument showing that regardless of how ingenious we are, how many marvelous gadgets and devices we invent, we simply won't be able to overcome the threat of the ultimate singularity if we continue business as usual.

Estimates from the theory suggest that we are due for another paradigm shift within the next twenty to thirty years' time. This is a little shorter than the estimates from fits to data by Johansen and Sornette, who suggest a number more like thirty-five years. The theory cannot, of course, tell us the nature of the change, so we can only make wild speculations as to its nature. It could be something as relatively mundane as driverless cars and associated smart devices or something as dramatic as the kind of science fiction fantasy of Kurzweil and the singularists. Most likely it's none of the above and, if we are able to make the paradigm shift, it'll be something totally unexpected. Perhaps more likely is that we can't make the shift, and that we will need to come to terms with the whole concept of open-ended growth and find some new way of defining "progress" or be content with what we've got and spend our energies raising the entire planet's standard of living to reflect a comparably high quality of life. Now, that would be a truly major paradigm shift!

Continuous growth and the consequent ever-increasing acceleration of the pace of life have profound consequences for the entire planet and, in particular, for cities, socioeconomic life, and the process of global urbanization. Until recent times, the time between major innovations far exceeded the productive life span of a human being. Even in my own lifetime it was unconsciously assumed that one would continue working in the same occupation using the same expertise throughout one's life. This is no longer true; a typical human being now lives significantly longer than the time between major innovations, especially in developing and developed countries. Nowadays young people entering the workforce can expect to see several major changes during their lifetime that will very likely disrupt the continuity of their careers.

This increasingly rapid rate of change induces serious stress on all facets of urban life. This is surely not sustainable, and, if nothing changes, we are heading for a major crash and a potential collapse of the entire socioeconomic fabric. The challenges are clear: Can we return to an analog of a more "ecological" phase from which we evolved and be satisfied with some version of sublinear scaling and its attendant natural limiting, or no-growth, stable configuration? Is this even possible? Can we have the kind of vibrant, innovative, creative society driven by ideas and wealth creation as manifested by the best of our world's cities and social organizations, or are we destined to a planet of urban slums and the ultimate specter of devastation raised by Cormac McCarthy's novel *The Road*?[8] Given the special, unique role of cities as the originators of many of our present problems and their continuing role as the superexponential driver toward potential disaster, understanding their dynamics, growth, and evolution in a scientifically predictable, quantitative framework is crucial to achieving long-term sustainability on the planet. Perhaps of even greater importance for the immediate future is to develop such a theory within the context of a *grand unified theory of sustainability* by bringing together the multiple studies, simulations, databases, models, theories, and speculations concerning global warming, the environment, financial markets, risk, economies, health care, social conflict, and the myriad other characteristics of man as a social being interacting with his environment.

AFTERWORD

---

1. SCIENCE FOR THE TWENTY-FIRST CENTURY

From the outset I have emphasized that the underlying philosophical framework that has guided many of the arguments presented in this book is based on a paradigm inspired by a physics perspective. Consequently, a major theme has been to explore the extent to which a quantitative, predictive understanding based on underlying generic principles that transcend the details of any particular system can be developed. A fundamental tenet of science is that the world around us is ultimately governed by universal principles, and it is in this context that scaling laws for highly complex systems such as organisms, cities, or companies should be viewed. As I have tried to demonstrate, scaling laws reflect systematic regularities that reveal underlying geometric and dynamical behaviors, suggesting that a possible quantitative science of such systems might be attainable. At the very least, they allow us to explore just how far such a paradigm can be pushed.

The quest for grand syntheses, for commonalities, regularities, ideas, and concepts that transcend the narrow confines of specific problems or disciplines, is one of the great inspirational drivers of science and scientists. Arguably, it is also a defining characteristic of *Homo sapiens*, manifested in our multitudinous creeds, religions, and mythologies that help us come to terms with the awesome mysteries of the universe. This search for synthesis and unification has been a major theme in science since its origins in early Greek thinking, which introduced concepts such as atoms and elements as fundamental building blocks from which everything else is constructed.

Among the classic grand syntheses in modern science are Newton's laws, which taught us that heavenly laws are no different from those on Earth; Maxwell's unification of electricity and magnetism, which brought the ephemeral ether into our lives and gave us electromagnetic waves; Darwin's theory of natural selection, which reminded us that we're just animals and plants after all; and the laws of thermodynamics, which suggest that we can't go on forever. Each of these has had profound consequences not only in changing the way we think about the world, but also in laying the foundations for technological advancements that have led to the standard of living many of us are privileged to enjoy. Nevertheless, they are all to varying degrees incomplete. Indeed, understanding the boundaries of their applicability, the limits to their predictive power, and the ongoing search for exceptions, violations, and failures have provoked even deeper questions and challenges, stimulating the continued progress of science and the unfolding of new ideas, techniques, and concepts.

One of the great ongoing scientific challenges that has dominated modern physics is the search for a *grand unified theory* of the elementary particles and their interactions, including its extension to understanding the cosmos and even the origin of space-time itself. Such an ambitious theory would be conceptually based on a parsimonious set of underlying mathematizable universal principles that integrate and explain all of the fundamental forces of nature from gravity and electromagnetism to the weak and strong nuclear forces, incorporating Newton's laws, quantum mechanics, and Einstein's general theory of relativity. Fundamental quantities like the speed of light, the four dimensions of space-time, and the masses of all of the elementary particles would all be explained, and the equations governing the origin and evolution of the universe through to the formation of galaxies and down to the planetary level including life itself would be derived. It is a truly remarkable and enormously ambitious quest that has occupied thousands of researchers for almost one hundred years at a cost of billions of dollars. Measured by almost any metric this ongoing quest, which is still far from its ultimate goal, has been enormously successful, leading, for example, to the discovery of quarks as the fundamental building blocks of matter, the Higgs particle as the origin of mass in the universe, and to black holes and the Big Bang . . . and to many Nobel Prizes.[1]

Emboldened by its great success, physicists endowed this fantastic vision with the grand title of the Theory of Everything. Demanding mathematical consistency between quantum mechanics and general relativity suggested that the basic building blocks of this universal theory might be microscopic vibrating strings rather than the traditional elementary point particles upon which Newton and all subsequent theoretical developments were based. Consequently this vision took on the more prosaic subtitle "string theory." Like the invention of gods and God, the concept of a Theory of Everything connotes the grandest vision of all, the inspiration of all inspirations, namely that we can encapsulate and understand the entirety of the universe in a small set of precepts, in this case, a concise set of mathematical equations from which literally everything follows. Like the concept of God, however, it is potentially misleading and intellectually dangerous.

Referring somewhat hyperbolically to a field of study as the Theory of Everything connotes a certain degree of intellectual arrogance. Is it really conceivable that there is a master equation that encapsulates *everything* about the universe? Everything? Where's life, where are animals and cells, brains and consciousness, cities and corporations, love and hate, your mortgage payments, this year's presidential elections, et cetera? How in fact does the extraordinary diversity, complexity, and messiness that we all participate in here on Earth arise? The simplistic answer is that these are inevitable outcomes of the interactions and dynamics encapsulated in this grand Theory of Everything. Even time itself is presumed to have emerged from the geometry and dynamics of these vibrating strings. Following the Big Bang the universe expanded and cooled and this resulted in the sequential hierarchy from quarks to nucleons, thence to atoms and molecules, and ultimately to the complexity of cells, brains, and emotions and all the rest of life and the cosmos to come tumbling out, a sort of deus ex machina. All of this metaphorically as a consequence of turning the crank of increasingly complicated equations and computations presumed, at least in principle, to be soluble to any sufficient degree of accuracy. Qualitatively, this extreme version of reductionism may indeed have some partial validity, though I'm not sure to what extent anyone actually believes it—but, in any case, *something* is missing.

The "something" includes many of the concepts and ideas implicit in a lot of the problems and questions considered in this book: concepts like information, emergence, accidents, historical contingency, adaptation, and selection, all characteristics of *complex adaptive systems* whether organisms, societies, ecosystems, or economies. These are composed of myriad individual constituents or agents that take on collective characteristics that are generally unpredictable in detail from their underlying components even if the dynamics of their interactions are known. Unlike the Newtonian paradigm upon which the Theory of Everything is based, the complete dynamics and structure of complex adaptive systems cannot be encoded in a small number of equations. Indeed, in most cases, probably not even in an infinite number. Furthermore, predictions to arbitrary degrees of accuracy are not possible, even in principle.

On the other hand, as I have tried to show throughout the book, scaling theory provides a powerful tool for forging a middle ground in which a quantitative framework can be developed for understanding and predicting the coarse-grained behavior of many broad aspects of such systems.

Perhaps, then, the most surprising consequence of a visionary Theory of Everything is that it implies that on the grand scale the universe, including its origins and evolution, though extremely complicated, is not *complex* but in fact is surprisingly *simple* because it can be encoded in a limited number of equations, conceivably even just a single master equation. This is in stark contrast to the situation here on Earth, where we are integral to some of the most diverse, complex, and messy phenomena that occur anywhere in the universe, and which require additional, possibly nonmathematizable concepts to understand. So while applauding and admiring the search for a grand unified theory of all the basic forces of nature, we should recognize that it cannot literally explain and predict *everything*.

Consequently, in parallel with the quest for the Theory of Everything, we need to embark on a similar quest for a grand unified theory of complexity. The challenge of developing a quantitative, analytic, principled, predictive framework for understanding complex adaptive systems is surely one of the grand challenges for twenty-first-century science. As a vital corollary to this and of greater urgency is the need to develop a grand unified theory of sustainability in order to come to terms with the extraordinary threats we now face. Like all grand syntheses, these will almost certainly remain incomplete, and very likely unattainable, but they will nevertheless inspire significant, possibly revolutionary new ideas, concepts, and techniques with implications for how we move forward and whether what we have thus far achieved can survive.

2. TRANSDISCIPLINARITY, COMPLEX SYSTEMS, AND THE SANTA FE INSTITUTE

Although such a vision may not be explicitly articulated in such grandiose terms, it does encapsulate what the Santa Fe Institute was founded to address. It's a remarkable place. Maybe not everybody's cup of tea, but for many of us who still harbor a naive, possibly romantic image of wanting to be part of an eclectic community of scholars searching for "truth and beauty"—and having been disappointed that we didn't find it in a classic university setting—SFI comes the nearest we're likely to get to realizing it. I feel extraordinarily fortunate and privileged to have been able to spend a number of very productive years in such a marvelous place, stimulated by like-minded colleagues from every possible corner of academia.

The ambience and character of SFI are perhaps best captured by the British science writer John Whitfield, who in 2007 wrote:

> The institute was intended to be truly multidisciplinary—it has no departments, only researchers. . . . Santa Fe and complexity theory have become almost synonymous . . . the institute, now situated on a hill on the town's outskirts, must be one of the most fun places to be a scientist. The researchers' offices, and the communal areas they spill into for lunch and impromptu seminars, have picture windows looking out across the mountains and desert. Hiking trails lead out of the car park. In the institute's kitchen, you can eavesdrop on a conversation between a paleontologist, an expert on quantum computing, and a physicist

who works on financial markets. A cat and a dog amble down the corridors and in and out of offices. *The atmosphere is like a cross between the senior common room of a Cambridge college and one of the West Coast temples of geekdom, such as Google or Pixar.*

The italics in the last sentence are mine; I added them because I think that Whitfield really got this singular combination of characteristics right: on the one hand, the ivory tower image of an Oxford or Cambridge college, a community of scholars devoted to the pursuit of knowledge and understanding for "its own sake" by following their proverbial noses wherever it might take them; on the other, the cutting-edge image of Silicon Valley grappling with problems of the "real" world, seeking innovative solutions and new ways of tackling the complexities of life. Although SFI is a classic basic research institute in that it is not driven by programmatic or applied agendas, the very nature of the problems it addresses necessarily bring many of us face-to-face with major societal issues. As a consequence, in addition to its academic network of scholars, the institute also has a very active business network (called the Applied Complexity Network) comprising diverse companies, some small and incipient, but many of them large and well-known corporations spanning the spectrum of business activities.

SFI occupies a unique place in the academic landscape. Its mission is to address fundamental problems and big questions across all scales at the cutting edge of science with a bias toward quantitative, analytic, mathematical, and computational thinking. There are no departments or formal groups, but rather a culture dedicated to facilitating long-term, creative, transdisciplinary research across all fields, ranging from the mathematical, physical, and biomedical sciences to the social and economic. It has a small resident faculty (but no tenure) and about one hundred external faculty whose major appointment is elsewhere and who spend varying periods of time ranging from a day or two to several weeks in residence. In addition, there are postdoctoral fellows, students, journalism fellows, and even writers. It supports lots of working groups and workshops, seminars and colloquia, and hosts a huge flux of visitors (several hundred a year). What a fantastic melting pot. There is almost no hierarchy, and its size is sufficiently small that everyone on-site can easily get to know everyone else; the archaeologist, economist, social scientist, ecologist, and physicist all freely interact on a daily basis to talk, speculate, bullshit, and seriously collaborate on questions big and small.

The philosophy of the institute stems from the underlying assumption that if you bring smart people together in a supportive, facilitative, dynamic environment that lets them freely interact, good things will inevitably result. The SFI culture is designed to create an open catalytic atmosphere where interactions and collaborations that are often difficult to promote within the traditional departmental structure of universities are strongly encouraged. Bringing together highly diverse minds prepared to engage in substantive, in-depth collaboration in the search for underlying principles, commonalities, simplicity, and order in highly complex phenomena is the hallmark of SFI science. In a curious sense the institute is an instantiation of the very thing it studies: a complex adaptive system.

The institute has been internationally recognized as "the formal birthplace of the interdisciplinary study of complex systems" and has played a central role in recognizing that many of the most challenging, exciting, and profound questions facing science and society lie at the boundaries between traditional disciplines. Among these are the origins of life; the generic principles of innovation, growth, evolution, and resilience whether of organisms, ecosystems, pandemics, or societies; network dynamics in nature and society; biologically inspired paradigms in medicine and computation; the interrelationship between information processing, energy, and dynamics in biology and society; the sustainability and fate of social organizations; and the dynamics of financial markets and political conflicts.

Having had the great privilege of serving as president of SFI for a few years, I obviously have a somewhat biased view regarding its philosophy, its standing, and its successes. So lest you think this is entirely hyperbole on my part, let me give you a couple of other comments and opinions regarding its character. Rogers Hollingsworth is a distinguished social scientist and historian at the University of Wisconsin, well known for his in-depth investigations into what the essential ingredients are that make research groups successful. In addressing a subcommittee of the National Science Board (it oversees the National Science Foundation) charged with reviewing "transformational" science, he remarked:

> *My colleagues and I have studied approximately 175 research organizations on both sides of the Atlantic, and in many respects the Santa Fe Institute is the ideal type of organization which facilitates creative thinking.*

And here's a quote from *Wired* magazine:

> *Since its founding in 1984, the nonprofit research center has united top minds from diverse fields to study cellular biology, computer networks, and other systems that underlie our lives. The patterns they've discovered have illuminated some of the most pressing issues of our time and, along the way, served as the basis for what's now called the science of complexity.*

The institute was originally conceived by a small group of distinguished scientists, including several Nobel laureates, most of whom had some association with Los Alamos National Laboratory. They were concerned that the academic landscape had become so dominated by disciplinary stovepiping and specialization that many of the big questions, and especially those that transcend disciplines or were perhaps of a societal nature, were being ignored. The reward system for obtaining an academic position, for gaining promotion or tenure, for securing grants from federal agencies or private foundations, and even for being elected to a national academy, was becoming more and more tied to demonstrating that you were *the* expert in some tiny corner of some narrow subdiscipline. The freedom to think or speculate about some of the bigger questions and broader issues, to take a risk or be a maverick, was not a luxury many could afford. It was not just "publish or perish," but increasingly it was also becoming "bring in the big bucks or perish." The process of the corporatization of universities had begun. Long gone were the halcyon days of polymaths and broad thinkers like Thomas Young or D'Arcy Thompson. Indeed, there were now scant few broad *intra*disciplinary thinkers, let alone *in-*

*ter*disciplinary ones, who were comfortable articulating ideas and concepts that transcended their own fields and potentially reach across to foreign territory. It was to combat this perceived trend that SFI was created.

The early discussions of what the actual scientific agenda of the institute might be centered on the burgeoning fields of computer science, computation, and nonlinear dynamics, areas where Los Alamos had played a seminal role. Enter the theoretical physicist Murray Gell-Mann. He realized that all of these suggestions revolved more around techniques rather than ideas and concepts and that if such an institute was to have a major impact on the course of science, its agenda would have to be broader and bolder and address some of the big questions. Whence arose the idea of complexity and complex adaptive systems as overarching themes, since they encompass almost all of the major challenges and big questions facing science and society today—and, furthermore, they invariably cross traditional disciplinary boundaries.

An interesting sign of the times, and I would argue a significant indicator of the impact that SFI has had, is that many institutions nowadays promote themselves as being multidisciplinary, transdisciplinary, cross-disciplinary, or interdisciplinary. Although such designations have to some extent been co-opted as buzzwords to describe any collaborative interaction across subfields *within* traditional disciplines, rather than being reserved for bold leaps *across* the vast divides that separate them, it does represent a significant change in image and attitude. This is infecting all of academia and is now almost taken for granted despite the reality that universities are to varying degrees as stovepiped as ever. Here's a quote from the Web pages of Stanford University as it rebrands itself in this image, even claiming that it has always operated in this mode:

> *Since its founding, Stanford University has been a pioneer in cross-disciplinary collaboration . . . producing innovative basic and applied research in all fields. . . . This naturally facilitates multidisciplinary collaboration.*

To give you a sense of the extraordinary shift in perception that has occurred over just the last twenty years, here's an anecdote from the early days of SFI.

Among its founding fathers were two other major figures of twentieth-century academia, both Nobel laureates: Philip Anderson, a condensed matter physicist from Princeton University who had worked on superconductivity and was an inventor, among many other things, of the mechanism of symmetry breaking that underlies the prediction of the Higgs particle; and Kenneth Arrow from Stanford University, whose many contributions to the fundamental underpinnings of economics, from social choice to endogenous growth theory, have been hugely influential. He was the youngest person ever to have been awarded the Nobel Memorial Prize for economics, which five of his students have also received. Anderson and Arrow together with David Pines, also a distinguished condensed matter physicist and founder of SFI, initiated the first major program that put SFI on the map. It was designed to address foundational questions in economics from this new complex systems perspective, asking, for instance, how ideas from nonlinear dynamics, statistical physics, and chaos theory might provide new insights into economic theory. After one of the early workshops in 1989, *Science* magazine wrote an article about the meeting titled "Strange Bedfellows."[2] It began:

> They make an odd couple, these two Nobel laureates. . . . Over the past 2 years, Anderson and Arrow have worked together in a venture that is one of the oddest couplings in the history of science—a marriage, or at least a serious affair, between economics and the physical sciences. . . . This ground-breaking venture is taking place under the auspices of the Santa Fe Institute.

How times have changed! These days collaborations between physicists and economists are hardly rare occurrences—witness the huge influx onto Wall Street of physicists and mathematicians, many of whom have since become absurdly rich—but just twenty-five years ago it was almost unheard of, especially between two such distinguished thinkers. It's still hard to believe that this was considered to be so rare and bizarre that it would be characterized as "one of the oddest couplings in the history of science." Maybe horizons are indeed expanding.

When I became president of SFI I came across some words of wisdom that strongly resonated with me from a man who had helped found and run an extraordinarily successful institute more than fifty years earlier. This was the crystallographer Max Perutz, who shared the Nobel Prize in Chemistry for discovering the structure of hemoglobin. X-ray crystallography, the technique used by Perutz, had been pioneered in the early twentieth century by the unique father-and-son team of William and Lawrence Bragg, who were jointly awarded the Nobel Prize in physics in 1915 when Lawrence, the son, was only twenty-five years old. He remains the youngest person ever to have received the prize in the sciences.

Lawrence Bragg had the great foresight to see that these techniques, which he had helped develop for exploring the crystalline structure of ordinary matter, could potentially prove to be a powerful tool for revealing the structure of the complex molecules that are the building blocks of life, such as hemoglobin and DNA. He strongly encouraged Perutz, who had been his student, to start a research program along these lines entirely devoted to unraveling the structural mysteries of life. Thus in 1947 was born one of the most successful enterprises in all of science, the Medical Research Council Unit (MRCU) within the famed Cavendish Laboratory at Cambridge, whose director was Lawrence Bragg. While under Perutz's guidance the MRCU produced in just a few short years no fewer than nine Nobel prizes, one of which was the famous discovery of the double-helix structure of DNA by James Watson and Francis Crick.

What was the secret to Perutz's extraordinary success? Is there some magic formula he had discovered for optimizing how research should be carried out? If so, how could we exploit it to ensure the future success of the Santa Fe Institute? These were questions that I naturally asked myself when I assumed the leadership of SFI. I learned that Perutz, while maintaining his own research program, gave his researchers independence and treated everyone equally, even turning down a knighthood because he thought it would separate him from younger researchers. He stayed fully conversant with everyone's work and made a point of sitting with different colleagues at coffee, lunch, or tea. Well, at least in spirit, though maybe not always in action, these were all things I was aspiring to do—except for the opportunity of turning down a knighthood in the highly unlikely event that it would have been offered to me.

But what really inspired me about Perutz was something I read about him in his obituary in the *Guardian*.[3] It read:

> *Impishly, whenever he was asked whether there are simple guidelines along which to organize research so that it would be highly creative, he would say: no politics, no committees, no reports, no referees, no interviews; just gifted, highly motivated people picked by a few men of good judgment. Certainly not the way research is usually run in our fuzzy democracy but, from a man of great gifts and of extremely good judgment, such a reply is not elitist. It is simply to be expected, for Max had practiced it and shown that this recipe is right for those who, in science, want to beat the world by getting the best in the world to beat a path to their door.*

So he did have a formula—and it had worked brilliantly. These days it's hard to believe it was for real: *no politics, no committees, no reports, no referees, no interviews,* "just" focus on excellence and use *extremely good judgment.* Well, at least in principle, that's what we were trying to do at SFI and, indeed, still are: find the best people, trust them, give them support, and don't hamper them with bullshit . . . and good things will happen. This was the spirit in which SFI had been founded and what all of its presidents from the visionary George Cowan to our wonderful current president, David Krakauer, have enthusiastically championed. It seemed so simple, so why wasn't everyone following Max's magic formula? Well, try suggesting this recipe to the funding agencies, the NSF, the DOE, the NIH, to the philanthropic foundations, or to the provosts and deans at universities or to your local congressman and you'll quickly discover the answer. The formula is, of course, simplistic, somewhat unrealistic, and easier said than done, harking back to an image of support for science and scholarship that probably never actually existed in its naive form. But perhaps that's its power. Aspiring to such lofty ideals and trying to create a spirit and culture where the development of ideas and the search for knowledge are unencumbered by the hegemony of quarterly reports, continual proposal writing and oversight committees, political intrigue and petty bureaucracy should supersede all other considerations. By example, Perutz had shown this to be a critical component of success. So every year at the conclusion of my annual report to our board of trustees, after boasting of our successes and bemoaning our financial situation and the difficulties of raising funds for our research activities, I would read that magic formula out loud as a mantra or aspiration to remind us to keep our priorities straight.

3. BIG DATA: PARADIGM 4.0 OR JUST 3.1?

Beginning with the quantitative observations of planetary motion by the Danish astronomer Tycho Brahe in the sixteenth century, measurement has played a central role in the development of our understanding of the entire universe around us. Data provide the basis for constructing, testing, and refining our theories and models whether they seek to explain the origins of the universe, the nature of evolutionary processes, or the growth of the economy.

Data are the very lifeblood of science, technology, and engineering, and in more recent years have begun to play an increasingly central role in economics, finance, politics, and business. Almost none of the problems I have addressed in this book could be analyzed without recourse to enormous amounts of data. Furthermore, we could not seriously think about developing anything approaching a theory of complex adaptive systems or a science of cities, companies, or sustainability without having access to the sorts of data that I have relied upon in earlier chapters. A good example is the billions of cell phone calls we used in our work to test predictions for the role of social networks and the movement of people in cities.

Critical in these more recent developments has been the IT revolution, not just in assembling and gathering data, but in analyzing and organizing the massive amounts being generated into a manageable form from which insights can be gained, regularities deduced, or predictions made and verified. The speed and capacity of even this thirteen-inch MacBook Air, which I am using to type this manuscript, is awesome and its power for analyzing and retrieving data, storing information, and making complicated calculations is truly extraordinary. My little iPad is more powerful than the Cray-2, the world's most powerful supercomputer just twenty-five years ago, which would have cost you around $15 million to buy. The amount of data now being amassed through the multiple devices that monitor almost everything around us from our bodies, social interactions, movements, and preferences to the weather and traffic conditions is mind-boggling.

The number of networked devices in the world is now more than double that of the entire global population and the total screen area of all such devices is now larger than one square foot per person. We have truly entered the era of big data. The amount of information currently being stored and exchanged continues to grow exponentially. And all of this has happened in just the last decade or so, yet another impressive manifestation of the accelerating pace of life. The advent of big data has been heralded with a fanfare of promises and hyperbole, suggesting that it will provide the panacea for solving any number of our impending challenges from health care to urbanization while simultaneously improving even further the quality of life. As long as we measure and monitor everything and shovel the data into mammoth computers that will magically produce all of the answers and solutions, then all of our problems and challenges will be overcome and life will be good for all. This evolving paradigm is aptly encapsulated in the flood of "smart" devices and methodologies that are increasingly dominating our lives. "Smart" has become a mandatory designation for almost any new product, whether it's smart cities, smart health care, smart thermostats, smart phones, smart cards, or even smart parcel boxes.

Data are good and more data are even better—this is the creed that most of us take for granted, especially those of us who are scientists. But this belief is implicitly based on the idea that more data lead to a deeper understanding of underlying mechanisms and principles so that credible predictions and further progress in constructing models and theories can be built upon a firm foundation subject to continual testing and refinement. Data for data's sake, or the mindless gathering of big data, without any conceptual framework for organizing and understanding it, may actually be bad or even dangerous. Just relying on data alone, or even mathematical fits to data, without having some deeper understanding of the underlying mechanism is potentially deceiving and may well lead to erroneous conclusions and unintended consequences.

This admonition is closely related to the classic warning that "correlation does not imply causation." Just because two sets of data are closely correlated does not imply that one is the cause of the other. There are many bizarre examples that illustrate this point.[4] For instance, over the eleven-year period from 1999 to 2010 the variation in the total spending on science, space, and technology in the United States almost exactly followed the variation in the number of suicides by hanging, strangulation, and suffocation. It's extremely unlikely that there is any causal connection between these two phenomena: the decrease in spending in science was surely not the cause of the decrease in how many people hanged themselves. However, in many situations such a clear-cut conclusion is not so clear. More generally, correlation is in fact often an important indication of a causal connection but usually it can only be established after further investigation and the development of a mechanistic model.

This is particularly important in medicine. For example, the level of high-density lipoproteins (HDL) in blood—often referred to as "good" cholesterol—is negatively correlated with the incidence of heart attacks, suggesting that taking medications to increase HDL should lower the probability of having a heart attack. However, the evidence supporting this strategy is inconclusive: cardiovascular health does not seem to be improved by artificially raising levels of HDL. This might be because other factors such as genes, diet, and exercise simultaneously affect both HDL levels and the incidence of heart attacks without there being any direct causal link between them. It's even possible that the causation is inverted so that that good cardiovascular health induces higher HDL levels. To determine what the predominant causes of heart attacks are clearly requires a broad research program involving the gathering of huge amounts of data coupled with the development of mechanistic models for how each factor—whether genetic, biochemical, diet, or the environment—contributes. And indeed, huge resources are devoted across the medical profession to carry out this strategy.

Big data should primarily be viewed within this context: the classic scientific method involving painstaking analyses, the development of models and concepts whose predictions can be tested and used for devising new therapies and strategies, can now be augmented with the additional power of "smart" devices for gathering huge amounts of relevant data. Central to this paradigm is that continual refinement guides what data are most important to measure, how much is needed, and how accurate it needs to be. The variables we choose to focus on and measure in order to obtain data are not arbitrary—they are guided by previous success and failure within the context of an evolving conceptual framework. Doing science is much more than a fishing expedition.

With the advent of big data this classic view is being challenged. In a highly provocative article published in *Wired* magazine in 2008 titled "The End of Theory: The Data Deluge Makes the Scientific Method Obsolete," its then editor, Chris Anderson, wrote:

> *The new availability of huge amounts of data, along with the statistical tools to crunch these numbers, offers a whole new way of understanding the world. Correlation supersedes causation, and science can advance even without coherent models, unified theories, or really any mechanistic explanation at all . . . faced with massive data, this approach to science— hypothesize, model, test—is becoming obsolete. . . . Out with every theory of human behavior, from linguistics to sociology. Forget taxonomy, ontology, and psychology. Who knows why people do what they do? The point is they do it, and we can track and measure it with unprecedented fidelity. With enough data, the numbers speak for themselves. . . . Today companies like Google, which have grown up in an era of massively abundant data, don't have to settle for wrong models. Indeed, they don't have to settle for models at all. . . . There's no reason to cling to our old ways. It's time to ask: What can science learn from Google?*

Well, I won't answer that question other than to say that this radical view is becoming fairly prevalent across Silicon Valley, the IT industry, and increasingly in the business community. In a less extreme version it is also rapidly gaining traction in academia. In the last few years almost every university has opened up a well-funded center or institute devoted to big data while at the same time paying due obeisance to the other buzzword, *interdisciplinary*. For example, Oxford University has just launched its Big Data Institute (BDI) in a new, sexy, "state-of-the-art building." Here's what they say: "this interdisciplinary research centre will focus on the analysis of large, complex, heterogeneous data sets for research into the causes and consequences, prevention and treatment of disease." Obviously an extremely worthy cause, despite there being no emphasis on theory or concept development.

A contrary view to this trend was forcibly expressed by the Nobel Prize–winning geneticist Sydney Brenner, whom I quoted in chapter 3 and who was coincidentally director of the famous institute in Cambridge founded by Max Perutz that I mentioned earlier: "Biological research is in crisis. . . . Technology gives us the tools to analyse organisms at all scales, but we are drowning in a sea of data and thirsting for some theoretical framework with which to understand it. Although many believe that 'more is better,' history tells us that 'least is best.' We need theory and a firm grasp on the nature of the objects we study to predict the rest."

Not long after the publication of Chris Anderson's article, Microsoft published a fascinating series of essays in a book titled *The Fourth Paradigm: Data-Intensive Scientific Discovery*. It was inspired by Jim Gray, a computer scientist at Microsoft who was sadly lost at sea in 2007. He envisioned the data revolution as a major paradigm shift in how science would advance in the twenty-first century and called it the fourth paradigm. He identified the first three as (1) empirical observation (pre-Galileo), (2) theory based on models and mathematics (post-Newtonian), and (3) computation and simulation. My impression is that, in contrast to Chris Anderson, Gray viewed this fourth paradigm as an integration of the previous three, namely as a unification of theory, experiment, and simulation, but with an added emphasis on data gathering and analysis. In that sense it's hard to disagree with him because this is pretty much the way science has progressed for the last couple of hundred years—the difference being primarily quantitative: the "data revolution" provides us with a much greater possibility for exploiting and enabling strategies we have been using for a very long time. In this sense this is more like paradigm 3.1 than paradigm 4.0.

But there is a new kid on the block that many feel promises more and, like Anderson, potentially subverts the need for the traditional scientific method. This invokes techniques and strategies with names like *machine learning, artificial intelligence*, and *data analytics*. There are many versions of these, but all of them are based on the idea that we can design and program computers and algorithms to evolve and adapt based on data input to solve problems, reveal insights, and make predictions. They all rely on iterative procedures for finding and building upon correlations in data without concern for why such relationships exist and implicitly presume that "correlation supersedes causation." This approach has become a huge area of interest and has already had a big impact on our lives. For instance, it is central to how search engines like Google operate, how strategies for investment or operating an organization are devised, and it provides the foundational basis for driverless cars.

It also brings up the classic philosophical question as to what extent these machines are "thinking." What, in fact, do we mean by that? Are they already smarter than we are? Will superintelligent robots eventually replace us? The specter of such science fiction fantasies seems to be rapidly encroaching on us. Indeed, we can readily appreciate why some like Ray Kurzweil believe that the next paradigm shift will involve the integration of humans with machines, or eventually lead to a world dominated by intelligent robots. As I expressed earlier I have a fairly jaundiced view of such futurist thinking, though the questions raised are fascinating and very challenging and need to be addressed. But the discussion needs to engage with another potential paradigm shift, driven by an impending finite time singularity associated with the accelerating pace of life and involves the challenge of global sustainability and the addition of four to five billion people who will shortly be joining us on our planet.

There is no question that big data will have a major influence across all aspects of life and, in addition, will be a huge aid in the scientific enterprise. Its success in terms of major discoveries and new ways in which we view the world will depend on the extent to which it is integrated with deeper conceptual thinking and traditional development of theory. The vision proposed by Anderson, and to a lesser extent by Gray, is the computer scientists' and statisticians' version of the Theory of Everything. It carries with it a similar arrogance and narcissism that this is *the* singular way to understand everything. How far it will truly reveal new science remains open to question. But when combined with the traditional scientific method, it surely will.

The discovery of the Higgs particle is a fascinating example of how Big Data can lead to important scientific discovery when integrated with traditional scientific methodology. First, I remind you that the Higgs is a crucial linchpin in the basic laws of physics. It permeates the universe, giving rise to the mass of all of the elementary particles of matter from electrons to quarks. Its existence was brilliantly predicted more than sixty years ago by a group of six theoretical physicists. This prediction didn't come out of the blue but was the end result of a traditional scientific process involving analyses of thousands of experiments carried out over many years iterated with mathematical theories and concepts that were developed to parsimoniously explain these observations, and thereby stimulate further experimentation to test predictions.

It took more than fifty years before the technology was sufficiently developed for a serious search for this elusive but crucial element of our unified theory of the fundamental forces of nature to be undertaken. Central to this was the construction of a giant particle accelerator in which protons move in opposite directions in a circular beam at almost the speed of light and collide with each other in a highly controlled interaction region. This machine, coined the Large Hadron Collider (LHC), was built at CERN in Geneva, Switzerland, at a cost of more than $6 billion. The scale of this gargantuan scientific instrument is huge: its circumference is about seventeen miles long and each of the two major detectors that actually observe and measure the collisions are about 150 feet long, 75 feet high, and 75 feet wide.

The entire project represents an unprecedented engineering achievement whose output is the mother of all big data—nothing comes close. There are about 600 million collisions per second monitored by about 150 million individual sensors in each detector. This produces about 150 million petabytes of data a year or 150 exabytes a day (a byte being the basic unit of information). Let me give you a sense of what the scale of this means. The Word document containing this entire book, including all of its illustrations, is less than 20 megabytes (20MB, meaning 20 million bytes). This MacBook Air can store 8 gigabytes of data (8GB, meaning 8 billion bytes). All of the films stored by Netflix amount to less than 4 petabytes—which is 4 million GB, or about half a million times larger than the capacity of this laptop. Now to the big one: each day the total amount of data produced by all of the world's computers and other IT devices taken together amounts to about 2.5 exabytes; an exabyte is 10^{18} bytes, or a billion gigabytes (GB).

This is awesome and is often touted as a measure of the big data revolution.

But here's what's truly awesome: it pales in comparison to the amount of data produced by the LHC. If every one of the 600 million collisions occurring each second were recorded it would amount to about 150 exabytes a day, which is about sixty times greater than the entire amount of data produced by all of the computational devices in the world added together. Obviously this means that the strategy of naively letting the data speak for themselves by devising machine-learning algorithms to search for correlations that would eventually lead to the discovery of the Higgs mechanism is futile. Even if the machine produced a million times less data it is extremely unlikely this strategy could succeed. How, then, did physicists discover the proverbial needle in this mammoth haystack?

The point is that we have a well-developed, well-understood, well-tested conceptual framework and mathematical theory that guides us in where to look. It tells us that almost all of the debris resulting from almost all of the collisions are actually uninteresting or irrelevant as far as searching for the Higgs particle is concerned. In fact, it tells us that out of the approximately 600 million collisions occurring every second, only about 100 are of interest, representing only about 0.00001 percent of the entire data stream. It was by devising sophisticated algorithms for focusing on only this very special tiny subset of the data that the Higgs was eventually discovered.

The lesson is clear: neither science nor data are democratic. Science is meritocratic and not all data are equal. Depending on what you are looking for or investigating, theory resulting from the traditional methodology of scientific investigation, whether highly developed and quantitative as in the case of fundamental physics, or relatively undeveloped and qualitative as in the case of much of social science, is an essential guide. It is a hugely powerful constraint in limiting search space, sharpening questions, and understanding answers. The more one can bring big data into the enterprise the better, provided it is constrained by a bigger-picture conceptual framework that, in particular, can be used to judge the relevance of correlations and their relationship to mechanistic causation. If we are not to "drown in a sea of data" we need a "theoretical framework with which to understand it . . . and a firm grasp on the nature of the objects we study to predict the rest."

One final point: The IT revolution is our most recent great paradigm shift, and like all previous ones it is driving us toward a "finite time singularity" whose nature I speculated about in chapter 9. It was enabled by the invention of a startling assortment of extraordinarily "smart" devices that are producing enormous amounts of data. And, like previous major paradigm shifts, it has predictably resulted in an increase in the pace of life. In addition, it has metaphorically brought the world closer together with instant communication anywhere across the globe at any time. It has also led to the possibility that we no longer need to live in an urban environment to participate in and benefit from the fruits of urban social networks and the dynamics of agglomeration, which are the very origin of superlinear scaling and open-ended growth. We can devolve to develop smaller, or even rural, communities that are just as plugged in as living in the heart of a great metropolis. Does this mean that we can avoid the pitfalls that lead to an ever-accelerating pace of life, finite time singularities, and the prospect of collapse? Have we somehow stumbled upon a way to avoid the ironic quandary that the very system that led to our great socioeconomic expansion of the past two hundred years may be leading to our ultimate demise, and that we can have our cake and eat it too?

This is clearly an open question. There are indeed signs that such a dynamic is beginning to develop but so far on an extremely small scale. In fact, the vast majority of people who could in principle de-urbanize and yet remain connected to the center of things choose not to. Even Silicon Valley, which was primarily suburban, has invaded downtown San Francisco, leading to tension between traditional commerce and the excesses of the high-tech lifestyle. I know of no high-tech geeks who are operating from high up in the mountain ranges of the California Sierra. The vast majority seem to prefer traditional urban living. Rather than depopulating, cities seem to be reviving and growing, partially because of the social attractiveness of real-time social contact.

Furthermore, we tend to think that nothing can compare to the changes wrought by the IT revolution with our iPhones, e-mail, text messages, Facebook, Twitter, and so forth. But think of what the railway brought in the nineteenth century or the telephone in the early twentieth. Before the coming of the railway most people didn't travel more than twenty miles from their home during their entire lifetime: suddenly Brighton was in relatively easy reach of London, and Chicago in reach of New York. Messages that took days, weeks, or even months to be communicated before the invention of the telephone could now be communicated instantaneously. The changes were fantastic. Relatively speaking, these had a greater impact on our lives and, in particular, in speeding up life and changing our visceral perception of space and time than our present IT revolution. But these didn't result in a de-urbanizing phenomenon or a contraction of our cities. On the contrary, they led to their exponential expansion and to the development of suburbs as an integral part of urban living. Whether the present paradigm continues this trend is open to question, though I suspect that life will continue to speed up and urbanization remain the dominant force as we head toward an impending singularity. How this plays itself out will determine much about the sustainability of the planet.

POSTSCRIPT AND
ACKNOWLEDGMENTS

Because this book covers such a vast and varied territory, an unexpected challenge in writing it was to settle on a suitable title that would capture its main message in just a few punchy words or even half a tweet. After floating some rather lame possibilities, such as *Size Really Matters, Scaling the Tree of Life*, and *The Measure of All Things*, I settled on the somewhat cryptic title of *Scale*, because this is indeed a unifying theme of the book. However, "scale" can mean a lot of different things to a lot of different people. For some it connotes maps and charts, for others music, for yet others weighing vegetables or meat, and for some deposits on a rough surface. These were decidedly not what the prime substance of the book was about, so calling it *Scale* simply deferred the challenge to finding a catchy subtitle that made its intended meaning much more explicit.

Hitting upon a grander image of scale, as in "the scale of the universe," I came up with the somewhat grandiose subtitle: *The Search for Simplicity and Unity in the Complexity of Life, from Cells to Cities, Companies to Ecosystems, Milliseconds to Millennia*. This at least captured some of the spirit of the book and in particular the crucial interplay between the big picture "cosmic" perspective and the more focused "real-world" problems that I was addressing. Although this suggestion was a bit of a mouthful, it still didn't capture many central aspects of the book, which my editor at Penguin Press, Scott Moyers, felt needed to be emphasized. Eventually, after considering several possibilities and variations suggested by Scott, Paul Murphy, my editor at Weidenfeld in the UK, my wife, Jacqueline, and my agent, John Brockman, I settled on what you see on the title page: *The Universal Laws of Growth, Innovation, Sustainability, and the Pace of Life, in Organisms, Cities, Economies, and Companies*. By far the most creative suggestion came from my son, Joshua, who is a professor in

Earth sciences at the University of Southern California in Los Angeles. He suggested the acronymic title *SCALE: Size Controls All of Life's Existence.*

It's pretty catchy, ridiculously hyperbolic, and rather clever and I wish I had had the chutzpah to actually use it. But had I done so, I'm sure it would have been vetoed by both Scott and Paul—and rightly so.

I have approached all of the problems addressed in the book primarily from the viewpoint of a theoretical physicist whose language is mathematics. Consequently, an underlying thread running throughout the book has been an emphasis on developing a more quantitative, computational, predictive understanding based on fundamental principles as a complement to the traditional, more qualitative, narrative arguments that tend to dominate the social, biological, medical, and business literature. Nevertheless, *there isn't a single equation in the book.* I took very seriously the admonition of Lord Ernest Rutherford, the famous discoverer of atomic nuclei—"the father of the nuclear age"—that *"a theory that you can't explain to a bartender is probably no damn good."* I'm not entirely convinced that he was right, but I did heed the spirit of what he said. I hope, therefore, that I have succeeded in some small way in keeping the arguments and explanations at an appropriately nontechnical level so that the proverbial "intelligent layperson" didn't have too much difficulty in following them. In doing so I have had to take a certain degree of poetic license in distilling the essence of complex technical or mathematical arguments to simple colloquial explanations and hope that my scientific colleagues will indulge and condone me for any oversimplifications, misrepresentations, or any subsequent lack of rigor.

The problems, questions, and explanations presented in the book are shamelessly addressed from my own personal perspective. As a result, the book is neither encyclopedic nor a comprehensive review of the enormous literature covering the many subjects and problems that are addressed. A major intent is to show that underlying the extraordinary complexity, diversity, and apparent messiness of the world we live in lies a surprising unity and simplicity when viewed through the lens of scale. Great tomes have been written by deep thinkers about almost everything that is considered in the book and it goes without saying that I am building on what many have already understood and analyzed. I have tried to give credit where appropriate, but I have in no way been comprehensive in referencing all those who have contributed to the development of the ideas and concepts that I explore. In so doing, I hope I have not offended too many people.

Many of the arguments and almost all of the examples used are based on an extensive body of work that I have been intensely engaged in over the past twenty years together with an extraordinarily talented group of colleagues. Not all of the grandiose themes or all of the specific problems that were attacked are given equal billing. Choices needed to be made and some were either neglected or given relatively short shrift. The specific topics and themes that were eventually chosen, and the depth to which they are explored, were determined partly by their conceptual significance, partly because they were judged to be an important topic of general interest, and partly out of my own idiosyncratic perspective. Throughout the book I veered toward emphasizing a bigger picture conceptual framework and explaining basic ideas rather than dwelling on details, though I tried not to shirk from drilling down deeper and presenting such details when I thought it necessary. Consequently, like the scientific enterprise itself, there are many loose ends and unanswered questions. However, the inquisitive reader should have little difficulty in further exploring the territory I have opened up that might be of particular interest by referring to the materials listed at the end of the book.

I interspersed the scientific narrative with occasional anecdotes about some of the key players in the development of various critical concepts that play an important role in the book. I concentrated primarily on a subset of a few remarkable men of broad intellect who changed the way we think about the world but who have not received the recognition they deserve even, in some cases, within the broader scientific community. Names that you might not have heard of, such as Adolphe Quetelet, Thomas Young, and William Froude. I also included a few personal anecdotes to illustrate how I came to think about some of these problems, and, in particular, how I transitioned from being obsessed with elementary particles, strings, dark matter, and the evolution of the universe to trying to understand cells and whales, life and death, cities and global sustainability, and why companies die.

A critical point in this transition was my meeting with the eminent ecologist and wonderful scientist Jim Brown. In chapter 3, I related the story of how this fortuitous encounter and my subsequent long-term engagement with the Santa Fe Institute came into being and how it led to an extraordinary collaborative relationship that changed my life, and I believe his, too. I also recounted the critical role played by Brian Enquist, then a student of Jim's and now a distinguished ecologist in his own right. Brian was the first of a small stream of outstanding young people who joined our little "scaling group" to work on many of the problems addressed in subsequent chapters: the ecologists Jamie Gillooly, Drew Allen, and Wenyun Zuo; the physicists Van Savage, Chen Hou, Alex Herman, and Chris Kempes; and the computer scientist Melanie Moses. An additional and very important member of the collaboration was the well-known biochemist Woody Woodruff, who has since retired to enjoy the hills of his native Tennessee.

In chapter 7, I recounted how the "cities group" evolved as a natural outgrowth of the scaling group. It actually began as part of a much larger social science project called ISCOM (Information Society as a Complex System), which was generously funded by the European Union. This was a collaboration with the Italian statistician/economist David Lane, the Dutch anthropologist Sander van der Leeuw, and the French urban geographer Denise Pumain, all of whom are senior leaders in their fields. Without their initial stimulus, enthusiasm, and support I doubt if any of this would have happened. The young researchers who did almost all of the analysis on the cities work, which is explained in chapters 7 and 8, were the physicists Luis Bettencourt, Hyejin Youn, and Dirk Helbing; the urban economists José Lobo and Debbie Strumsky; the anthropologist Marcus Hamilton; the mathematician Madeleine Daepp; and the engineer Markus Schlapfer. Other collaborators who were intermittently involved but nevertheless made important contributions and influenced my thinking include the ecologist Ric Charnov; the systems biologist Aviv Bergman; the physicists Henrik Jensen, Michelle Girvan, and Christian Kuhnert; the investment analyst Eduardo Viegas; and the architect Carlo Ratti, whom I talked about in chapter 8.

I have been truly blessed in having each of these as my collaborators and am deeply indebted to all of them. I have deliberately made their individual disciplinary backgrounds explicit so as to highlight the broad transdisciplinary nature of the collaborations that were needed to seriously address the sorts of topics and problems that constitute this book. Their individual and collective commitments and passion for conceptual understanding and for attacking important problems were hallmarks of our ongoing meetings and interactions. Their probing questions and insights, their technical and conceptual contributions, and their willingness to engage in intense group discussions were crucial ingredients to our success. I am sure that some of them may have misgivings about how I have presented the results of some of our work and apologize in advance for any embarrassment or concern that this might cause. I take full responsibility for any mistakes or misrepresentations.

I am delighted to say that all of the young researchers moved on to successful careers at excellent universities, where, among other things, they established their own brand of this kind of science. Two who were particularly important in terms of my own interactions were Van Savage and Luis Bettencourt. This is very likely because both were trained in theoretical physics, so we spoke the same language. Luis, who is now a colleague at the Santa Fe Institute, played a central role in the development of the work on cities, a story that is recounted in some detail in chapter 7. Van, who had originally been hired as a postdoctoral fellow at SFI, eventually left for Harvard before moving on to UCLA, where he has established himself as a leading theoretical ecologist. Among the many problems we had great fun working on together, I want to mention two that are not discussed in the detail they deserve in the book even though both are fascinating, challenging, and very important. One is the development of a quantitative theory of sleep, showing, for example, why whales sleep for just a couple of hours, mice for fifteen, and we for about eight. Together with Van's bright young student Junyu Cao we recently extended this to understand sleep patterns in babies and children and showed how this framework provides important insights into early brain development. The other problem, carried out in collaboration with Alex Herman, was to develop the first quantitative theory for understanding the growth, metabolic rate, and vascular structure of tumors that we hope will stimulate new therapeutic strategies for attacking cancer.

I've probably been a little remiss in not drawing attention to the fact that some of the work in biology discussed in chapters 3 and 4 has not been without its critics. This in spite of, or possibly because of, the significant impact it has had, as evidenced by the numerous citations to it in the literature and the broad attention it has received across the scientific and popular press from the *Financial Times* to the *New York Times*. There have been many feature articles about it in high profile media outlets worldwide, including TV channels ranging from the National Geographic to the BBC. It was hyperbolically referred to in *Nature* as a "biological theory of everything" and "as potentially important to biology as Newton's contributions are to physics"—though very flattering, this is clearly a highly inflated characterization. In a separate article in *Nature* it was also stated that ". . . this theory explains so much with so little. It is breathtaking in its ambition and scope. Any new theory that is apparently so omniscient will attract as many grumbles of doubt as gasps of admiration. . . . No comparable idea yet matches it, despite its inevitable limitations."

In writing the book I made a strategic decision not to directly address the "grumbles of doubt" but rather concentrate on getting the big message across. A major reason for this was that from our biased point of view none of the criticisms was convincing. Some were simply incorrect and many typically rested on a single technical issue in some specific system for which there were often at least equally supportable alternative explanations. Furthermore, almost all of the concerns were focused solely with mammalian metabolic rate, failing to appreciate the great breadth of the framework and that it offers a single parsimonious explanation rooted in basic principles of biology, physics, and geometry for an enormous variety of empirical scaling relations. Needless to say, such criticisms have been dealt with in the scientific literature and can be accessed through references cited below.

It goes without saying that many other colleagues and friends have been enormously important in providing me with the enthusiastic moral and intellectual support and encouragement needed to complete such a book, especially at times when my own enthusiasm waned. The Santa Fe Institute provided exactly the right ambience and cultural mix of colleagues needed for developing most of the ideas that are articulated in the preceding chapters. A few anecdotes about SFI are recounted throughout the book, and part of my afterword is dedicated to extolling its virtues and why I believe that its mission represents an important harbinger of science for the twenty-first century. I am particularly indebted to the wonderfully effervescent Ellen Goldberg, a past president of SFI, for persuading me to join the institute—a move that reset my intellectual clock and gave me a new lease on life. Being exposed to a never-ending flux of extraordinary individuals at various stages of their careers, from students to Nobel Prize winners across a breathtaking spectrum of intellectual and cultural pursuits, was like letting a child loose in a candy store.

In that context I would also like to thank the extended SFI community, both individually and collectively, for broadening my scientific horizons and helping me begin to understand some of the subtleties and challenges inherent in the study of complex adaptive systems. I would particularly like to mention Pablo Marquet, John Miller, Murray Gell-Mann, Juan Perez-Mercader, David Krakauer, Cormac McCarthy, and Bill Miller and Michael Mauboussin, past and present chairmen, respectively, of SFI's board of trustees, all of whom have given me their unwavering enthusiastic support and encouragement over many years. I am deeply grateful and indebted to all of them. I am particularly grateful to Cormac for painstakingly reading and editing the manuscript in excruciating detail, providing extensive feedback, which helped greatly in improving the final product. Although I accepted most of his advice about grammar and sentence construction, I continue to argue with him over his *total* aversion to semicolons and exclamation marks; and his insistence on the Oxford comma.

In addition to my close associates I owe a measure of gratitude to an eclectic group of nonscientists who felt that I had something of broad interest to say and enthusiastically encouraged me to write a book for a general audience. It was their feedback that persuaded me to change gears and write a nontechnical "popular" book rather than one directed to my scientific colleagues. Among these are the historian Niall Ferguson; the art curator and critic Hans Ulrich Obrist; the writer-actor Sam Shepard; the founder of Amazon, Jeff Bezos; and the founder of Salesforce, Marc Benioff. I was really touched when Marc sent me a large painting of the Sephirot—the traditional Kabbalistic image representing the spiritual unity of life—suggesting that I meditate upon it each day. I can't say that I religiously followed his advice but it did inspire me to stay connected to the big picture when the going got tough. In this context I owe particular thanks to the amazing Richard Wurman, the original founder of TED, who has been indefatigable in his enthusiastic appreciation of my work.

Even though theoretical research requires only a pencil and paper—at least metaphorically—it can no longer be done without substantial financial support. I have been very fortunate in receiving funding from several diverse sources in both the public and private sectors for support of much of the research that forms the basis of this book. I am deeply grateful to Los Alamos National Laboratory and the Department of Energy for supporting my exploratory foray into biology while I was still leading the high energy physics program at the lab. At that crucial incipient stage the physics division of the National Science Foundation awarded me a modest grant to pursue research on scaling in biology. I am indebted to Bob Eisenstein, the then leader of the division, and to Rolf Sinclair, the program manager, for sticking their necks out in supporting this line of research, which at the time was rather unfashionable. Over the years the NSF continued its support of the work in biology, extending it to some of our early work on cities. This is in no small part due to the vision of the irrepressible Krastan Blagoev, who later initiated, and still runs, a dedicated program called the "The Physics of Living Systems," whose aim is to address important problems at the interface between traditional disciplines.

Significant support has also come from nongovernmental sources, including the Hewlett Foundation, the Rockefeller Foundation, the Bryan and June Zwan Foundation, and, in particular, the Eugene and Clare Thaw Charitable Trust. Gene Thaw has been extraordinarily generous in providing support for both the research and, equally important, for the writing of this book. The successive directors of the trust, beginning with the visionary Susan Herter through Sherry Thompson to Katie Flanagan, facilitated a very special relationship. Gene is a remarkable man, a gentleman of the old school who sports cravats and tweed jackets, a man of great culture who truly cares about the world. He is a well-known collector, critic, and dealer who has vigorously supported the arts for many years. He will turn ninety next year and the Morgan Library & Museum in New York will be entirely turned over to exhibit his remarkable collection of drawings ranging from Piranesi and Rembrandt to Cézanne and Picasso, while the Metropolitan Museum of Art will be showing his unequaled collection of Native American art and artifacts. Gene's passion for opera and the arts is equaled only by his passion for the environment and the challenge of global sustainability and it is in this context that he volunteered to support our research. He is the nearest embodiment I know of to a traditional patron—his support for my research agenda gave me the freedom to explore wherever my imagination and curiosity took me as I embarked on writing this book. I take great pleasure in thanking him for his generosity and patience.

In addition to the ongoing support of the Thaw trust it is also true that the book would not have been written without the urging and cajoling of my agent, the inexorable John Brockman. I'm still not entirely sure why but he was determined that I write a book. It's been a long haul during which John, and now his son, Max, has been there with me and I am hugely grateful for their support. John gently bullied me into writing the original proposal for the book, which was eventually completed at the spectacular Rockefeller Foundation retreat in Bellagio, Italy. This was exactly the right setting and I am extremely grateful to the foundation for hosting my wife, Jacqueline, and me there for a month—it was extremely productive. Even though the Rockefeller Foundation does not generally support basic research, it was also very generous in providing major funding for our work on cities. The president of the foundation, Judith Rodin, was very supportive, but we have to thank our program officer at the time, Benjamin de la Pena, for fighting hard on our behalf.

The book would not have been completed, and it certainly would have been much less coherent, had it not been for my wonderful editor at Penguin Press, Scott Moyers. He worked relentlessly on my behalf, always encouraging, always thoughtful, and always gentle even when being critical . . . and always remarkably patient and understanding. He must have been horrified as he saw this book grow from its originally projected modest size to something gargantuan, taking twice as long to complete as anticipated. His meticulously detailed editing of the manuscript, his probing questions, and his wise advice were invaluable. Scott, I cannot thank you enough. In addition to Scott, the entire team at Penguin Press was marvelous: Christopher Richards and Kiara Barrow aided by Thea Traff at the *New Yorker* were critical in getting all of my garbled illustrations and edits straight.

Finally, I take enormous joy in thanking my family for its tremendous support and patience throughout this long process. Our wonderful children, Joshua and Devorah, have been cheering me on from the sidelines, providing encouragement each time I dropped the ball and wildly celebrating the occasional touchdown. I'm sure they will be relieved now that this book is behind me. My deepest appreciation is owed to my extraordinary wife, Jacqueline, who has been my moral, spiritual, and intellectual companion not just in the writing of this book but throughout our remarkable journey together that has now lasted for almost fifty-five years—and what a journey it has been! Her honesty, intelligence, and profound love have been the mainstay of our life together, providing a depth to the meaning of life complemented only by the eternal search for understanding.

NOTES

1. THE BIG PICTURE

1. This is notoriously difficult to estimate, with numbers ranging from around 5 million to up to a trillion. The latest estimates give a number of 8.7 million. See Camilo Mora, et al., "How Many Species Are There on Earth and in the Ocean?" *PLOS Biology* 9 (8) (Aug. 23, 2011): e1001127.

2. Although everyone is familiar with the unit of watts (W), there is considerable confusion as to its meaning. Unfortunately, it is often perceived as a unit of energy, whereas it is in fact a unit for the *rate* of energy use or generation per unit time. The unit of energy is the joule (J); 1 watt is simply 1 joule per second. As there are 3,600 seconds in an hour, a 100W lightbulb uses 360,000J in an hour. Electricity bills typically express the electrical energy used over the previous month in kilowatt-hours (a kilowatt is 1,000 watts). Thus a 100W lightbulb left on for an hour uses 0.1 kilowatt-hours of energy.

3. The scaling of metabolic rate was first proposed by Max Kleiber. M. Kleiber, "Body Size and Metabolism," *Hilgardia* (1932); 6: 315–51. The graph shown in the figure is based on data from F.G. Benedict, *Vital Energetics: A Study in Comparative Basal Metabolism*. Washington, DC: Carnegie Institute of Washington, 1938.

4. H. J. Levine, "Rest Heart Rate and Life Expectancy." *Journal of the American College of Cardiology* 30 (4) (1997): 1104–6.

5. L.M.A. Bettencourt, J. Lobo and D. Strumsky, "Invention in the City: Increasing Returns to Patenting as a Scaling Function of Metropolitan Size," *Research Policy* 36 (2007): 107–120.

6. L.M.A. Bettencourt and G. B. West, based on data supplied by Professor F. Schweizer of the Swiss Federal Institute of Technology (ETH) in Zurich, Switzerland. Each point represents an average over a number of companies of approximately the same size. A more detailed version of this graph including almost 30,000 publicly traded companies in the United States is shown in Figures 60–63 in chapter 9.

7. As I shall discuss in chapter 3, humans obeyed this approximate general rule until relatively recently. In the developed world our life spans have increased by almost a factor of two over the last 150 years, so that now we can expect our hearts to beat roughly three billion times during our lives.

8. A good source for detailed statistics on cities and urbanization are reports from the United Nations. See, for instance, their publication "World Urbanization Prospects," https://esa.un.org/unpd/wup/Publications/Files/WUP2014-Highlights.pdf.

9. And, by the way, to several Nobel Prizes.

10. Stephen Hawking quoted in an interview, "Unified Theory Is Getting Closer, Hawking Predicts," San Jose *Mercury News*, Jan. 23, 2000; www.mercurycenter.com/resources/search.

11. There are a number of popular books devoted to elucidating the new science of complexity. Among them are: M. Mitchell, *Complexity: A Guided Tour* (New York: Oxford University Press, 2008); M.

M. Waldrop, *Complexity: The Emerging Science at the Edge of Order and Chaos* (New York: Simon & Schuster, 1993); J. Gleick, *Chaos: Making a New Science* (New York: Viking Penguin, 1987); S. A. Kauffman, *At Home in the Universe: The Search for the Laws of Self-Organization and Complexity* (Oxford, UK: Oxford University Press, 1995); J. H. Miller, *A Crude Look at the Whole: The Science of Complex Systems in Business, Life, and Society* (New York: Basic Books, 2016).

12. Those familiar with the mathematics of power laws will be aware that a ¾ power scaling law means that strictly speaking the increase in metabolic rate when the size is doubled is by a factor of $2^{3/4}$, which is 1.68, namely, an enhancement of 68 percent and therefore slightly less than the 75 percent increase quoted. For ease of presentation I shall ignore this difference throughout the book when presenting pedagogical examples such as this.

13. There are several excellent texts summarizing the various allometric scaling laws in biology. Among them are: W. A. Calder, *Size, Function and Life History* (Cambridge, MA: Harvard University Press, 1984); E. L. Charnov, *Life History Invariants* (Oxford, UK: Oxford University Press, 1993); T. A. McMahon and J. T. Bonner, *On Size and Life* (New York: Scientific American Library, 1983); R. H. Peters, *The Ecological Implications of Body Size* (Cambridge, UK: Cambridge University Press, 1986); K. Schmidt-Nielsen, *Why Is Animal Size So Important?* (Cambridge, UK: Cambridge University Press, 1984).

14. These ideas were originally proposed in G. B. West, J. H. Brown, and B. J. Enquist, "A General Model for the Origin of Allometric Scaling Laws in Biology," *Science* 276 (1997): 122–26. Nonmathematical reviews summarizing the general theory and its implications can be found in G. B. West and J. H. Brown, "The Origin of Allometric Scaling Laws in Biology from Genomes to Ecosystems: Towards a Quantitative Unifying Theory of Biological Structure and Organization," *Journal of Experimental Biology* 208 (2005): 1575–92; and G. B. West and J. H. Brown, "Life's Universal Scaling Laws," *Physics Today* 57 (2004): 36–42. The various technical papers devoted to specific elaborations and ramifications of this framework will be cited in the appropriate places in later chapters.

15. The seminal paper detailing these results is L. M. A. Bettencourt, et al., "Growth, Innovation, Scaling, and the Pace of Life in Cities," *Proceedings of the National Academy of Science USA* 104 (2007): 7301–6. Subsequent papers dealing with specific subtopics will be cited in the appropriate places in later chapters. Brief overviews can be found in L.M.A. Bettencourt and G. B. West, "A Unified Theory of Urban Living," *Nature* 467 (2010): 912–13, and "Bigger Cities Do More with Less," *Scientific American* (September 2011): 52–53.

16. M.I.G. Daepp, et al., "The Mortality of Companies," *Journal of the Royal Society Interface* 12 (2015): 20150120.

2. THE MEASURE OF ALL THINGS

1. The title of the book is often shortened to *Dialogues Concerning Two New Sciences*. The classic source in English is the 1914 translation by Henry Crew and Alfonso de Salvio, originally published by Macmillan (1914) but reissued in 1954 by Dover Publications Inc., New York.

2. The full quote from Einstein is worth repeating because it emphasizes a central dictum of science: "Propositions arrived at by purely logical means are completely empty as regards reality. Because Galileo saw this, and particularly because he drummed it into the scientific world, he is the father of modern physics—indeed, of modern science altogether." Taken from Einstein's "On the Method of Theoretical Physics," in *Essays in Science* (New York: Dover, 2009), 12–21.

3. J. Shuster and J. Siegel, *Superman*, Action Comics 1 (1938).

4. For those who are mathematically inclined this comes about because $(10^1)^{3/2} = 31.6$ and $(10^2)^{3/2} = 1,000$.

5. M. H. Lietzke, "Relation Between Weightlifting Totals and Body Weight," *Science* 124 (1956): 486.

6. L. J. West, C. M. Pierce, and W. D. Thomas, "Lysergic Acid Diethylamide: Its Effects on a Male Asiatic Elephant," *Science* 138 (1962): 1100–1102.

7. Children's dosage guide for Tylenol can be found at www.tylenol.com/children-infants/safety /dosage-charts (accessed September 25, 2016). For babies, see www.babycenter.com/0_acetamin ophen-dosage-chart_11886.bc (accessed September 25, 2016).

8. See, for example: Alex Pentland, *Social Physics: How Good Ideas Spread—The Lessons from a New Science* (New York: Penguin Press, 2014).

9. There are many readily accessible BMI calculators on the Web from which you can easily determine your BMI. Here's one from the NIH: www.nhlbi.nih.gov/health/educational/lose_wt/BMI/bmi calc.htm.

10. See, for instance, T. Samaras, *Human Body Size and the Laws of Scaling* (New York: Nova Science Publishers, 2007).
11. G. B. West, "The Importance of Quantitative Systemic Thinking in Medicine," *Lancet* 379, no. 9825 (2012): 1551–59.
12. An absorbing overview of the development of the steamship in the nineteenth century, including Brunel's seminal role, can be found in Stephen Fox, *The Ocean Railway* (New York: Harper-Collins, 2004).
13. Barry Pickthall, *A History of Sailing in 100 Objects* (London: Bloomsbury Press, 2016).
14. The remarkable story of the *Vasa* from her conception through her disastrous launching to her miraculous resurrection is brilliantly displayed in a specially built museum in the center of Stockholm close to the site where she went down. She has been cleaned up and restored to her original magnificent condition and is alive and well. It's a wonderful museum and a must-see for any visitor to the city, where it has become Sweden's top tourist attraction.
15. R. Feynman, in the wonderful set of books R. Feynman, R. B. Leighton, and M. Sands, *The Feynman Lectures on Physics* (Boston: Addison-Wesley, 1964), where there is an excellent though technical discussion of the Navier-Stokes equation.
16. Lord Rayleigh, "The Principle of Similitude," *Nature* 95 (1915): 66–68.

3. THE SIMPLICITY, UNITY, AND COMPLEXITY OF LIFE

1. John Horgan, *The End of Science: Facing the Limits of Science in the Twilight of the Scientific Age* (New York: Broadway Books, 1996).
2. Erwin Schrödinger, *What Is Life?* (Cambridge, UK: Cambridge University Press, 1944).
3. Commencement address delivered by Steve Jobs at the graduation ceremony at Stanford University on June 12, 2005.
4. The version most often referred to is the abridged version: D'A. W. Thompson, *On Growth and Form* (Cambridge, UK: Cambridge University Press, 1961).
5. M. Kleiber, "Body Size and Metabolism," *Hilgardia* 6 (1932): 315–51.
6. In addition to the references cited earlier, see G. B. West, J. H. Brown, and W. H. Woodruff, "Allometric Scaling of Metabolism from Molecules and Mitochondria to Cells and Mammal," *Proceedings of the National Academy of Science* 99 (2002): 2473; V. M. Savage, et al., "The Predominance of Quarter Power Scaling in Biology," *Functional Ecology* 18 (2004): 257–82.
7. Huxley's classic, originally published in 1932, has been reproduced: Julian Huxley, *Problems of Relative Growth* (New York: Dover, 1972). J.B.S. Haldane published a famous essay, "On Being the Right Size," in the March 1926 edition of *Harper's Magazine*. It can be found at http://irl.cs.ucla.edu/papers/right-size.html.
8. See the references cited above.
9. J. H. Brown, *Macroecology* (Chicago: University of Chicago Press, 1995).
10. S. Brenner, "Life's Code Script," *Nature* 482 (2012): 461.
11. Two recent discussions urging a greater integration of a more theoretical approach into biology and ecology are contained in P. A. Marquet, et al., "On Theory in Ecology," *Bioscience* 64 (2014): 701; D. C. Krakauer, et al., "The Challenges and Scope of Theoretical Biology," *Journal of Theoretical Biology* 276 (2011): 269–76.
12. The first paper detailing this approach is G. B. West, J. H. Brown, and B. J. Enquist, "A General Model for the Origin of Allometric Scaling Laws in Biology," *Science* 276 (1997): 122. Relatively nontechnical overviews can be found in G. B. West and J. H. Brown, "The Origin of Allometric Scaling Laws in Biology from Genomes to Ecosystems: Towards a Quantitative Unifying Theory of Biological Structure and Organization," *Journal of Experimental Biology* 208 (2005): 1575–92; G. B. West and J. H. Brown, "Life's Universal Scaling Laws," *Physics Today* 57 (2004): 36–42; J. H. Brown, et al., "Toward a Metabolic Theory of Ecology," *Ecology* 85 (2004): 1771–89.
13. Physiologists deconstruct the aorta into several subcomponents (the ascending aorta, the aortic arch, the thoracic aorta, and so on).
14. G. B. West, J. H. Brown, and B. J. Enquist, "A General Model for the Structure and Allometry of Plant Vascular Systems," *Nature* 400 (1999): 664–67.
15. Traditional technical overviews of the physiology of the circulatory system can be found in C. G. Caro, et al., *The Mechanics of Circulation* (Oxford, UK: Oxford University Press, 1978); Y. C. Fung, *Biodynamics: Circulation* (New York: Springer-Verlag, 1984).
16. There is, however, the subtlety that some of the tree is deadwood that does not participate in the

hydrodynamics of fluid flow through its branches, even though it plays a crucial role in its biomechanics. The theory shows that this does not change the result that the volume of the active network scales linearly with the total mass of the tree.

17. B. B. Mandelbrot, *The Fractal Geometry of Nature* (San Francisco: W. H. Freeman, 1982).

18. An excellent summary of Richardson's endeavors, including appropriate references, can be found in Anatol Rapaport, *Lewis F. Richardson's Mathematical Theory of War*, University of Michigan Library; it can be downloaded at https://deepblue.lib.umich.edu/bitstream/handle/2027.42/67679/10.1177_002200275700100301.pdf?sequence=2.

19. L. F. Richardson, *Statistics of Deadly Quarrels*, ed. Q. Wright and C. C. Lienau (Pittsburgh: Boxwood Press, 1960).

20. See, for instance, A. Clauset, M. Young, and K. S. Cleditsch, "On the Frequency of Severe Terrorist Events," *Journal of Conflict Resolution* 51 (1) (2007): 58–87.

21. L. F. Richardson, in *General Systems Yearbook* 6 (1961): 139.

22. Benoit Mandelbrot, "How Long Is the Coast of Britain? Statistical Self-Similarity and Fractional Dimension," *Science* 156 (1967): 636–38.

23. See, for example, Rosario N. Mantegna and H. Eugene Stanley, *An Introduction to Econophysics: Correlations and Complexity in Finance* (Cambridge, UK: Cambridge University Press, 1999).

24. See, for example, J. B. Bassingthwaighte, L. S. Liebovitch, and B. J. West, *Fractal Physiology* (New York: Oxford University Press, 1994).

25. Mandelbrot, *The Fractal Geometry of Nature*.

26. See, for instance, Manfred Schroeder, *Fractals, Chaos, Power Laws: Minutes from an Infinite Paradise* (New York: W. H. Freeman, 1991).

4. THE FOURTH DIMENSION OF LIFE

1. G. B. West, J. H. Brown, and B. J. Enquist, "The Fourth Dimension of Life: Fractal Geometry and Allometric Scaling of Organisms," *Science* 284 (1999): 1677–79.

2. M.A.F. Gomes, "Fractal Geometry in Crumpled Paper Balls," *American Journal of Physics* 55 (1987): 649–50.

3. G. B. West, W. H. Woodruff, and J. H. Brown, "Allometric Scaling of Metabolic Rate from Molecules and Mitochondria to Cells and Mammals," *Proceedings of the National Academy of Science* 99 (2002): 2473–78.

4. G. B. West, J. H. Brown, and B. J. Enquist, "A General Model for Ontogenetic Growth," *Nature* 413 (2001): 628–31.

5. G. B. West, J. H. Brown, and B. J. Enquist, "A General Quantitative Theory of Forest Structure and Dynamics," *Proceedings of the National Academy of Science* 106 (2009): 7040; B. J. Enquist, G. B. West, and J. H. Brown, "Extensions and Evaluations of a General Quantitative Theory of Forest Structure and Dynamics," *Proceedings of the National Academy of Science* 106 (2009): 7040.

6. C. Hou, et al., "Energetic Basis of Colonial Living in Social Insects," *Proceedings of the National Academy of Science* 107(8) (2010): 3634–38.

7. A. B. Herman, V. M. Savage, and G. B. West, "A Quantitative Theory of Solid Tumor Growth, Metabolic Rate and Vascularization," *PLoS ONE* 6 (2011): e22973.

8. Van M. Savage, Alexander B. Herman, Geoffrey B. West, and Kevin Leu, "Using Fractal Geometry and Universal Growth Curves as Diagnostics for Comparing Tumor Vasculature and Metabolic Rate with Healthy Tissue and for Predicting Responses to Drug Therapies, Discrete Continuous," *Dynamical Systems Series B* 18(4) (2013).

9. G. B. West, J. H. Brown, and B. J. Enquist, "A General Model for the Structure and Allometry of Plant Vascular Systems," *Nature* 400 (1999): 664–67; B. J. Enquist, et al., "Allometric Scaling of Production and Life-History Variation in Vascular Plants," *Nature* 401 (1999): 907–11.

10. Quoted in Max Jammer, *Einstein and Religion* (Princeton, NJ: Princeton University Press, 1999).

11. J. F. Gillooly, et al., "Effects of Size and Temperature on Metabolic Rate," *Science* 293 (2001): 2248–51; J. F. Gillooly, et al., "Effects of Size and Temperature on Developmental Time," *Nature* 417 (2002): 70–73.

12. Taken from the opening to Ingmar Bergman's 1968 film *The Hour of the Wolf*.

13. See, for example, Claudia Dreifus, "A Conversation with Nir Barzilai: It's Not the Yogurt; Looking for Longevity Genes," *New York Times*, February 24, 2004.

14. T. B. Kirkwood, "A Systematic Look at an Old Problem," *Nature* 451 (2008): 644–47; Geoffrey B. West and Aviv Bergman, "Toward a Systems Biology Framework for Understanding Aging and Health Span," *Journal of Gerontology* 64 (2009): 2.

15. H. Bafitis and F. Sargent, "Human Physiological Adaptability Through the Life Sequence," *Journal of Gerontology* 32(4) (1977): 210, 402.
16. H. J. Levine, "Rest Heart Rate and Life Expectancy," *Journal of American College of Cardiology* 30(4) (Oct. 1997): 1104–6. See also M. Y. Azbel, "Universal Biological Scaling and Mortality," *Proceedings of the National Academy of Science* 91 (1994): 12453–57.
17. A. T. Atanasov, "The Linear Allometric Relationship Between Total Metabolic Energy per Life Span and Body Mass of Mammals," *Bulgarian Journal of Veterinary Medicine* 9(3) (2006): 159–74.
18. T. McMahon and J. T. Bonner, *On Size and Life* (New York: Scientific American Books—W. H. Freeman & Co., 1983).
19. J. F. Gillooly, et al., "Effects of Size and Temperature on Metabolic Rate," *Science* 293 (2001): 2248–51; J. F. Gillooly, et al., "Effects of Size and Temperature on Developmental Time," *Nature* 417 (2002): 70–73.
20. R. L. Walford, *Maximum Life Span* (New York: W. W. Norton, 1983); R. L. Walford, *The 120-Year Diet* (New York: Simon & Schuster, 1986).

5. FROM THE ANTHROPOCENE TO THE URBANOCENE

1. E. Glaeser, *The Triumph of the City* (New York: Penguin Books, 2012).
2. L.M.A. Bettencourt and G. B. West, "A Unified Theory of Urban Living," *Nature* 467 (2010): 21, 912.
3. Two excellent books that provide detailed background are G. Clark, *A Farewell to Alms: A Brief Economic History of the World* (Princeton, NJ: Princeton University Press, 2008); and I. Morris, *The Measure of Civilization: How Social Development Decides the Fate of Nations* (Princeton, NJ: Princeton University Press, 2013). Both are provocative and somewhat controversial.
4. P. Ehrlich, *The Population Bomb* (New York: Ballantine Books, 1968).
5. D. Meadows, et al., *The Limits to Growth* (New York: Universe Books, 1972).
6. J. Simon, *The Ultimate Resource* (Princeton, NJ: Princeton University Press, 1981).
7. P. M. Romer, "The Origins of Endogenous Growth," *Journal of Economic Perspectives* 8(1) (1994): 3–22.

6. PRELUDE TO A SCIENCE OF CITIES

1. J. Moore, "Predators and Prey: A New Ecology of Competition," *Harvard Business Review* 71(3) (1993): 75.
2. The results of the program are summarized in the volume edited by D. Lane, et al., *Complexity Perspectives in Innovation and Social Change* (Berlin: Springer-Verlag, 2009).
3. Jane Jacobs, *The Death and Life of Great American Cities* (New York: Random House, 1961).
4. Interviewed by Bill Steigerwald in the June 2001 issue of *Reason* magazine.
5. B. Barber, *If Mayors Ruled the World: Dysfunctional Nations, Rising Cities* (New Haven, CT: Yale University Press, 2013).
6. B. Bryson, *Down Under* (New York: Doubleday, 2000).
7. See, for instance, L. Mumford, *The City in History: Its Origins, Its Transformations, and Its Prospects* (New York: Harcourt, Brace & World, 1961).

7. TOWARD A SCIENCE OF CITIES

1. C. Kuhnert, D. Helbing, and G. B. West, "Scaling Laws in Urban Supply Networks," *Physica A* 363 (1) 2006: 96–103.
2. An important issue that has plagued all discussions of cities is the definition of what a city actually is. We all have an intuitive understanding, but in order to develop a quantitative understanding we need something a little more precise. Generally speaking, a city in the sense that I am using it here is *not* to be identified with its political or administrative definition. For example, if you look up the population of San Francisco it's only about 850,000, whereas the contiguous metropolitan area has a population of about 4.6 million. In terms of its dynamics, growth, and socioeconomic structure, it is clearly this latter agglomeration that defines "San Francisco" or any other city. It generally includes suburbs and other communities with their own specific names but that are functionally part of the greater urban network. This is universally recognized by most urbanists, administrators, and

governments by introducing broader categories to accommodate this more realistic representation of the "city." For example, these functional agglomerations are called *metropolitan statistical areas* (MSA) in the United States, *metropolitan areas* in Japan, and *large urban zones* (LUZ) in Europe. Unfortunately, there is no common definition, so some caution needs to be exercised when comparing across different countries. Almost all of the data used in constructing scaling graphs are based on these operational definitions of cities.

3. L.M.A. Bettencourt, et al., "Growth, Innovation, Scaling, and the Pace of Life in Cities," *Proceedings of the National Academy of Science* 104 (2007): 7301–6.

4. L.M.A. Bettencourt, J. Lobo, and D. Strumsky, "Invention in the City: Increasing Returns to Patenting as a Scaling Function of Metropolitan Size," *Research Policy* 36 (2007): 107–20.

5. See, for example, B. Wellman and S. D. Berkowitz, *Social Structures: A Network Approach Sciences* (Cambridge, UK: Cambridge University Press, 1988); M. Granovetter, "The Strength of Weak Ties: A Network Theory Revisited," *Sociological Theory* 1 (1983): 201–33, in P. V. Marsden and N. Lin, eds., *Social Structure and Network Analysis* (Thousand Oaks, CA: Sage, 1982); Claude Fischer, *To Dwell Among Friends: Personal Networks in Town and City* (Chicago: University of Chicago Press, 1982); R. Sampson, "Local Friendship Ties and Community Attachment in Mass Society: A Multilevel Systemic Model," *American Sociological Review* (1988).

6. M. Batty and P. Longley, *Fractal Cities: A Geometry of Form and Function* (Cambridge, MA: Academic Press, 1994); M. Batty, *Cities and Complexity* (Cambridge, MA: MIT Press, 2005).

7. M. Batty, *The New Science of Cities* (Cambridge, MA: MIT Press, 2014).

8. See, for instance, A.-L. Barabási, *Linked: The New Science of Networks* (New York: Perseus Books Group, 2002); M.E.J. Newman, *Networks: An Introduction* (Oxford, UK: Oxford University Press, 2010).

9. Stanley Milgram, *The Individual in a Social World: Essays and Experiments* (London: Pinter & Martin, 1997).

10. D. J. Watts, *Six Degrees: The Science of a Connected Age* (New York: W. W. Norton, 2004).

11. S. H. Strogatz, et al., "Theoretical Mechanics: Crowd Synchrony on the Millennium Bridge," *Nature* 438 (2005): 43–44.

12. This is fun: S. H. Strogatz, *The Joy of X: A Guided Tour of Mathematics, from One to Infinity* (New York: Houghton Mifflin Harcourt, 2013).

13. P. Zimbardo, *The Lucifer Effect: Understanding How Good People Turn Evil* (New York: Random House, 2007).

14. S. Milgram, "The Experience of Living in Cities," *Science* 167 (1970): 1461–68.

15. R.I.M. Dunbar, *How Many Friends Does One Person Need?: Dunbar's Number and Other Evolutionary Quirks* (London: Faber & Faber, 2010).

16. R.I.M. Dunbar and S. Shultz, "Evolution in the Social Brain," *Science* 317 (2007): 1344–47.

17. G. K. Zipf, *Human Behavior and the Principle of Least Effort* (Boston: Addison-Wesley, 1949).

18. Combining these arguments with the concepts inherent in the fourth dimension of life as discussed in chapter 4 and that underlie ¼ power scaling in biology (see the reference of note 1), Luis Bettencourt has suggested that the 0.15 observed in urban phenomena is actually an approximation of ⅙: L.M.A. Bettencourt, "The Origins of Scaling in Cities," *Science* 340 (2013): 1438–41.

8. CONSEQUENCES AND PREDICTIONS

1. This quote is taken from the wonderful collection of letters between Goethe and the composer Carl Friedrich Zelter. A. D. Coleridge, trans., *Goethe's Letters to Zelter* (London: George Bell & Sons, 1887). Although well known in his day, Zelter is nowadays remembered primarily for his relationship with Goethe. I am indebted to my friend the Goethe scholar David Levine for bringing this quote to my attention.

2. Though originally published in 1914, it is now back in print: J. G. Bartholomew, *An Atlas of Economic Geography* (London: Forgotten Books, 2015).

3. C. Marchetti, "Anthropological Invariants in Travel Behavior," *Technological Forecasting and Social Change* 47(1) (1994): 88.

4. G. B. West, "Big Data Needs a Big Theory to Go with It," *Scientific American* 308 (2013): 14; originally published as "Wisdom in Numbers."

5. M. Schläpfer, et al., "The Scaling of Human Interactions with City Size," *Journal of the Royal Society Interface* 11 (2014): 20130789.

6. Examples of such rankings are *The Economist*, which can be found at www.economist.com/blogs

/graphicdetail/2016/08/daily-chart-14, and *Forbes*, which can be found at www.forbes.com/sites
/iese/2016/07/06/the-worlds-smartest-cities/#7f9bee254899.

7. L.M.A. Bettencourt, et al., "Urban Scaling and Its Deviations: Revealing the Structure of Wealth,
Innovation and Crime Across Cities," *PLoS ONE* 5(11) (2010): e13541.

8. San Jose profited early on from the location of the first IBM research establishment on the West
Coast of the United States in 1956.

9. Here's the link to NAICS through the U.S. census: www.census.gov/eos/www/naics/.

10. H. Youn, et al., "Scaling and Universality in Urban Economic Diversification," *Journal of the Royal
Society Interface* 13 (2016): 20150937.

11. G. U. Yule, "A Mathematical Theory of Evolution, Based on the Conclusions of Dr. J. C. Willis,
F.R.S.," *Philosophical Transactions of the Royal Society B* 213 (402–10) (1925): 21–87; H. A. Simon,
"On a Class of Skew Distribution Functions," *Biometrika* 42 (3–4) (1955): 425–40. *Preferential attachment* was popularized in the modern network context by A.-L. Barabási and R. Albert, "Emergence of Scaling in Random Networks," *Science* 286 (5439) (1999): 509–12.

12. The actual number 0.4 arises from a subtle interplay between the scaling exponent for the number
of lawyers as a function of city size (the approximate 1.15) and the Zipfian scaling of diversity of
businesses shown in Figures 52 and 53.

13. L. Mumford, *The City in History* (New York: Harcourt, Brace & World, 1961).

14. In the much more restrictive engineering thermodynamic sense, estimates of urban metabolism
have been made by A. Wolman, "The Metabolism of Cities," *Scientific American* 213(3) (1965):
179–90, and more recently by C. Kennedy, S. Pincetl, and P. Bunje, "The Study of Urban Metabolism and Its Applications to Urban Planning and Design," *Environmental Pollution* 159 (2011):
1965–73.

9. TOWARD A SCIENCE OF COMPANIES

1. R. L. Axtell, "Zipf Distribution of U.S. Firm Sizes," *Science* 293 (5536) (2001): 1818–20.

2. Good overviews of traditional views of companies are provided by G. R. Carroll and M. T. Hannan,
The Demography of Corporations and Industries (Princeton, NJ: Princeton University Press, 2000);
and R. H. Coase, *The Firm, the Market, and the Law* (Chicago: University of Chicago Press, 1988).

3. See, for instance, J. H. Miller and S. E. Page, *Complex Adaptive Systems: An Introduction to Computational Models of Social Life* (Princeton, NJ: Princeton University Press, 2007).

4. J. D. Farmer and D. Foley, "The Economy Needs Agent-Based Modeling," *Nature* 460 (2009):
685–86.

5. N. N. Taleb, *The Black Swan: The Impact of the Highly Improbable* (New York: Random House,
2007).

6. M.I.G. Daepp, et al., "The Mortality of Companies," *Journal of the Royal Society Interface*,
12:20150120.

7. E. L. Kaplan and P. Meier, "Nonparametric Estimation from Incomplete Observations," *Journal of
American Statistical Association* 53 (1958): 457–81; R. Elandt-Johnson and N. Johnson, *Survival
Models and Data Analysis* (New York: John Wiley & Sons, 1999).

8. R. Foster and S. Kaplan, *Creative Destruction: Why Companies That Are Built to Last Underperform
the Market—and How to Successfully Transform Them* (New York: Doubleday, 2001).

9. For a discussion of mergers and acquisitions and how they can be understood in the framework of
their heredity, see E. Viegas, et al., "The Dynamics of Mergers and Acquisitions: Ancestry as the
Seminal Determinant," *Proceedings of the Royal Society A* 470 (2014): 20140370.

10. THE VISION OF A GRAND UNIFIED THEORY OF SUSTAINABILITY

1. This was first suggested in G. B. West, "Integrated Sustainability and the Underlying Threat of Urbanization," in *Global Sustainability: A Nobel Cause*, ed. H. J. Schellnhuber (Cambridge, UK: Cambridge University Press, 2010).

2. A. Johansen and D. Sornette, "Finite-Time Singularity in the Dynamics of the World Population,
Economic and Financial Indices," *Physica A* 294 (3–4) (2001): 465–502.

3. Nor, by the way, does the situation remain fixed in between major innovations. However, these
changes are relatively small and smooth compared with the dramatic, almost discontinuous,
changes that occur at the transition or tipping point when a major innovation occurs.

4. See, for instance, W. B. Arthur, *The Nature of Technology: What It Is and How It Evolves* (New York: Free Press, 2009); H. Youn, et al., "Invention as a Combinatorial Process: Evidence from U.S. Patents," *Journal of the Royal Society Interface* 12 (2015): 20150272.
5. R. Kurzweil, *The Singularity Is Near: When Humans Transcend Biology* (New York: Viking, 2005).
6. V. Vinge, "The Coming Technological Singularity: How to Survive in the Post-Human Era," *Whole Earth Review* (1993).
7. This is quoted by the great mathematician Stanislaw Ulam in a eulogy to von Neumann following his death in 1957: "Tribute to John von Neumann," *Bulletin of the American Mathematical Society* 5(3), part 2 (1958): 64.
8. C. McCarthy, *The Road* (New York: Alfred A. Knopf, 2006).

AFTERWORD

1. Two popular nontechnical books that present a broad overview of the enormously exciting quest for the fundamental constituents of matter and a grand unified theory for understanding their interactions, including its extension to the evolution of the cosmos and the origin of space-time itself, are S. Carroll, *The Particle at the End of the Universe* (New York: Dutton, 2012); and L. Randall, *Warped Passages* (New York: Harper Perennial, 2006).
2. "Strange Bedfellows," *Science* 245 (1989): 700–703.
3. A. Tucker, "Max Perutz," *Guardian*, Feb. 7, 2002; www.theguardian.com/news/2002/feb/07/guardianobituaries.obituaries.
4. See, for example, www.fastcodesign.com/3030529/infographic-of-the-day/hilarious-graphs-prove-that-correlation-isnt-causation.

INDEX

Numbers in *italics* indicate pages with figures or illustrations.

LIST OF ILLUSTRATIONS

Graph art by Jeffrey L. Ward